WunderCAD/CAMmer
A Digital Mirabilia's Compendium in Contemporary Material Science and Bioinspired Design

Massimo Lumini
Science communicator in the field of biomimetics and design; founder of BIONIKONLab/FABNAT14 – Laboratory of Applied Bionics & Biomimetics Design, Iglesias - South Sardinia - Italy

CRC Press
Taylor & Francis Group
Boca Raton London New York

CRC Press is an imprint of the
Taylor & Francis Group, an **informa** business
A SCIENCE PUBLISHERS BOOK

Cover credit: Massimo Lumini—created through ChatGPT® (OpenAI®), 2025.

First edition published 2026
by CRC Press
2385 NW Executive Center Drive, Suite 320, Boca Raton FL 33431

and by CRC Press
4 Park Square, Milton Park, Abingdon, Oxon, OX14 4RN

Library of Congress Cataloging-in-Publication Data (applied for)

ISBN: 978-1-032-57333-5 (hbk)
ISBN: 978-1-032-57337-3 (pbk)
ISBN: 978-1-003-43894-6 (ebk)

DOI: 10.1201/9781003438946

Typeset in Times New Roman
by Prime Publishing Services

"Nella mia esperienza la spinta a scrivere
è sempre legata alla mancanza di qualcosa
che si vorrebbe conoscere e possedere,
qualcosa che ci sfugge."

Italo Calvino[1]

[1] Calvino, I. (2010). *Mondo scritto e mondo non scritto*. Mondadori.
"In my experience, the urge to write is always linked to the lack of something one wishes to know and possess, something that eludes us."

Preface

The opening *incipit* of this concept book dates to 2019, but it was only in 2022 that a publishing partnership made the vision concrete. In the meantime, rapid advances in CAD/CAM—particularly AI and quantum computing—have continuously reshaped the technological landscape, compelling successive updates to the book's structure and content.

The Renaissance *Wunderkammer*—a *cabinet of curiosities* blending *naturalia* (fossils, shells, exotic specimens) with *artificialia* crafted by human ingenuity—offers a fitting metaphor for our digital *mirabilia*. Today's *WunderCAD/CAMmer* extends that tradition into the digital realm, housing not only rare and unseen objects, but also complex algorithms, artificial intelligence, and the latest design-engineering tools. Here, creativity meets computation, and design becomes both exploration and innovation catalyst.

Modern digital design—now intertwined with AI and machine deep learning—is in constant flux, making yesterday's information obsolete almost as soon as it's published. This presents not merely a challenge, but an opportunity to imagine and anticipate a future in which the line between natural and artificial dissolves and today's marvels become tomorrow's commonplace.

Early CAD (Computer-Aided-Design) systems once handled only simple electronic drafting; today's platforms have evolved into parametric design environments capable of generating intricate geometries on demand. With the integration of CAE (Computer-Aided Engineering) and CAM (Computer-Aided Manufacturing), these tools simulate engineering processes and new material behaviors, revolutionizing fields from structural engineering to medicine and industrial design.

These technologies are no longer passive instruments for producing shapes or technical drawings. They have become dynamic, predictive systems—powerful oracles in the design process. Through virtual simulations, engineers, designers, and scientists can test the performance of structures, medical devices, and products before they exist in the physical world. This capability unlocks entirely new realms of innovation and paves the way for smarter, more efficient solutions that transform industries and improve lives.

In this book, we embark on a tour of that evolving terrain, exploring how creativity, technology, and biology converge. We examine how cutting-edge tools in design and engineering not only solve today's challenges but also help shape a future where the unimaginable becomes reality.

Writing this book has meant immersing ourselves in a vast ecosystem of algorithms and innovations. The sheer volume of global research—from AI breakthroughs to nanotechnology—has been impossible to ignore. Hundreds of references and footnotes bear witness to the field's interconnectedness and complexity: each citation is a thread in a tapestry of breakthroughs that continually raise new questions and possibilities.

These pages do more than sketch a vision of the future—they strive to define it. Design, technology, and nature collide in unexpected ways. The future is no longer a distant horizon but a blank canvas; the tools we wield today are the brushstrokes that will color it.

Acknowledgments

The seeds of this book were sown in my early years, when as a child I wandered—with my father, Julius—among the dusty 19th-century display cases filled with extinct and uncanny creatures in the expansive halls of the Museum of Natural Sciences in Milan. Those mysterious galleries, and the ancient writings of Camille Flammarion on natural evolution, shaped my aesthetic and scientific imagination.

My deepest gratitude goes to Prof. Carmelo Di Bartolo of Milan—Italian designer, founder and former director of the Istituto Europeo di Design, CRIED and Design Innovation—whose mentorship since 1988 introduced me to the wonders of bionics, biomimetic design, and the philosophy of biophilia. I also thank Prof. Amilton J. V. Arruda of the Federal University of Pernambuco for his friendship, esteem, and scholarly guidance—and for inviting me to collaborate on his editorial projects—and Professors Felipe Luis Palombini and Subramanian Senthilkannan Muthu for bringing my work on bionics and sustainability to an international audience.

I am grateful to Simone Lumini for his invaluable advice and for keeping me up to date on emerging technologies, which have helped keep this manuscript at the cutting edge.

A special thank-you to Dr. Prashant K. Singh at Vanderbilt University for granting me access to his *WunderCAD/CAMmer*, where he explores the mysterious realm of bacterial molecular motors, and for sharing invaluable insights into contemporary CAD/CAM innovations in structural biology.

I am also grateful to Dr. Xiaoyang Zheng at the University of Tokyo for generously providing study materials and imagery on auxetic materials. My thanks extend to Professors Arthur Mamou-Mani and Toby Burgess of the University of Westminster for permission to reproduce images and text from the *Living Fabric Tower* project.

A heartfelt thanks to Kelly Egorova of OXMAN Studio for her attentive support with my editorial requests.

I would like to thank my highly esteemed friend Antonio Arcadu, who, from Sardinia, has forged an outstanding international reputation in the motorcycle-industry CAD/CAM sector by adapting the industrial-clay technique to 3D printing. With Antonio, I have shared many projects on digital innovation in digital manufacturing and FAB_LABs, and he has kindly provided me with several info and photos from his portfolio.

Sincere appreciation goes to Jorge Llorente-Bousquets, Emeritus Professor at UNAM, for his invaluable research resources in biology imaging. I also thank the Editorial Committee at CRC Press/Taylor & Francis for their assistance throughout the manuscript's preparation.

Finally, this book is dedicated to my wife, Maristella, and to my children, Simone and Sara Agnese—the bright stars of my family life's constellation.

Contents

Introduction

The Brilliant Ghost of the *Wunderkammer*:[1] The Evolution of Biology and the Advent of Contemporary Digital *Mirabilia*

Throughout history, there have been moments when the boundaries between science, art, and imagination have blurred, giving birth to new knowledge and even new ways of perceiving the world. From the late fifteenth century onward, the *Wunderkammer*—those early *Cabinets of Curiosities*—embodied this convergence. These collections, often theatrical in nature, brought together natural wonders and human ingenuity, creating spaces designed not just to inform but to inspire awe and contemplation. Today, amidst the rise of biotechnology, artificial intelligence, and parametric design, the spirit of wonder captured by the Wunderkammer has not vanished—it has merely evolved. From those secret rooms—guardians of natural mirabilia—to the digital esoterism of parametric design, which translates algorithms and organic geometries into structures of formal exuberance, the dialogue between wonder and method is renewed, celebrating the very marvel of Nature's and Design exuberance (Lumini 2018). No longer confined to wooden cabinets or intricate displays, this sense of *mirabilia* now thrives in the form of high-resolution screens, dynamic simulations, and synthetic organisms. The *brilliant ghost* of the old rooms returns, embedded in the very molecular architectures of synthetic life, and reflected in the digital laboratories and algorithmic models that are transforming our understanding of life and creativity. The sensibility of the *Nature's cabinet of wonder*—its thirst for the unknown, its theatrical fusion of art and science—finds unexpected resonance in contemporary biology and computational design. Just as the Renaissance *wunderkammers* brought

[1] The iconic image of the dusty and mysterious Renaissance *Wunderkammer* is reimagined through the linguistic twist of *WunderCAD/CAMmer*, serving as a metaphor, narrative avatar, and leitmotif to introduce an up-to-date exploration of the *mirabilia* emerging from today's most advanced laboratories and workshops. A particular focus is placed on cutting-edge CAD/CAM protocols for bio-inspired design and manufacturing. This work outlines a landscape of ongoing research and applications—highly interdisciplinary and deeply interconnected—drawing inspiration from the remarkable lessons of molecular material design, the foundation of biological life on Earth. Through the evolution of technological bio-inspiration, unprecedented intersections between nature, technology, and art are continuously reshaping the future of civilization. Within the digital *cabinets of wonders*, powered by state-of-the-art CAD/CAM suites, 3D virtualizations, and quantum supercomputers, every aspect of design is undergoing radical transformation. Driving this epochal shift are Machine Learning and Deep Learning—the first steps in the rapid rise of AGI, the Artificial General Intelligences set to redefine human creativity and artificial design innovation.

together the marvels of the natural world and human-made artifacts, today's *digital mirabilia* combine the boundaries of biology and technology (McCall et al. 2016). This thread of fascination, extending from the past to the present, reveals how the intellectual spirit of curiosity and exploration continues to haunt, shape, and inspire the technological imagination of the present. The evolution of biology from a descriptive natural science to a dynamic, data-driven discipline has profoundly reshaped our understanding of life itself. Once focused on the classification of life, contemporary biology—enriched by molecular genetics, systems theory, and synthetic biology—has now expanded into a field that seeks not only to decode life's mechanisms but to simulate, predict, and even reengineer living systems. In this new paradigm, life is no longer a passive object of study but an active, manipulable system capable of being designed and fabricated. It is precisely within the forge of this deeply bioinspired design approach that the contours of future science and technology are being cast. Far from being mere branches of contemporary research, these emerging fields serve as potent conceptual frameworks—dissolving the boundaries between nature and invention, the organic and the computational. They are transforming biology into an engineering science capable of designing novel forms of living matter. Here, in the convergence of code and cell, algorithm and organism, the future of knowledge and innovation is quietly being redefined. And this long arc of transformation was launched only a few centuries past. As Barsanti (2003) notes, "between the 18th and 19th centuries, biology emerged as an autonomous science, following a long process of erosion of the mechanical philosophy (as advocated by Robert Boyle), driven by the persistent difficulty in explaining vital phenomena solely through the laws of physics." The mechanical worldview, which saw living organisms as complex machines, became increasingly inadequate as new experimental methods revealed the extraordinary diversity and complexity of life. The inadequacy of this philosophy becomes particularly evident starting from the first half of the 18th century, when the widespread introduction of the microscope and the shift from a descriptive to an experimental approach began to reveal the extraordinary variety and complexity of the structure of living nature. The microscope, coupled with experimental methods, exposed a world of biological complexity that could not be explained by mechanical philosophy alone. The microscope, Cartwright (2022) observes, was among the most important instruments of the Scientific Revolution, as it allowed scientists to examine phenomena not visible to the naked eye. The first biologists observe that this is not a system reducible to simple general laws, as it is a realm populated by *unusual* phenomena and true *exceptions* to the norms and theories commonly accepted. Furthermore, even beyond these anomalies, properties of nature emerge that are far more intricate than had been imagined until then (Beekman and Jochemsen 2022). Early magnifying optical instruments unveiled an unseen realm of tiny living organisms, opening the door to entire worlds previously invisible to the human eye (Pouyan 2016). What emerged was a landscape teeming with incredible, unusual, and bewildering phenomena—extraordinary and astonishing, at times paradoxical, and often deemed embarrassing precisely because, as someone aptly noted, they are simply too complex to be fully understood.

The phenomenology of life thus presented itself to the eyes of modern biology as a two-faced *Janus*. In some respects, it continued to follow the line of continuity with the theories related to the order of physical-chemical phenomena, but in many other elusive aspects it manifested, this new science took on the characteristics of a qualitative leap towards new dimensions of reality. Having abandoned the eighteenth-century idea of a *vital force*, of an unknown *quid* that could no longer serve as an explanatory principle, two contrasting approaches emerged from the most representative physiologists and naturalists in the medical faculties and scientific circles of Germany, France, Italy, and England. The majority view was the *physicalistic* or *mechanistic* conception, inspired by the scientific culture of the time, according to which nothing happens that cannot be reconciled with known physical-chemical laws. The minority view, though no less combative, was the *neovitalistic* conception, which does not deny the explanatory role of physical and chemical causality, but nevertheless hopes for the advent of a *complete biology* capable of explaining the final organic complexity of the psychic phenomena of the human mind, considered the pinnacle of biological creation. This research process would begin along a path still leading to the discovery, in the general manifestation of life, of something irreducible to pure physical-chemical causality: a peculiarity that suggests hypotheses of a different physiology. Fully nineteenth-century biology, in particular, in an attempt to define its own autonomous domain within the map of classical scientific disciplines, thus affirms that:

"The science of living bodies must be treated in a completely different way from the way in which the sciences of inorganic bodies are treated."[2]

The autonomy of research will lead biologists to limit the interpretative influences derived from the application of physics and chemistry theories, in the experimental effort to uncover autonomous laws capable of deciphering the dynamics that govern the mysterious processes enabling the manifestation of life. This separation between biology, physics, and chemistry is formalized by theories that establish that living bodies, as Pierre-Jean-Georges Cabanis (1757–1808) confirms in his *Rapports du physique et du moral de l'homme*, possess "forces exclusively their own," which means that hasty applications of purely mechanical theories lead to results that are "vague, uncertain, insufficient, contradictory, and even contrary to the facts." According to Corbellini (2004), from the late 19th century onward, life sciences gained increasing methodological and conceptual independence. This development paved the way for the emergence of coherent, detailed explanations of biological phenomena, particularly regarding functional and evolutionary dynamics, as well as the factors influencing health and disease. The project of emancipating biology, aimed at clearly distinguishing the study of organic matter from that of inorganic matter,

[2] Marie-François-Xavier Bichat (1771–1802) argued that "the science of living bodies must be treated in a completely different way from the science of inorganic bodies" and that "a different language should be employed, so to speak; since most of the terms we transfer from the physical sciences to those of animal or plant economy evoke ideas that are completely incompatible with the phenomena of this science." In addition, he considers: "Physics, chemistry, etc., intersect because the same laws govern their phenomena; but an immense gap separates them from the science of organic bodies, because there is a vast difference between those laws and the laws of life from *L'Anatomie générale* (1801)."

turned out to be short-lived. The extraordinary discoveries in physics and chemistry, starting from the early decades of the 20th century and deepened until today, have in fact redefined the so-called raw matter, revealing an unexpected, enigmatic, and surprisingly complex face. MDC (2016) notes that Max Delbrück (1906–1981), a pioneering biologist, played a key role in the establishment of molecular biology and was awarded the Nobel Prize in Physiology or Medicine in 1969 for his work on bacteriophages. He remarked, "The more one closely observes these performances of matter in living organisms, the more the spectacle becomes impressive. The humblest living cell becomes a magic box full of elaborate and changing molecules." The observation and analysis at the nanometric scale of the physical and chemical phenomena that dominate the atomic and subatomic universe are indeed opening up and revealing to our senses the contents of this *magic box*, showing that within it, worlds teem with a sort of vitality and *molecular intelligence* once thought impossible. These observations have the consequence of resurrecting old neovitalist ghosts and of increasingly bringing together and blending the opposing categories of *organic* and *inorganic*. Contemporary biology is achieving extraordinary results in the analysis and modeling of cellular dynamics through the advanced application of the laws of physics and chemistry, which describe the emergence of life via algorithms capable of processing vast data flows. This is further enhanced by continuous innovations in bioimaging techniques enabled by computing, the increasingly optimized and fast performance of 3D design and modeling CAD/CAM (Computer-Aided-Design and Computer-Aided-Manufacturing) suites, and the now omnipresent integration with AI algorithms, machine and deep Learning, and the extraordinary technological advancements in emerging quantum computing. The *bottom of the box* has opened up to new dimensions, crossing the invisible boundaries of matter, almost touching the mysterious quantum horizon. Davies (2019) proposes a radical theory suggesting that life is defined not solely by the ability to replicate, but by a dynamic balance between order and disorder, which enables organisms to adapt and evolve in complex environments. In the cold and rational rooms of contemporary science, an ancient presence resurfaces: the reflection of a dormant unease that rekindles wonder and amazement. These feelings, as we will see, have driven emperors and scholars since the late 1400s in Europe to gather, collect, and publicly display in the *Wunderkammer* the most mysterious relics of nature alongside the marvels of human Art and Technology of their time. In this work, the evocative topos of the *Kunstkammer*—the room of art, a true *Theater of Scientific Wonders* (*Theatrum Natura Mirandum* Fig. 1) and an archetype of 17th-century scientific culture—has been chosen to evoke a sense of continuity with that wonder which still resonates in contemporary science and digital technology.

Galambosova (2024) argues that *Wunderkammern*, or *Cabinets of Curiosities*, were intentionally designed to provoke wonder in the viewer. These spaces blended science and art, assembling rare and unusual natural specimens (*naturalia*) alongside man-made artifacts (*artificialia*), creating eclectic collections that embodied both curiosity and astonishment. Today, that spirit lingers like a oblique presence in the cold corridors of laboratories, among sophisticated super microscopes, high-resolution monitors, and 3D printers housed in advanced research centers. It is an ancient wraith that now dances with the digital forms of protean and synthetic creatures, animated

Fig. 1. Levinus Vincent, Wondertooneel der Nature (1706–1715).

within the models and simulations generated by the most advanced parametric and generative programs. According to Dassault Systèmes (2025), parametric modeling is a powerful approach for developing flexible and adaptive designs, enabling users to manipulate the parameters that define a model's geometry. This method has revolutionized product development by allowing for precision combined with rapid adjustability. By establishing relationships between components, parametric modeling ensures that any modification to a parameter automatically updates the associated geometry, streamlining the revision process without the need for manual reconfiguration. Its true strength lies in facilitating design iteration and optimization. Designers can easily explore variations—adjusting dimensions, angles, and other defining features—to refine their work toward more efficient, functional, or aesthetically compelling outcomes. This flexibility proves especially valuable in complex fields such as automotive, aerospace, and architecture, where project requirements frequently evolve throughout the design cycle. Furthermore, parametric modeling fosters collaboration among multidisciplinary teams. In a typical design environment, engineers, architects, and manufacturers may all have input on the final product, and with parametric modeling, changes made by one team member are automatically reflected in the work of others. This results in more streamlined workflows, reduced errors, and an overall more efficient design process. Additionally, parametric design tools often integrate seamlessly with CAM and Computer-Aided Engineering (CAE) software, allowing for a smooth transition from digital design to physical production. The integration of parametric modeling with artificial intelligence (AI) and machine learning algorithms further enhances its capabilities. These technologies can help predict optimal design solutions by analyzing vast datasets and providing insights based on patterns and trends. This can lead to smarter design decisions that improve performance, reduce costs, and accelerate time-to-market for new products. The ongoing advancements in computational power, coupled with the increasing sophistication of AI algorithms, are likely to expand the potential of parametric modeling, allowing for even more innovative and complex designs. As digital fabrication technologies, such as 3D printing, continue to evolve towards the nanoworld, the role of parametric modeling becomes even more critical. With digital manufacturing, designers can create sophisticated physical prototypes directly from parametric models, enabling rapid prototyping and testing. This direct link between digital and physical realms opens up new possibilities for customization and personalization in product design, as well as creating complex geometries that were previously unattainable through traditional manufacturing methods. In summary, parametric modeling serves as a cornerstone for modern design and manufacturing processes, offering unprecedented flexibility, efficiency, and accuracy. By leveraging advanced computational tools, AI, and emerging fabrication techniques, parametric modeling is helping shape the future of design across industries, allowing for the creation of more complex, sustainable, and innovative products.The editorial concept of this *compendium*[3] is primarily technical, but its perspective lies at the

[3] The term comes from the Latin *compendium*, meaning a short or summary collection, derived from *com-*(together) and *pendĕre* (to weigh). It suggests a collection that condenses and balances the essential elements of a topic. In the context of the *WunderCAD-CAMmer project*, this word refers to a collection of emerging knowledge—both technical and theoretical—that aims to encompass a broad and multifaceted panorama. Although difficult to define due to the rapid evolution of the field and its numerous interconnections, a compendium serves as a concise yet comprehensive resource. It gathers key concepts, facts, or knowledge in a way that allows for quick understanding while preserving the completeness of the subject. Often used to provide an overview of complex fields, a compendium distills the main aspects without delving into excessive detail. Additionally, it can be seen as a curated selection of works or materials related to a common theme, offering a broad perspective on key trends, innovations, or discoveries.

intersection of the scientific mind and aesthetic and sentimental sensitivity. It seeks subtle connections between the psychic encyclopedism of the 17th century, represented in the *Wunderkammern*, which historically served as spaces dedicated to exploration and awe. Vaiani (2022) highlights how these collections were not only places of curiosity but also of artistic and scientific contemplation, where the marvels of the natural world were staged to evoke amazement and reflection. This reasoned collection of design possibilities for the near future also focuses on the dissemination and popularization of techno-scientific advances, particularly in the context of CAD/CAM and computational biosciences applied to highly innovative sectors like the research and development of new materials and technological processes inspired by biomimicry, bionics, and bio-inspired design. Given that these fields remain largely outside the complete public domain, even within expert circles, the book seeks to fill this editorial gap by offering a reference that bridges the scientific and humanistic realms. Its aim is to uncover the traces of the near future, providing a primary guide and a basic dissemination tool for an international audience, while contributing to the historicization and systemic analysis of research results and innovative applications. Historiographical study helps us understand how the digital present is deeply rooted in the intellectual, cultural, and economic foundations of the past, as well as in its deepest psychological motivations. The *Wunderkammer* is a striking example: extraordinary spaces that emerged during a transitional era, where new forces and ideas challenged medieval dogmas. It was a time when human intellect, rediscovering classical sources, once again observed forms, materials, and natural phenomena with a liberated gaze, exploring their secrets and technological potential. This period, which laid the foundations of the Modern Age, was marked by the convergence of Art, Science, Technology, and *Humanitas*, but also by wars, invasions, famines, and plagues that devastated entire populations. It was amid the ruins of these events that, in the secluded elegance of rooms imbued with alchemical symbolism, enlightened nobles and intellectuals refined the mental tools that would shape the future of Western science and technology. Buzard (2013) explains that from the mid-15th century onward, the first *Theaters of the World* began to emerge in Europe—secret, aristocratic spaces where the wonders of both the artificial and natural worlds were displayed. Alongside the well-documented cabinets of rulers and aristocrats, members of the merchant class and early practitioners of science also assembled collections that served as precursors to modern scientific museums (Fig. 2).

Buzard (2013) explains that from the mid-15th century onward, the first *Theaters of the World* began to emerge in Europe—secret, aristocratic spaces where the wonders of both the artificial and natural worlds were displayed. Alongside the well-documented cabinets of rulers and aristocrats, members of the merchant class and early practitioners of science also assembled collections that served as precursors to modern scientific museums. These collections were heirs to the thesauri and medieval repositories of knowledge, safeguarded in monastic scriptoriums. True arks of civilization, they preserved ancient wisdom through time. Among these mythical spaces were the *studiolo*—a small humanistic chamber for art collections—and the European *Wunderkammer* of the 16th and 17th centuries, an expression of the interplay between art and early science, which eventually influenced the 19th-

Fig. 2. Ole Worm, Museum Wormianum. Leiden: Johannes Elzevier, 1655. This edition showcases the natural history cabinet of Danish physician and scholar Ole Worm (1588–1654), Professor of Philosophy and Medicine at the University of Copenhagen, who in 1609 also studied at Padua, Italy.

century concept of the scientific museum. In the context of the *Wunderkammer*, the term *mirabilia* refers to extraordinary objects, marvels, and curiosities that were the heart of these collections. Derived from the Latin *mirabĭlis* (meaning *extraordinary* or *worthy of admiration*), and related to the verb *mīror* (*to admire* or *to wonder at*), *mirabilia* were the objects that triggered awe and wonder. These collections were designed not only to showcase rare and fascinating items but also to engage visitors in an intellectual exploration of the natural world, pushing the boundaries of what was known and understood at the time. The *mirabilia* were not limited to rare natural specimens but also included artificial curiosities—man-made objects that demonstrated the ingenuity of human creativity. The root *mir-* is related to the word *mirum*, meaning a prodigy or wonder, and is the origin of the English word *mirror* and the French *miroir*. This etymology reflects how objects in these unseen collections served as mirrors, reflecting the mysteries of the natural world and offering both intellectual and visual puzzles for the beholder. The disturbing organic specimens of abnormal human or animal births, partially fused twins, and grotesque botanical deformities were preserved in large glass jars. Their sight evoked a mix of awe and horror in visitors, suggesting hidden meanings and inscrutable properties of the natural world. They were not merely objects to be contemplated, but visual metaphors of the mysteries of reality. The *Wunderkammer*, initially created by

Italian scientists and rulers, was later adopted by Austrian, German, and Bohemian princes. These were authentic cabinets de curiosités: private, sometimes esoteric spaces, open only to a select audience. Here, scholars and intellectuals engaged in a spectacular cosmic workshop, where aristocratic and notable owners accumulated rare globes, scientific instruments, bizarre stuffed animals—sharks, devilfish, giant alligators—alongside narwhal tusks, platypus skeletons, ivory, coral, and supposed unicorn horns. These collections, built upon a fascination with both the natural world and human-made wonders, served as microcosms of knowledge that blended the realms of science, art, and curiosity. Today, digital technologies and an increasing understanding of the hidden laws of nature are leading us into an era where the boundary between Nature and Artifice is becoming increasingly blurred. Just as the *cabinets of curiosities* once merged the natural and the man-made in ways that sparked intellectual wonder, modern advancements continue to redefine and reshape our understanding of what constitutes the natural and the artificial. The hybridization of biology and technology has become a hallmark of contemporary design. Bio-inspired materials, modeled on molecular architectures honed by millions of years of evolution, open new avenues for eco-compatible products and prompt us to rethink our relationship with the environment. At the same time, this convergence feeds into today's cyberculture, where natural and artificial systems are viewed as parts of a single, programmable continuum. Recent research by Vuylsteke et al. (2023) highlights the need for a more precise vocabulary in the context of biologically related materials within design practice. Rather than relying on the broad and often ambiguous term *biomaterial*, the authors advocate for more specific classifications such as *biobased*, *biodegradable*, and *biomanufactured*, in order to enhance clarity and communication across industrial and academic settings. Their study suggests that the core of innovation in this domain lies not merely in the materials themselves, but in the design processes they enable. Drawing on four analytical perspectives—origin, production, use, and end-of-life—the authors propose a conceptual model of *biodesign* aimed at reestablishing a symbiotic relationship between human systems and natural ecosystems. A second, ongoing phase of the study seeks to further investigate the intentions and methodologies that drive contemporary biodesign practices, with the ultimate goal of identifying barriers to its broader implementation in professional contexts. The culture of technological design is appropriating fundamental ethical paradigms to ensure a possible survival of the planet and to establish a *New Alliance* with Nature. Material design plays a crucial role in this transformation, promoting sustainable industrial practices and production cycles inspired by biomimicry. Researchers and designers are called upon to rethink how we design and consume, placing the well-being of the Earth and future generations at the heart of the process. Today, the genius of those ancient *rooms of wonders* resurface in a new form. Advanced optical tools like SEM microscopes reveal biological geographies never explored before, allowing us to observe up close the *Invisible Factories* of living matter: microscopic workshops where nature, with its molecular intelligence, assembles extraordinary materials, surpassing even our most advanced engineering (Lumini 2021). The sense of wonder once kindled by ancient *cabinets of curiosity* now manifests through parametric design, virtual reality, artificial intelligence, and advanced scientific imaging technologies. In today's *WunderCAD/CAMmer*[4] environments, engineers and researchers manipulate nanocrystals, bioplastics, and organic tissue scaffolds, modeling bold cellular metropolises and intricate molecular architectures. Generative and Computational Design mark the evolution of digital design tools, ushering in a new era of *Emerging Objects* and *Material Ecology*®—a realm where human artifacts engage in a dialogue with the intelligence of Nature. Fascinated and inspired by these images—illuminated from the darkness of virtual *cabinets des merveilles* on TFT (Thin Film Transistor) screens or projected as holographic illusions in augmented reality via CADMATIC™ and HoloLens™ technologies—the algorithmic alchemy of digital pulses generates exuberant chimeras (Fig. 3).[5]

Once 3D printed, these hybrids begin their material lives suspended between Nature and Artifice. These connections between the symbolic mental spaces of the pre-scientific past and the high-tech domains of contemporary design echo the philosophical duality proposed by Blaise Pascal (1623–1662) in his posthumous work *Pensées* (1670): The *Esprit de Géométrie*—the spirit of rigorous reasoning—and the *esprit de finesse*—the spirit of intuitive subtlety.[6] They justify this comparison and offer food for thought on the evolution of contemporary technology, including: the questioning of the boundaries between the natural and artificial worlds, the

---

[4] The iconic image of the dusty and mysterious Renaissance *Wunderkammer* is reimagined through the linguistic twist of *WunderCAD-CAMmer*, serving as a metaphor, narrative avatar, and leitmotif to introduce an up-to-date exploration of the *mirabilia* emerging from today's most advanced laboratories and workshops. A particular focus is placed on cutting-edge CAD-CAM protocols for bio-inspired design and manufacturing. This work outlines a landscape of ongoing research and applications—highly interdisciplinary and deeply interconnected—drawing inspiration from the remarkable lessons of molecular material design, the foundation of biological life on Earth. Through the evolution of technological bio-inspiration, unprecedented intersections between nature, technology, and art are continuously reshaping the future of civilization. Within the digital *cabinets of wonders*, powered by state-of-the-art CAD-CAM suites, 3D virtualizations, and quantum supercomputers, every aspect of design is undergoing radical transformation. Driving this epochal shift are Machine Learning and Deep Learning—the first steps in the rapid rise of AGI, the Artificial General Intelligences set to redefine human creativity and artificial design innovation.

[5] CADMATIC. (n.d.). *eShare for HoloLens*. CADMATIC. From https://www.cadmatic.com/it/products/cadmatic-eshare/eshare-for-process-and-industry/eshare-for-hololens/.

[6] The opposition between these two terms can be related—albeit with due caution—to the modern concept of functional specialization in the brain's hemispheres. *Esprit de géométrie* aligns with functions typically attributed to the left hemisphere, such as analytical, logical, sequential, and mathematical thinking. This side of the brain is traditionally associated with language, rationality, and systematic thought—traits that resonate with Pascal's notion of a spirit that proceeds through rigorous deduction based on clear principles. *Esprit de finesse*, by contrast, reflects the capabilities often linked to the right hemisphere, which is more engaged in intuition, holistic perception, creativity, and the interpretation of emotional nuance. This hemisphere plays a greater role in grasping complex contexts, artistic sensitivity, and interpersonal understanding—all qualities that mirror Pascal's description of the spirit of finesse. However, it is important to note that the notion of a strict division between a rational left hemisphere and an intuitive right one is an oversimplification in light of contemporary neuroscience. Today, we know that the brain's hemispheres function in close collaboration and that cognitive processes are not strictly compartmentalized. Nonetheless, Pascal's distinction endures as a powerful metaphor for two modes of understanding: one grounded in logic and mathematical clarity, the other in sensitivity and intuition. Ultimately, as Pascal himself implied, a profound grasp of the world likely requires both.

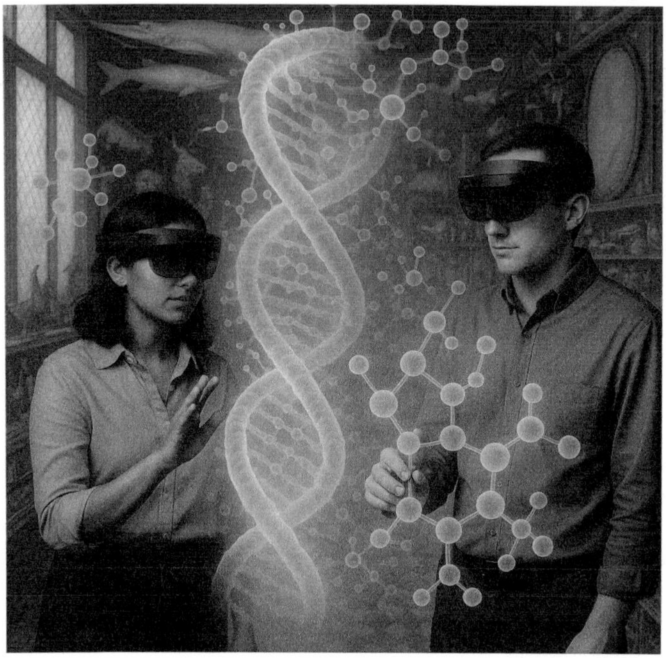

Fig. 3. VR-WunderCAD/CAMmer.

reformulation of the model of knowledge of reality, and the role of images in science. As Giallombardo (2013) argues about the *Encyclopedic Palaces*—spaces that symbolize the pursuit of total knowledge and the forms of possession of such wisdom—they represent an aspiration to encompass all human understanding, while also revealing the limitations and boundaries inherent in such an ambition. Pascal clearly distinguished two paths to understanding things: the *geometric path,* which is demonstrative and resides in reason (*raison*), and the *intuitive path*, which is the ability to grasp the essential, residing in the heart (*coeur*). To fully grasp the complexity of the world and what truly matters in life, both reason and heart are necessary. However, the way of the heart can precede and transcend reason, or at least guide it toward subtler and deeper considerations that go beyond the driest and most sterile materialism. Very often, the study and research needed to draft this compendium have been guided and inspired more by rhythm and the *reasons of finesse*, enabling us to break through the difficult barriers of formulas, data, charts, and statistics, and transcend the coldness of algorithms and digital bits. This might seem like an oxymoron, akin to a *lucid madness*, yet it is precisely algorithms and digital bits that, like delicate scalpels, penetrate and reveal the fabric of reality itself at the nanometric scale, unveiling the deepest secrets of living matter and allowing us to cultivate the conceptual seeds of a renewed and extraordinary contemporary *digital mirabilia.* According to an article by *Nature Research Custom Media and The Catholic University of America* (2023), the *joy of scientific discovery* not only drives research progress but also contributes to the well-being of researchers. The *experience of wonder* and *beauty in scientific work* provides enrichment that goes beyond the mere pursuit of results, fostering personal growth and the fulfillment of professional aspirations. In this context, the passion for knowledge and the emotional connection to science appear to be essential for the psychological and professional well-being of scientists. In this way, the boundaries between the scientific and the spiritual, the rational and the emotional, begin to blur, revealing that the pursuit of knowledge is not just an intellectual endeavor, but a deeply human one—rooted in both the mind and the heart. As we continue to explore the wonders of the universe through the lens of science and technology, we are reminded that the mysteries we uncover are not only of the world around us, but also of the very essence of what it means to be human, in a *sort of a sublime state of the soul, both aesthetic and scientific*.

In recent discussions of the relationship between science and aesthetics, Louis (2023) highlights the fascinating role of beauty in theoretical physics, suggesting it is a subject worthy of deeper exploration. Arcangeli and Dokic (2023), on their part, explore the role of aesthetic experiences in science, highlighting that scientists are often guided by sensory considerations when evaluating and constructing scientific theories and other artefacts. The authors propose expanding the scientific aesthetic debate by including the sublime, which, unlike beauty, evokes ambivalent feelings: both disturbing and enlightening. In this context, they argue that both beauty and sublimity can contribute to judgments of truth and understanding, with the sublime, in particular, being tied to relational experiences that foster deep judgments about the limits of human cognition.

The year 2019 marked the 75th anniversary of the publication of *What Is Life? The Physical Aspect of the Living Cell* by the great physicist Erwin Schrödinger (1887–1961), a brief book described by Roger Penrose in his preface to a reprint of this classic as "one of the most influential writings of the 20th century" (Phillips 2021). The book is based on a series of public lectures given by the distinguished physicist in February 1943, under the auspices of the Dublin Institute for Advanced Studies, where he was the director of theoretical physics at Trinity College, Dublin. In this text, the author, attempting to open the way to a new theoretical synthesis, reflects on how physics and chemistry can help us understand "the events in space and time that occur within the spatial boundaries of a living organism", calling for the advent of "other laws of physics". It is also worth noting that Schrödinger was one of the founders of *quantum mechanics*, and his reflections have often been praised for their influence on some of the giants who founded molecular biology, such as James Watson and Francis Crick (1916–2004), as well as their critical reception by others like Linus Pauling (1901–1994) and Max Perutz (1914–2002). *What Is Life?* aims to bring a fundamental question to the forefront of the scientific materialism debate, addressing the mysterious living world that encompasses and surrounds us. It reminds us how harmful it is to separate science and consciousness when confronting the great challenges that, in the study of living matter and humanity itself, remain unsolved by the science of the 21st century. The structures of natural organisms emerge hierarchically, scaling the abyssal depths of the space-time domains of matter, where atoms and particles dance the dance of life. The renewed study of Nature through Structural Biology, as well as Biomimetics and other emerging Digital Life Sciences, fosters theories that seek to identify the nature of the *sèmina rerum*, or the *seeds of things*, paraphrasing Lucretius' atomistic definition, retracing this process backward (Lucretius 1761). These disciplines immerse the human gaze into the very core of living matter, into tissues and cells, greatly amplifying our limited senses. It is precisely through these technological marvels that the eyes and minds of researchers expand, exploring the realms of the tiniest structures: from crystals to molecules, atoms, atomic nuclei, to being astonished by the fleeting dance of subatomic particles, and finally reaching the contemplation of the incredible and bewildering vastness of the *quantum vacuum*. Despite sudden political shifts and backslides in recent times, for the past few decades, companies, technologists, and scientists have shown a renewed interest in the structures of life and its technologies. In an era where sustainability is at the heart of global discussions, innovative approaches and reflections on green thinking are becoming increasingly relevant. As highlighted by Richardson (2023), *Green Thinking* represents a fundamental approach to addressing global environmental challenges, offering solutions ranging from education to integrated environmental policies. The economy and industry, for their part, are increasingly inclined to embrace design and management philosophies oriented towards this eco-thinking, envisioning futuristic scenarios for creating concrete and feasible conditions that facilitate the advent of a more sustainable human-made material universe, one that is more respectful of the logic of Nature and its ecosystems. This perspective is based on the tangible possibility of possessing a solid body of technological know-how, increasingly futuristic experimental methodologies, and the experiential maturity needed to bring about the advent of a new phase in human evolution. On the horizon, dark shadows seem to emerge, cast upon Earth by the dramatic advent of the Anthropocene. Yet, there appears a phase of virtuous design, capable of developing models of the artificial that can slow down and, in the medium-to-long term, decisively reverse the destructive and predatory course of the *Great Acceleration*—the exponential anthropological trend that has characterized the ongoing economic and productive escalation since 1750 in relation to the *Earth System Trend*s (Steffen et al. 2015). The green-oriented design is characterized by an extraordinary and unprecedented fusion of technoscientific creativity, bioinspired thinking, and computational power, and it is in this domain that ultimate CAD/CAE programs dedicated to biomimetic design have transformed and evolved (Duarte Poblete et al. 2025). A combination of factors is quickly propelling the entire Earth's economic, social, and ecological system into yet another exponential evolutionary leap, with ethical and ecological contours that remain underdeveloped, uncertain, but constantly evolving and transforming (Ethology Ecology & Evolution 2025). The sectors that make up today's techno-sciences are vast, intricate, and tightly interwoven. They cover every branch of engineering—from mechanics to integrated medicine—as well as material science, architecture, and the many facets of industrial design. All of these fields are being reshaped by a growing, highly specialized interest in biology and the life sciences. Researchers now look to the countless technological *patents* perfected by living organisms, and this bio-inspired focus is transforming the entire landscape of innovation (Euchner 2021). The viral surge of *biorevival*, spreading across the dynamic global networks, is not only reshaping scientific and technological discourse but also fueling fascinating socio-cultural reflections in the arts. This phenomenon is unprecedented in the last five centuries of human technological and industrial civilization. According to *BioArtLab* (n.d.), the laboratory is dedicated to the intersection of biology and art, fostering interdisciplinary research and creative exploration in *bioart*. This collaboration between scientific and artistic disciplines opens new possibilities for experimentation and understanding of life forms in art. According to the American National Standards Institute (ANSI 2024), the agents of this immense global revolution signal the emergence, within the current framework of Industry 4.0, of the more advanced and ecologically oriented production paradigm of Industry 5.0—at least in terms of the system's stated intentions. The rapid proliferation of these innovation catalysts is driven by the convergence of multiple, increasingly integrated factors through the application of innovative cooperative protocols. This evolution aligns with the principles of the circular economy (Rejeb et al. 2025). Among the most important *design enzymes* that stimulate the vitality of this complex contemporary digital-organic process, it can be argued that:

- The rapid evolution and widespread adoption of increasingly high-performance digital protocols across all design sectors, such as CAD/CAM, have progressively expanded to include CAE (Computer-Aided Engineering), CAO (Computer-Aided Optimization), CADD (Computer-Aided Drug Design), and ECAD (Electronic Computer-Aided Design). For a detailed overview of the developments and consequences in the evolution of CAD technology, it is recommended to consult Dergano (2024), who delves into the main advancements and implications of the field in the current context, as well as Scan2CAD (2017) and NobleCAD (2023), which examine the major historical developments and transformations in this industry domain.
- The assertion at a global level of four *V* fundamental pillars: *Variety*, *Veracity*, *Volume*, and *Velocity*, which form the essence of the analysis of so-called Big Data, the *cornucopia* of information that increasingly constitutes the thesaurus and the business core of current industrial, commercial, corporate, financial strategies, and technological innovations.
- The proliferation of a multifaceted digital ecosystem made up of a myriad of software, scientific applications, and other subfamilies of tools and computer hardware, within which algorithms operate that are increasingly fast, powerful, and user-friendly, designed to support operational methods and tools at the nanoscale. Cryo-electron microscopy (cryo-EM), electron crystallography and electron diffraction of microcrystals, or the emerging technology of lattice light sheet microscopy (LLSM), are instruments used in the laboratories of the most recent and advanced branches of scientific research that employ advanced computational approaches in the study of cellular structures, such as structural biology, molecular modeling, quantum chemistry, and bioinformatics (e.g., the 2024 Nobel Prize in Chemistry awarded to the developers of predictive and protein design systems managed by AI: AlphaFold® and Rosetta®). The ArchWiki's *List of applications/Science* page offers a comprehensive compilation of scientific software available on Arch Linux. It categorizes applications across various scientific disciplines, including physics, chemistry, biology, and engineering. This resource is particularly valuable for researchers and developers seeking open-source tools tailored to scientific computing and

analysis. For detailed information on Visual Molecular Dynamics (VMD), a powerful tool for molecular simulations, it is recommended to consult the Theoretical and Computational Biophysics Group at the University of Illinois at Urbana-Champaign (Theoretical and Computational Biophysics Group, n.d.). For comprehensive insights into structural biology and biomolecular modeling, OpenEye Scientific offers a suite of validated tools designed to enhance the preparation, search, and assessment of protein structural complexes. Their platform provides automated solutions for protein preparation, access to extensive protein structure databases, and advanced methods for comparing protein binding sites, making it a valuable resource for researchers in the field.

- The very recent, explosive, and virally globalized advent—and consequent intrusion and establishment as an operational standard in the digital-informatic system dedicated to design sectors—of futuristic AI (Artificial Intelligence) and AGI (Artificial General Intelligence) systems, along with related machine learning and deep learning protocols. These systems are shattering existing structures at breathtaking speed. They are reshaping the cognitive frameworks that have long guided design, prediction, data processing, and decision-making in industry, commerce, finance, and defense. Their explosive growth means that skills once thought exclusively human—research methods, hypothesis testing,
- Strategic problem-solving—are losing their privileged status. One fragment at a time, this know-how is being broken down and fed to voracious digital algorithms. Not to mention the intrusive and ominous *weaponization* of AI in the domain of human arts, including their most creative expressions like painting, music and movie making. For a quick yet comprehensive understanding of the distinctions and interrelations between artificial intelligence (AI), machine learning (ML), and deep learning (DL), it is recommended to consult Coursera's guide (Coursera Staff 2024). The article clarifies that *deep learning* is a subset of *machine learning*, which is itself a subset of AI. It highlights key differences—such as data requirements, training processes, and computational demands—offering valuable insights for anyone seeking to grasp the foundational concepts in these fields.
- Armies of ultra-powerful machines, concentrated in super servers, operate in the cloud day and night, processing colossal volumes of data at ever-more staggering speeds to assist humans in their intellectual and operational endeavors. These systems learn and assimilate knowledge in an almost compulsive, encyclopedic manner, imitating and *personally* reinterpreting every image, sound, voice, word, number, theory, and nuance of thought produced by humans and circulating online. At this rate of evolution, it is estimated that AGI systems could soon surpass the biological standard of human cognitive evolution in both speed and efficiency. In her article, Gordon (2023) explores the transformative potential of General AI, emphasizing its capacity to revolutionize various sectors. She discusses the ongoing advancements in AI research and development, highlighting the efforts to create more autonomous and adaptable systems. Gordon also addresses the implications of General AI on industries, economies, and societal structures, suggesting that its widespread adoption could lead to

significant changes in how businesses operate and how individuals interact with technology. This could lead to a radical upheaval, potentially compromising the very foundations of human knowledge and civilization. The socio-economic, ethical-cultural, and psycho-anthropological repercussions of such a shift are profoundly unsettling, keeping entire fields of scholars and experts awake at night as they grapple with the implications of this unprecedented transformation (Molek 2023).

- The rapid surge in studies, research, and experimental applications—collectively grouped under the broad term *bioinspired*—encompasses a diverse array of techno-scientific themes, methodologies, and characteristics. These include *Bionics*, which focuses on translating biological functions into technological solutions; *Biomimetics*, which aims to replicate nature's engineering principles to enhance human-made systems; and *Biomimicry*, which goes beyond mere imitation, advocating for a deeper integration of nature's strategies into sustainable design and innovation. This interdisciplinary approach is revolutionizing fields as varied as materials science, robotics, medicine, and architecture, fostering an unprecedented synergy between biology and technology (Iouguina et al. 2014). Biomimetics, notably, beyond being a structured and methodologized scientific discipline, is a mindset—an intrinsic attitude of biophilia—that draws inspiration from biological systems to develop innovative technological solutions (Garbuio et al. 2017). As Einstein famously stated: "Everything you can imagine, nature has already created." This approach allows the application of principles and mechanisms found in living organisms to human design, aiming to enhance performance, adaptability, and sustainability in products and processes. By observing and analyzing the evolutionary strategies adopted by nature, biomimetics studies the optimization of cellular structures, materials, biological functions, and adaptation systems in living organisms at various dimensional scales and levels of complexity. In recent years, this field has expanded exponentially thanks to advancements in high-resolution microscopy, programming languages, digital modeling, and 3D printing applied to life sciences. Biomimetics represents a convergence between biology, technology, and design: the former provides optimized models refined by evolution, while the latter offers the tools to replicate its principles and processes in artificial materials and systems. One of the main challenges of bio-inspired research is the deep understanding of natural systems, ranging from microscopic organisms to plants and complex animals. Since geometric representation and dynamic visualization of morphological and functional details are essential in this field, 3D digital modeling plays a crucial role by enabling high-resolution simulations and analyses. As a discipline that bridges biology and engineering, biomimetics integrates specialized knowledge with methodologies typical of computer-aided design and manufacturing (CAD/CAM). However, an interesting reverse approach also exists: although less common, the transfer of engineering concepts and models to biological systems can provide new insights into nature, leveraging advanced fabrication processes. This iterative exchange between biology and engineering not only fosters new scientific discoveries but also paves the way for groundbreaking technological innovations. The extraordinary diversity of characteristics found in nature offers an inexhaustible reservoir of solutions for creating materials with unique properties—already tested and perfected by evolution. For decades, scientists from various disciplines have studied how to replicate the behavior of natural materials and apply them to fields such as materials science, architecture, design, security, and medicine. Biomimetics is, therefore, not just a bridge between science and technology but a true revolution in the way we conceive innovation (Bar-Cohen 2006).

- The accelerated development driven by the explosion of maker culture and the advent of Industry 4.0 and Industry 5.0 has significantly advanced the verticalized evolution of digital manufacturing technologies based on CNC (Computerized Numerical Control) protocols, mechatronic technology, and measurement and control systems applied to machine tools, robotics, and next-generation 3D printing. These technologies span from the architectural scale down to the nanoscale, as seen in applications such as advanced medical robotics.

Asimer Group® is one of many examples of companies embracing the Fourth Industrial Revolution (Industry 4.0) to revolutionize traditional manufacturing processes. Through the integration of cutting-edge technologies such as automation, real-time data analytics, and IoT (Internet of Things), companies like Asimer® are leveraging CNC machining to create smarter, more efficient, and customizable production systems. In this context, Asimer® represents how the application of these technologies in industries such as automotive, aerospace, and energy can greatly enhance precision, reduce downtime, and improve operational efficiency (Asimer n.d.). This is just one of the numerous companies utilizing the advances of Industry 4.0 to drive innovation and maintain competitiveness in the rapidly evolving global market.

Most likely, alongside this list of *essentials*, it is necessary to include and consider many other emerging factors—economic, financial, cultural, social, political, anthropological, psychological—that equally contribute to stimulating and accelerating the processes of techno-scientific innovation. In the following chapters, these aspects will be further detailed, but the previous *magnificent seven* undoubtedly form the backbone that supports the development of the entire organism of the design of the new artificial world, with human-made scientific material and structural engineering (Conway 2023). A *plethora* of hypotheses is emerging from the natural vital source, contributing to the advent of a salvific industrial revolution capable of outlining the guidelines for a truly human-centered and sustainable development for the future of humanity and the planet Earth. This is demonstrated by the hundreds of studies and publications that are published and shared annually on the global stage, at an exponential rate. The scenario is one whose characteristic traits change and transform almost instantaneously before our eyes, leaving non-experts increasingly disoriented in trying to understand, assimilate, and internalize, on the scale of their own daily existential experience, such a vast amount of data, information, theories, innovations, and projects. The Pew Research Center report titled *Themes: The most harmful or menacing changes in digital life that are likely by 2035* (2023) presents insights from over 300 experts about the future of digital technologies. While some remain hopeful, a significant portion express concerns about key emerging threats:

- **Human-Centered Tech Is at Risk** – Technology may continue to prioritize profit and power over public interest, leading to increased surveillance, data exploitation, and social inequality.
- **Threats to Human Rights** – Privacy erosion, biometric tracking, deepfakes, and growing digital divides could undermine individual freedoms and democratic structures.
- **Erosion of Trusted Knowledge** – Misinformation and disinformation may dominate online spaces, weakening public institutions and making facts harder to verify.
- **Declining Mental and Physical Well-being** – Digital overexposure, AI-driven manipulation, and loss of autonomy could intensify anxiety, depression, and overall societal stress.

The report calls for ethical tech development, transparent governance, and digital literacy to counteract these risks.

The rapid obsolescence of information and the ever-evolving professional skills required to navigate continuous innovation processes, along with the real or potential applications emerging from contemporary techno-scientific research aimed at the new economic and industrial protocol, is so significant that it can be truly frustrating. This has been the central theme, the constant underlying rhythm that, from the creation of the editorial concept to its development, has accompanied me in my effort to encapsulate, in as comprehensive and organic a manner as possible, the complex, dynamic, and proliferating character of a technological era like our own—one that is nervous, fluid, unstable, exhilarating, utopian, prophetic, yet at times deeply dystopian and unsettling. It has truly felt like a ride on the rollercoaster of technological and scientific innovation, as well as the Research and Development efforts of universities, research centers, industrial districts, companies, factories, visionary minds, and emerging, spontaneous seeds of futurity that shape and drive the evolution of the global economy. However, a red thread of Ariadne, weaving through the warm and challenging points of this labyrinthine artificial *multiverse*, as this unfolding might be understood, seemed to unravel and offer me guidance, clearly signifying its meaning. The path it has traced offers a historical perspective in which human civilization is undergoing a profound transformation, marked by an intense struggle played out not only in the realm of fierce competition but also through the establishment of deep alliances between the realms of Nature and Human Artifice. In the late 1970s, *La Nouvelle Alliance. Métamorphose de la Science, was* published in Europe, a work translated into English as *Order out of chaos: Man's new dialogue with nature*, edited by Nobel Prize winner Ilya Prigogine (1917–2003) and Belgian philosopher Isabelle Stengers (Prigogine and Stengers 1984, Bertrand 2023). The work is hard to place in a specific genre, as it is neither a popular science book nor a classic text on the philosophy of science. It discusses scientific themes in a way that is accessible even to those without expertise, although a basic understanding of classical physics is helpful. However, the authors move beyond a mere scientific exposition and provide a philosophical reflection on the results of the physics and chemistry of the time. The *New Alliance* referred to in the text attempts to create a connection between the natural sciences and the humanities, inspired by breakthroughs in physics and the natural sciences that emerged during those years, such as thermodynamics, cybernetics, and chaos theory. This view suggests that the modern scientific approach—one that acknowledges the complexity, unpredictability, and *creativity* of natural systems (such as thermodynamic phenomena and complex systems)—can be connected to human freedom and self-determination, which lie at the heart of the humanities. Here, the *creativity* of Nature refers to the formation of new, complex structures through random fluctuations in systems far from equilibrium, while *human freedom* refers to the individual's capacity to make choices and act beyond the laws governing the observable world. The New Alliance, then, signifies a link between these two domains—scientific and human—but without confusing or merging them. In other words, it strives to integrate our understanding of complex natural processes with that of human freedom and culture, while respecting the distinct ontological nature of physical and human phenomena. Building on this, we found it worthwhile to revisit the idea of a New Alliance, a pact that could be extended and urgently signed between Nature and Artifice. Beyond simply analyzing and compiling specific sectors, this concept has been a constant guide in the editorial project and the present work. The goal has been, within the constraints of possibility, to extract from the intricate and multifaceted contemporary landscape—spanning economic, industrial, political, socio-political, military, cultural, psychological, and environmental aspects, all of which are as complex and layered as the dynamics of a rainforest ecosystem—the research trends most capable of advancing the development of a human-made system that respects the balance of our planet. The vast contemporary artificial system, brimming with technological diversity, exuberant, invasive, and ecologically devastating in many ways, is driven by a multitude of competitive corporate, research, industrial, and business entities, all involved in the relentless struggle for economic and financial survival (Forster 2012). The system that has shaped the governing laws of human material existence so far is one that is now showing signs of radical transformation. It is the legacy of an evolutionary approach that emerged from 18th-century utilitarian economic theories, later bolstered by biological theories developed between the First and Second Industrial Revolutions, such as the *struggle for life* and Evolutionism, which were ultimately transposed into 19th-century social Darwinism. The consequences of this aggressive, predatory model of development are now becoming a heavy burden for both human and natural planetary communities, yet it appears that no one is willing to settle the costs. While the term *ecology* was originally coined in relation to emerging Darwinism, with a meaning quite distinct from its contemporary use, the *New Ecology* that has developed since the post-World War II era has introduced alternative models for describing the dynamics of interactions among living organisms. This approach has brought new concepts to the scientific discussion, such as interconnection, cooperation, and symbiosis, all encapsulated within the overarching metaphor of the network (Watts et al. 2019):

" By *ecology* (Ökologie), we mean the science of the economy, the household of animal organisms. This must study the entirety of an animal's relationships with both its inorganic and organic environment, particularly the benign and hostile interactions with the plants and animals it comes into direct contact with; or, to be concise, all the intricate interrelations that Darwin calls the *struggle for existence* (Haeckel 1866).

These new interpretations of natural dynamics have inspired reflections and debates on the search for alternative models of human growth and development, such as those related to the productive theories of the so-called *Circular Economy*. As discussed in recent sustainable development models, the Circular Economy offers a transformative approach to resource management (Ellen MacArthur Foundation n.d.). It has been suggested (Tazzi 2017) that most biological design concepts can be directly applied to business management to improve efficiency, effectiveness, and sustainability. This is because living organisms, through the continuous refinements of evolution, serve as exceptional models for achieving significant improvements in the use of organizational resources. Even pedagogy and basic education, which aim to make known and understand the extraordinary logic with which Nature solves problems, have led, in just a few decades, to a greater general awareness, contributing to the birth of an environmental democracy. This is especially evident in the realm of energy decisions and their impact on the ecosystems of the affected areas. This increased awareness can lead to the development of innovative economic and production models, but also of social behavior and mental attitudes, that foresee more environmentally respectful processes, promoting, by ecological analogy, greater interconnectedness and cooperation between businesses, politics, and human communities. By now, various social sectors are calling for the implementation of feasible and effective alternative models for the production and consumption of material goods, capable of counterbalancing the dynamics of consumption, aggression, and rampant economic competition, in which the agents are inextricably bound by lethal neo-Darwinian relationships of predator-victim. According to Planet Labs PBC (2024), the Nature Tech Alliance aims to transform how companies measure and report on their biodiversity impact through advanced technological solutions. For these reasons, the emerging *green economy* preaches well that, in order to ensure the survival of humanity and the planet, the human artificial economic system must necessarily ally itself with the other vast system operating on the planet, that of the communities of natural organisms. It is increasingly acknowledged that humanity stands at a pivotal moment in history, where the need to measure, monitor, and evaluate biodiversity has become urgent. Technological advancements—ranging from satellite observation to artificial intelligence and large-scale data infrastructures—now make such efforts not only possible but actionable. The challenge ahead lies in fostering genuine collaboration across disciplines and tools to effectively drive this transformation. In this context, research by Italian plant neurobiologist Stefano Mancuso, internationally recognized for his contributions, draws attention to the long-term environmental impact of human activity. Since the dawn of agriculture—and most significantly over the past two centuries—approximately three trillion trees have been lost, effectively halving the planet's tree population. Furthermore, temperate primary forests that once extended across much of Europe have disappeared entirely, a striking indicator of the scale and permanence of ecological change. Moreover, Mancuso also argues that approximately four billion and four hundred million tons of concrete, the most extracted terrestrial resource after water and oil, are produced every year. In just a few decades, the human economy has generated such an enormous amount of concrete that, by 2020, the weight of the artificial *material body* created by humans surpassed the dry weight of all natural plant biomass. Only a century ago, this ratio stood at about one percent. At the root of these urgent considerations, possibilities are emerging for the advent of a new human material identity, as argued in the work aimed at creating a brilliant, sustainable, symbiotic artificial society (Calvano et al. 2025). It is an epochal project that must, however, accelerate its course, as we are all marching at full speed toward the brink of an environmental apocalypse of monumental proportions. To find effective remedies, an unprecedented metamorphosis of technoscience is underway—a profound revision that could radically transform the still dramatically contradictory and conflictual aspects that govern both human relations and, above all, the environmental interface between Nature and Artifice. A 360-degree, state-of-the-art overview of the design corpus that drives innovation has been compiled to capture the spirit of this intricate process, tracing the evolution of design-to-production frameworks in contemporary research and engineering of bio-inspired materials and technological solutions. The aim has been to make this radiography as organic and comprehensive as possible, capturing the continuous processes of transformation and evolution that unfold at an accelerated pace within the technological Research and Development system. It was intriguing to compare the ancient, disarming fascination evoked by the esoteric *cabinets* of the past. In those *Wunderkammern*, the sight of the strangest and most extraordinary artefacts (*mirabilia*) stirred the imagination of curious researchers and daring scholars. By contrast, a modern sense of estrangement now emerges when we witness the nanometric wonders made possible by computational design. As digital fabrication methods evolve and grow more accessible, the marvels once restricted to cutting-edge laboratories are now crossing into the visible, tangible realm of everyday human-scale needs. They are materially embodied in response to the functional *requirements* of the body, offering practical, context-sensitive solutions—for instance, designs that tackle the daily challenges faced by people with disabilities. A notable example is provided by *Modern Machine Shop* (2023), which recounts the work of Devin Hamilton and his Raw-RapAdapt FabLab/*WunderCAD/CAMmer*, where advanced fabrication technologies are harnessed to develop adaptive tools that enhance autonomy and improve quality of life. *Studioli*, *Wunderkammern* e *Cabinets de Curiosités*, have been fundamental institutions for the evolution of human civilization, from which the idea of the modern scientific museum developed. In these equipped spaces, where the challenge of understanding matter was welcomed and nurtured—proudly launched toward the unknown by the emerging experimental science, as well as by the early techniques and technologies—a modern mentality of the researcher was formed. The transition from the Renaissance *Wunderkammer* to the modern museum reflects a deeper cultural shift—from personal, often esoteric systems of categorization to institutionalized, standardized frameworks for organizing knowledge. As highlighted by Grinnell College (2013), early collectors—scholars, kings, and priests—sought to assemble fragments of the world into a "personally coherent whole," driven by curiosity and a desire to understand the unknown. Today, this urge is mirrored in digital environments where CAD/CAM systems and data-driven design protocols serve as new spaces where *naturalia* and *artificialia* converge, not in cabinets, but within design software, code, and fabrication labs. That sense of *mirabilis*, which over the centuries was overshadowed by the arrogance of positivist materialism, has returned in the form of a ghost from the past, evoked by the (virtual) visit to the *digital rooms* of today's most sophisticated *WunderCAD/CAMmers*. These ultratechnological laboratories use cutting-edge cryo-electron microscopy (cryo-EM) and advanced instrumental procedures to push the human eye to the thresholds of the infinitely small, revealing wonders such as the tiny molecular motor that has been tirelessly powering the flagellum of bacteria for billions of years. Graphic workstations equipped with the most advanced molecular software, including Relion®, CryoSarc®, Scipion®, Eman®, AlphaFold®, LightDock®, and others, allow researchers to process and visualize enormous flows of big data that describe the chemical-physical states of matter. These data can be manipulated and visualized in 3D through CAD suites such as Blender®, ChimeraX®, or PyMOL®. In these environments, images of nanometric biomechanical structures are meticulously processed, offering new insights into the intricacies of life at a molecular level. They explore, scan, disassemble, and reassemble at the atomic level, the astonishing biotechnologies that nature has devised to bring inorganic matter to life, and through CAM protocols, they send these to next-generation 3D printers, displaying functional models at the human perceptual scale. The eye and the lay mind are astonished in *Make Way for Wonder* at TEDMED 2020 in Boston, by contemplating the images produced by the digital wonder machines of bioimaging pioneers like Professor Gokul Upadhyayula, Biophysicist Faculty Scientist at Lawrence Berkeley National Laboratory, and scientific director at the Advanced BioImaging Center (ABC) at UC Berkeley. His work shows us how biological imaging of *life in action* and the microscopic universe inside each cell change our perception of Nature and the future possibilities of scientific challenges. Although the conventional CAD/CAM protocol is not explicitly involved in its digital bioimaging technology, this *mirabilis WunderCAD/CAMmer* represents experiments that interface with the broader field of computational systems. The Advanced Bioimaging Center (ABC) exemplifies the frontier where cutting-edge bioimaging merges with computational design. As highlighted in his institutional profile (Lawrence Berkeley National Laboratory n.d.), Upadhyayula's research focuses on developing next-generation optical systems and algorithms to visualize biological processes in real time and three dimensions. His contributions demonstrate how today's *Wunderkammer* of scientific tools includes not only rare artifacts, but dynamic, immersive windows into the microscopic world—transforming both biomedical research and our epistemological relationship with the invisible.

These systems, developed historically in computer graphics and digital rendering, are used for the extreme representation and control of digital 3D bio-modeling processes. The ABC is an international project that brings together advanced, pre-commercial imaging technologies, including the next-generation multifocal adaptive optical microscope. This initiative will allow imaging across a wide range of scales, both spatially and temporally, enabling the analysis of samples that can range from a few millimeters to several days in duration.

An evolution is underway that is rapidly transforming all the computer-assisted design protocols known so far from the ground up. CAD software, originally designed as *simple* digital drafting tools to support the design routines of mechanical draftsmen, civil engineers, and architects, has rapidly evolved, over the course of just a few decades, into highly sophisticated computational tools. Current algorithms are capable of producing and managing vast amounts of incredibly complex data flows in 3D, enabling the creation, simulation, and manipulation of intricate models that replicate real-world structures and systems in a highly interactive and dynamic environment. The greatest challenge faced by designers and researchers is the ability to visualize, analyze, and understand this vast amount of data in order to extract biologically and technologically significant information. Therefore, projects as innovative and visionary as ABC are not only to provide cutting-edge microscopy but also to offer human and hardware resources capable of handling projects at scales ranging from terabytes to petabytes and developing open-source computational workflows that allow biologists to easily extract meaningful data. These experiments leverage the powerful combination of artificial intelligence tools and state-of-the-art multimodal imaging systems to accelerate the pace of discoveries in both fundamental and translational sciences. This continuous evolution of digital microscopy tools, connected to bioimaging systems, where the manipulation suites of massive digital image data and molecular CAD play a fundamental role, is revolutionizing the systems of biological observation *in life*, allowing for an increasingly competent understanding of the fundamentals of life, such as cells and their complex components. Gokul refers to himself as a *cell's paparazzo*, offering a humorous, transgressive yet effective image of his role as a *celebrity cellular stalker*. Science and imagination seem to have no limits or boundaries in exploring the secrets of life (Upadhyayula 2024):

> "There is no other area of science that will allow you to observe the dynamic nature of life at the level of detail that bioimaging can offer, and it will allow you to see things that you have only been able to put together in your imagination."

The goal of these researchers, who literally *spy on cell*s in their *molecular private life*, is to develop technologies that allow them to observe while trying to preserve as much as possible their agitated, chaotic, random, and mysterious vital organic dynamics. What is interesting about Gokul's work, his team, and other researchers at the Howard Hughes Medical Institute in Virginia, working in collaboration with Boston Children's Hospital and Harvard Medical School, is the combination of technologies that produces remarkably sharp results for high-resolution imaging of cellular behaviors in real-time *in vivo* (Liu et al. 2018).

One key aspect of their approach is the transfer of solutions from contemporary telescopy to instrumental practices, as one of the core challenges in advanced microscopy lies in the use of light. Despite the enormous progress made by contemporary biosciences, many details of the overall functioning of the cell and its components, such as proteins, are still not fully understood. This lack of understanding obviously impacts sectors like drug design, where the creation of increasingly effective protocols for understanding and curing diseases, such as cancer or the thousands of other pathologies linked to deep organic dysfunctions of the human body, is now delegated to sophisticated computational algorithms and big data management suites. The dialogue between the ancient and new *rooms of wonder* has accompanied us throughout the creation of this text, in the search and exploration of the vanguards that are erasing the boundaries between science and technology, redrawing the very contours of the artificial universe, and charting new paths for the future of humanity, not just in the material sense.

It was believed that this type of technoscientific approach, blended with humanistic and literary language and references, which is somewhat unorthodox in the dissemination of specialized scientific knowledge, could offer an inspiring editorial proposal. At the same time, the aim was to provide a toolbox of fundamental knowledge, particularly for readers who may not be familiar with these specialized subjects, helping them navigate the complex and rapidly evolving landscape of contemporary technoscientific, economic, and industrial competition in the broader field of information technologies (IT), and specifically in the evolving areas of CAD/CAM (Chen et al. 2021b). Meanwhile, using arguments that go beyond the simple compilation of the lights and shadows of research and development realities that still unfold under the dictates of an industrial development model that remains firmly neodarwinist, the aim is to present examples of pure wonder between *biophilia* and *technophilia*. Certain initiatives and projects demonstrate how it is possible to restore an ancient sense of intimate, magical, and aesthetic engagement with the mysteries of the world to which we belong, regenerating what was once the most genuine, refined, and enlightening cultural perspective of European Humanism and the Renaissance. This approach not only seeks to bridge the gap between the past and the present but also explores how modern science, technology, and artistic expression can reinvigorate our connection with the mysteries of nature, offering a new understanding of our place in the world. Merging innovation and tradition, many of the projects analyzed embody a synthesis of exquisitely technoscientific intellectual curiosity and almost artistic-inspired exploration, capturing the essence of wonder and awe for discovery, aspects that sustain the very spirit of research. Through this fusion, we rediscover not only the complexity of scientific inquiry but also the profound beauty that accompanies it, inviting both mind and spirit to engage with the universe in a more holistic and deeply resonant way (Meadow 2022). A refined and enlightening cultural perspective inspired the brightest minds of artists, philosophers, and scientists of the era, guiding them in the creation of a new Terrestrial Utopia founded on Harmony and Beauty. Although history has often produced realities in contrast to these ideals, Humanism and the Renaissance had the merit of attempting an unprecedented integration of Aesthetics, Technique, Science, and Nature. This synthesis laid the foundations of modern technoscientific and artistic thought, and despite its contradictions, it reintroduced to the artificial world the thirst for knowledge, existential doubt, and the most genuine aspirations of *Homo Faber*—archaic, archetypal aspirations of craft intelligence. Through the technological mediation of CAD/CAM, these aspirations are evolving into a new synthesis: HOMO ROBO FABER. This project, developed within the Master in Robotics and Advanced Construction (MRAC) at IAAC, explores the interaction between traditional craftsmanship and advanced robotics. The research team divided the craft process into three main phases: *Think, Sense* and *Do*, aiming to transfer the tacit knowledge of the craftsman to a robotic arm. Using motion sensors and force detectors, the researchers analyzed the movements and forces applied by the craftsman during the creative process. The collected data was used to train a machine learning model, which enabled the robot to replicate and interpret the craftsman's actions in real-time. This approach aims to preserve and transmit craft knowledge, integrating human intelligence with the precision and production capabilities of robotics (Chen et al. 2021a).

The figure of the man who not only creates objects but also shapes destinies is one that is, in a sense, reawakened in the personality and technical mindset of the contemporary maker within FabLabs. These global institutions have become the true digital ateliers of a technological neotech-Renaissance, which has permeated the daily lives of millions, spreading the constructive philosophy that bridges the digital bit to the physical atom, historically inspired by the ideas of Neil Gershenfeld and the CBA at MIT. The *Homo Faber* is a gigantic figure of our culture, shaped in the heart of the new Era, and central to all subsequent Industrial Revolutions from the 18th century onward. This technical mythology occupies almost the entirety of the psychological development of Western thought, and today brings together the community of the diverse CAD/CAM universe, bioimaging, structural biology, molecular biology, and bioinspired design. Within it converge the aspirations of a humanity that dares to recreate and manipulate the energy and matter of the world through its ingenuity, labor, and technology. In this figure, with Luciferian traits and deeply rooted in archaic myths, lies the very essence of the desire to challenge and dominate nature and its mysteries—the psychology of a man-magician who acts within the Universe, refusing to limit himself to mere passive contemplation (Philosophy Kitchen 2024). In this image, we recognize the renewed symbol of a humanity that refuses to merely observe, to spy on reality through the narrow keyholes of ideological and metaphysical constraints that, for centuries, have hindered its thirst for knowledge from taking shape. It is the forceful sign of a new psychology, of an epochal challenge; of a mind that, above all, seeks to act, modify, design, and shape the world in its own image and likeness—a true rebellion against the Father, against universally recognized Authority, as exemplified by the case of Galileo Galilei. Within the context of cultural evolution, *Homo Faber* embodies the deep-seated will to transcend the very limits of the nature that created it, in order to generate an anthropized, civilized, technological, and artificial space. On one hand, this figure celebrates the Greek *technè*—art, technique, and craftsmanship. On the other, it psychologically represents a constant tension between the aspiration for knowledge and the risk of losing touch with the sense of sacredness and mystery that the Universe embodies. The trajectory that extends from the ashes of medieval mysticism, through five epic Industrial Revolutions, has now reached the shores of the Anthropocene—the era of the rise and global affirmation of *Homo Digitalis* (Geerts 2022). The contemporary also contemplates the arrival of a disturbing yet stimulating metamorphic transcendence of *Homo Faber* towards a *transhumanism* thought that celebrates the unsettling traits of an *augmented humanity*, pushing the limits of its traditional and evolutionary biological and inventive potential (Saracco 2019).

Today, these ideals continue to resonate in the most advanced laboratories and design studios, where art, science, and technology merge to redefine our relationship with both the artificial and natural worlds. This is not just a conceptual legacy but an active paradigm in cutting-edge projects such as those by Soma Bioworks® (Soma Biowork 2024) and Festo® (Festo n.d), where bionics and biomimetics drive the development of innovative solutions, from robotic engineering inspired by bird flight to industrial systems replicating plant strategies. Similarly, studios like Zaha Hadid Architects and Neri Oxman integrate generative algorithms and bio-inspired materials in a continuous dialogue between the artificial and the natural, demonstrating how aesthetics and functionality can coexist in an organic balance.In the biomedical field, the Centre for Computational Imaging and Simulation Technologies in Biomedicine (CISTIB) at the School of Computing & School of Medicine - University of Leeds– focuses on computational imaging and modeling with applications in biomedical research (CISTIB, n.d.). This center merges materials science, advanced imaging, and computational modeling to uncover hidden patterns in biological and evolutionary processes. Its work is centered on developing multiparadigm diagnostic methods, supporting medical intervention planning and guidance, optimizing devices and implants through image-based simulations, enabling high-throughput biomedical data analysis for pharmaceutical applications, and enhancing biomedical information management. One of its key technologies is GIMIAS (Graphical Interface for Medical Image Analysis and Simulation), an open-source platform that integrates tools for medical imaging, computational modeling, and computer graphics. This platform facilitates the development of clinical prototypes and accelerates the transfer of technology to clinical and industrial contexts. CISTIB also emphasizes Precision Imaging, an approach that combines mechanistic and phenomenological modeling to build descriptive, predictive, and integrative representations of biological systems. This enhances our understanding of complex physiological dynamics and fosters innovation in personalized medicine (CISTIB n.d.). Meanwhile, companies like Organovo® are developing 3D-printed biological tissues, demonstrating that the synergy between technology and biology can lead to solutions that are not only functional but also aesthetically and harmonically refined (Flaherty 2012, Organovo Holdings Inc. 2025). Even art is exploring these territories, as exemplified by the works of Refik Anadol, who transforms data streams into immersive visual experiences. This proves that artificial intelligence can serve as an expressive tool capable of shaping a new digital aesthetic. These projects are not merely technological innovations; they are fragments of a renewed Wunderkammer—a modern cabinet of curiosities where the pale ghost of Renaissance collections comes back to life in new forms. The boundaries between Art, Science, and Technology dissolve, creating a new space for exploration where the sense of wonder becomes the driving force of knowledge. Just as in Renaissance thinking, the desire to understand and shape the world is not only an exercise in technical skill but a profound inquiry into our very nature. In an era of artificial intelligence, advanced digital simulations, and programmable biotech materials, our connection with life is redefined through Utopia and Innovation, revealing the ancient and mysterious dialogue between humanity and the organic world (Martinuzzi 2020). Although the historical course of human events has unfolded in ways that have often contradicted these ideologies,

Humanism and the Renaissance undeniably had the value of attempting to integrate, within an unprecedented human project, Aesthetics, Technique, Science, and Nature. In doing so, they laid the foundations of modern technoscientific, artistic, and design thought, both in its positive and negative aspects. Despite political and social failures and contradictions, the Renaissance Project had the historical merit of placing at the center of the artificial world the renewed intentions, thirst for knowledge, doubt, and the most genuine and archetypal existential anxiety of Homo Faber (Çalışkan 2024).

The unlocking of knowledge potential through naturalistic and anatomical analysis and drawing, combined with the birth and development of optical microscopy, allowed researchers and new scientists to push the limits of the human perceptive horizon and embark on that extraordinary, fantastical journey inside the structure of infinitesimal matter, a journey that continues today, enormously amplified by extraordinary nanometric-scale exploratory instruments and refined digital and computational methodologies. That time also produced a sort of nostalgic *spleen*,[7] enchanting us still today with the harmony and incomparable aesthetics of its architecture and art, which emanate the charm of an ideal of beauty and harmony, perhaps now unattainable and lost forever. It is precisely by allowing ourselves to be fascinated by the ghosts that still wander through the halls of science and by the intellectual and refined atmospheres of that era, which permeated the early study rooms of the Italian Humanist princes and intellectuals and the ancient, bizarre collections of Renaissance Wunderkammern, that this cultural analogy is proposed through the pages of this book. In doing so, it casts light upon and heralds the dawn of the extraordinary technoscientific wonders of the contemporary and near digital future. The *WunderCAD/CAMmers* of our time are ultratechnological laboratories where unprecedented design possibilities are brought to life, which, within a few decades, could catapult humanity and its planet into a future shaped by a revolutionary neo-biodigital artificial ecosystem. Human actions can create worlds of unprecedented and never-before-seen possibilities, yet they also show how, in a very short time, the same actions can bring life on Earth to the brink of annihilation and self-destruction as a biological species, triggering the so-called sixth mass extinction. The advent and evolutionary affirmation of *Homo Digitalis* has not been painless for the entire terrestrial ecosystem. Its feverish material activity, concentrated in just a few decades and marked by the term Anthropocene, has caused and continues to produce epochal environmental damage. It has brought with it fundamental questions and issues regarding the relationship that must exist between the limits of human creative power and the resulting ethical, social, and ecological responsibilities (Ellis 2024). The ability to manipulate matter and life is reaching levels never before attained, and with

[7] The term *spleen* (from the Greek *σπλήν* "spleen," whose black bile, according to Hippocratic medicine, caused hypochondria) refers to a state of mind characterized by melancholy, dissatisfaction, boredom, and irritability with everything, without a precise cause. In literary and philosophical contexts, *spleen* refers not just to the anatomical organ, but to a deep, melancholic state of mind often associated with existential ennui, dissatisfaction, and creative introspection. The term became especially prominent in 19th-century Romantic and Symbolist literature, notably in the poetry of Charles Baudelaire, who used *spleen* to evoke a sense of inner turmoil, weariness with the world, and the burden of modern life. Far from simple sadness, *spleen* embodies a complex emotional and intellectual malaise—at once depressive, reflective, and artistically fertile.

it comes the responsibility regarding the limits of human technological intervention (Team EMB 2024). The modern technoscientific challenge is taking place on unstable ground. The act of creation is crossing into domination, with the risk that technique and artifice may replace nature, sacrificing something essential for humanity and the entire planetary ecological system. History stands in a precarious balance, as never before. The choices affecting billions of human beings are now determined by a delicate balance between two opposing directions. On one side lies the pursuit of a possible equilibrium, attempting to mediate the omnipotent and blind transformative power of globalized economic and financial strategies pursued at any cost; on the other stands a growing awareness of the consequences of our actions within the delicate—and in many respects still unknown—ecosystemic balances, ideally rediscovering the value of all those mysterious and elusive aspects of reality that cannot be fully understood or dominated by rationality, technique, and materialistic economy. An intellectual gaze cast toward the past guides this search for historical inspiration. Yet our feet remain firmly planted in the fertile soil of the present. From that stance, we seek the keys that might unlock the digital cabinets of wonder awaiting us in the near future. The reflection is complex. It concerns the bonds linking creation, Technique, Nature, and Humanity. Its urgency grows as biomimetics, artificial intelligence, and digital design converge. In this hybrid terrain, the boundary between the natural and the artificial becomes ever more intricate—and ever more subtle. The present-day *Digital Homo Faber* has not only become a powerful builder of synthetic worlds—often artificial paradises[8]—but also an acute interpreter of complex systems: a kind of *digital alchemist* who, through CAD/CAM and bioengineering, seeks to merge the precision of technique with the freedom of imaginative creativity.

Besides the current introduction, the volume opens with a clear structure: ten chapters that provide a 360-degree view of emerging trends in CAD/CAM and industrial bio-design, weaving together theory, case studies, and design insights. Each section begins with the question "*How to make* ..."—a deliberate choice that immediately turns concepts into operational tools, reveals concrete impacts on today's production chains, and proposes solutions that can be replicated in different contexts. Alongside this pragmatic yet broadly cultural approach, every chapter looks beyond the present horizon, outlining the key trends that will shape artefact production in the near future—scenarios of futurability that consider the evolution of bio-inspired materials, the integration of artificial intelligence and digital twins, the logics of circular sustainability, and the prospects of distributed manufacturing. In this way, the *compendium* spans the most significant idea-to-prototype-to-market cycles, offering readers not only a snapshot of the state of the art in bio-inspired materials science but also a compass for navigating the innovations to come:

- On the prospects of hybridization between Nature and Artifice in the effort to create a new Material Ecology and new generations of ELMs (Engineered Living Materials) (1.0);

---

8 The word *paradise* comes from the ancient Persian *piridaeza*, meaning garden or park. Translated by the Greeks, it became *paradeisos*, and thus the Garden of Eden was renamed in the Septuagint Bible, which is the Greek translation from Hebrew made in the 3rd century BC.

- On design ideas aimed at creating new materials capable of combining high-tech performance with maximum eco-compatibility (2.0);
- On the biomimetics creation of lighter and more resistant materials capable of overcoming the so-called *gravitational paradox* (3.0);
- On the possibilities of creating self-cleaning materials inspired by the properties of the purity and cleaning in Nature (4.0);
- On efforts to create increasingly smart materials, objects, and cybernetic systems endowed with movement or decision-making autonomy (5.0);
- On the possibilities of design in creating ever greater connectivity (6.0);
- On the design of smaller and more performant materials that challenge and compete with the nanometric technologies of living matter (7.0);
- On the creation of *wild materials* capable of simulating increasingly organic morphological and functional characteristics, inspired by extraordinary solutions found in natural technology (8.0);
- On the advent and effects of new *esoteric* digital oracles born from Deep Learning, now capable of massively amplifying the predictive capabilities of digital algorithms (9.0);
- On frontier research that is imagining and creating science-fiction systems and materials made possible by the evolution of CAD/CAM connected with AI and Quantum Design (10.0).

References

American National Standards Institute (ANSI) (2024). The Evolution of Industry 5.0: What's to Come. Available at: https://blog.ansi.org/the-evolution-of-industry-5-0-whats-to-come/ [Accessed: 7 May 025].

Arcangeli, M., and Dokic, J. (2023). A plea for the sublime in science. Published in The Aesthetics of Science: Beauty, Imagination and Understanding, eds. S. French & M. Ivanova, 104–124, Routledge. Available at: [link] [Accessed: 7 May 2025].

Asimer Group (n.d.). Home. Available at: https://www.asimergroup.com/en/home-en/ [Accessed: 21May 2025].

Bar-Cohen, Y. (2006). Biomimetics – Using nature to inspire human innovation. Bioinspiration & Biomimetics, 1(1): 1–12.

Barsanti, G. (2003). L'Ottocento: biologia. Dal Settecento all'Ottocento: la nascita della biologia come scienza autonoma. Treccani. Available at: https://www.treccani.it/enciclopedia/l-ottocento-biologia-dal-settecento-all-ottocento-la-nascita-della-biologia-come-scienza-autonoma_(Storia-della-Scienza)/ [Accessed: 7 May 2025].

Beekman, W., and Jochemsen, H. (2022). The Kantian account of mechanical explanation of natural ends in eighteenth and nineteenth century biology. History and Philosophy of the Life Sciences, 44(1): 10.

Bertrand, E. (2023). A controversy about chance and the origins of life: Thermodynamicist Ilya Prigogine replies to molecular biologist Jacques Monod. History and Philosophy of the Life Sciences, 45.

BioArtLab (n.d.). Laboratory. Available at: https://bioartlab.com/laboratory [Accessed: 7 May 2025].

Buzard, K. (2013). The First Museums: Kunst-und Wunderkammer. Travel To Eat. Available at: https://traveltoeat.com/the-first-museums-kunst-und-wunderkammer/ [Accessed: 7 May 2025].

Çalışkan, A. S. (2024). Homo Faber and Homo Economicus in the Scientific Revolution (136 pages, 20 b/w illustrations). Routledge.

Calvano, M., Curci, A., Desolda, G., Esposito, A., Lanzilotti, R., and Piccinno, A. (2025). Building symbiotic AI: Reviewing the AI Act for a human-centred, principle-based framework. ArXiv.

Cartwright, M. (2022). The Microscope & the Scientific Revolution. World History Encyclopedia. Available at: https://www.worldhistory.org/article/2271/the-microscope--the-scientific-revolution/ [Accessed: 7 May 2025].

Chen, C., Abelson, S., Dubor, A., Sollazzo, A., Garivani, S., Foroughi, A. et al. (2021a). HOMO ROBO FABER. Institute for Advanced Architecture of Catalonia (IAAC). Available at: https://www.iaacblog.com/programs/homo-robo-faber/ [Accessed: 8 May 2025].

Chen, T., Cheng, H. K., Jin, Y., Li, S., and Qiu, L. (2021b). Impact of competition on innovations of IT industry: An empirical investigation. Journal of Management Information Systems, 38(3): 647–666.

CISTIB (n.d.). Software. Centre for Computational Imaging & Simulation Technologies in Biomedicine. Available at: https://www.cistib.org/software/en/software [Accessed: 8 May 2025].

Conway, E. (2023). Material world: A substantial story of our past and future. WH Allen.

Corbellini, G. (2004). La seconda rivoluzione scientifica: introduzione. Le specificità della materia vivente. Treccani. Available at: https://www.treccani.it/enciclopedia/la-seconda-rivoluzione-scientifica-introduzione-le-specificita-della-materia-vivente_(Storia-della-Scienza)/ [Accessed: 7 May 2025].

Coursera Staff. (2024). How Does AI Work? Basics to Know. Coursera. Available at: https://www.coursera.org/articles/how-does-ai-work [Accessed: 21 May 2025].

Dassault Systèmes. (n.d.). Parametric Modeling. 3DS Store. Available at: https://www.3ds.com/store/cad/parametric-modeling [Accessed: 07 May 2025].

Davies, P. (2019). Life's secret ingredient: a radical theory of what makes things alive. New Scientist, 241(3215). Available at: https://www.newscientist.com/article/mg24132150-100-lifes-secret-ingredient-a-radical-theory-of-what-makes-things-alive/ [Accessed: 7 May 2025].

Dergano, F. (2024). The Evolution of CAD Technology in 2024: Advances and Consequences. Medium. Available at: https://raccomandino.medium.com/the-evolution-of-cad-technology-in-2024-advances-and-consequences-34a12287f0c1 [Accessed: 7 May 2025].

Duarte Poblete, S. S., Romani, A., and Rognoli, V. (2024). Emerging materials for transition: A taxonomy proposal from a design perspective. Sustainable Futures.

Ellen MacArthur Foundation. (n.d.). Home. Available at: https://www.ellenmacarthurfoundation.org/ [Accessed: 21 May 2025].

Ellis, E. C. (2024, March 5). The Anthropocene is not an epoch, but the age of humans is most definitely underway. The Conversation. Available at: https://theconversation.com/the-anthropocene-is-not-an-epoch-but-the-age-of-humans-is-most-definitely-underway-224495 [Accessed: 8 May 2025].

Ethology Ecology and Evolution. (2025). Ethology Ecology & Evolution: Aims and Scope. Available at: https://www.tandfonline.com/journals/teee20/about-this-journal#aims-and-scope [Accessed: 7 May 2025].

Euchner, J. (2021). Is the evolution of technology inevitable? Research-Technology Management. 64(4): 11–13.

Festo. (n.d.). Bionic Projects by Festo. Available at: https://stem.festo.com/us/en/our-approach/bionics/bionic-projects-by-festo/index.html [Accessed: 8 May 2025].

Flaherty, J. (2012). Wired. Autodesk Developing CAD Software to Design, 3-D Print Living Tissue. Available: https://www.wired.com/2012/12/autodesk-3-d-print-tissue/ Accessed: May 27: 2025.

Forster, B. A., Forster, J. D., and Forster, K. E. (2012). The struggle for human and economic development in sub-Saharan Africa. North American Business Press.

Galambosova, C. (2024). What Is a Wunderkammer? Best Cabinets of Curiosities. DailyArt Magazine, 23 September. Available at: https://www.dailyartmagazine.com/cabinets-of-curiosities/ [Accessed: 7 May 2025].

Geerts, T. (2022). Homo Digitalis: How Digitalisation Is Making Us More Human. Lannoo Publishers.

Giallombardo, F. (2013). I Palazzi enciclopedici: le passioni della conoscenza e le forme del possesso. roots§routes. Available at: https://www.roots-routes.org/speciale-55-biennale-di-venezia-i-palazzi-enciclopedici-le-passioni-della-conoscenza-e-le-forme-del-possesso-di-floriana-giallombardo/ [Accessed: 7 May 2025].

Gordon, C. (2023, September 30). How General AI Will Eventually Reshape Everything. Forbes. Available at: https://www.forbes.com/sites/cindygordon/2023/09/30/how-general-ai-will-eventually-reshape-everything/ [Accessed: May 8, 2025].

Grinnell College. (2013, October 1). From Wunderkammer to the Modern Museum. Grinnell College. Available at: https://www.grinnell.edu/news/wunderkammer-modern-museum [Accessed: 8 May 2025].

Haeckel, E. (1866). Generelle Morphologie der Organismen: allgemeine Grundzüge der organischen Formen-Wissenschaft, mechanisch begründet durch die von Charles Darwin reformirte Descendenz-Theorie. Internet Archive. Available at: https://archive.org/details/b22651007_0001/page/n17/mode/2up [Accessed: 11 May 2025].

Homo Faber. (2024) Homo Faber. Michelangelo Foundation for Creativity and Craftsmanship. Available at: https://www.homofaber.com/ [Accessed: 8 May 2025].

Iouguina, A., Dawson, J., Hallgrimsson, B., and Smart, G. (2014). Biologically informed disciplines: A comparative analysis of bionics, biomimetics, biomimicry, and bio-inspiration among others. International Journal of Design & Nature and Ecodynamics, 9(3): 197–205.

Liu, T. L., Upadhyayula, S., Milkie, D. E., Singh, V., Wang, K., Swinburne, I. A. et al. (2018). Observing the cell in its native state: Imaging subcellular dynamics in multicellular organisms. Science, 360, eaaq1392.

Logan, L. (2016). Early microscopes revealed a new world of tiny living things. Smithsonian Magazine. Available at: https://www.smithsonianmag.com/science-nature/early-microscopes-revealed-new-world-tiny-living-things-180958912/ [Accessed: 7 May 2025].

Louis, A. (2023). Beauty and the sublime in physics. Theos Think Tank. Available at: https://www.theosthinktank.co.uk/comment/2023/02/02/beauty-and-the-sublime-in-physics [Accessed: 7 May 2025].

Lucretius Carus, T. (1761). De rerum natura. Translated by A. Marchetti. Internet Archive. Available at: https://archive.org/details/bim_eighteenth-century_de-rerum-natura-italia_lucretius-carus-titus_1761_1 [Accessed: 11 May 2025].

Lumini, M. (2018). About Biomorphic Exuberance and Digital Rococo in Design and Parametric Contemporary Architecture. *In*: Pau, F., and Vargiu, L. (eds.). Following Forms, Following Functions: Practices and Disciplines in Dialogue (pp. 123-149). Newcastle upon Tyne, UK: Cambridge Scholars Publishing.

Lumini, M. (2021). The Invisible Factories: Nature's technologies and design of artificial innovation. Cuadernos del Centro de Estudios en Diseño y Comunicación, (149): 145–166.

McCall, C., Hildebrandt, L. K., Hartmann, R., Baczkowski, B. M., and Singer, T. (2016). Introducing the Wunderkammer as a tool for emotion research: Unconstrained gaze and movement patterns in three emotionally evocative virtual worlds. Computers in Human Behavior, 59: 93–107.

McNeill, J. R., and Engelke, P. (2016). The Great Acceleration: An Environmental History of the Anthropocene Since 1945. Harvard University Press.

Mancuso, S. (2019). The Nation of Plants: A History of a Silent Revolution. The Overlook Press.

Martinuzzi, C. (2020). Utopia in Renaissance Philosophy. *In*: Sgarbi, M. (ed.). Encyclopedia of Renaissance Philosophy. Springer.

Max Delbrück Center (2016, 4 September). Who was Max Delbrück? Max Delbrück Center for Molecular Medicine. Available at: https://www.mdc-berlin.de/news/news/who-was-max-delbruck [Accessed: 11 May 2025].

Meadow, M. A. (2022). Kunstkammer. In Sgarbi, M. (Ed.), Encyclopedia of Renaissance Philosophy (pp. 1–2). Springer.

Modern Machine Shop (2023, 16 October). How Integrated CAD/CAM Transforms Inventions into Products. Modern Machine Shop. Available at: https://www.mmsonline.com/articles/how-integrated-cadcam-transforms-inventions-into-products [Accessed: 8 May 2025].

Molek, N. (2023). AI and organizational transformation: Anthropological insights into higher education. Izzivi prihodnosti, 8: 148–177.

Nature Research Custom Media & The Catholic University of America (2023). Wonder at work: The joy of discovery helps scientists flourish. Available at: https://www.nature.com/articles/d42473-023-00357-5 [Accessed: 7 May 2025].

Organovo Holdings Inc. (2025). Home. Organovo. Available at: https://organovo.com/ [Accessed: 8 May 2025].

Phillips, R. (2021, June 16). Schrödinger's What Is Life? at 75. Cell Systems, 12(6): 465–476. Elsevier.
Philosophy Kitchen. (2024). Dietro le quinte della magia rinascimentale: un profilo del mago erudito. Available at: https://philosophykitchen.com/2024/10/dietro-le-quinte-della-magia-rinascimentale-un-profilo-del-mago/ [Accessed: 8 May 2025].
Planet Labs PBC (2024, 14 May). The Nature Tech Alliance: Planet, ERM, Salesforce, and NatureMetrics launch initiative to measure biodiversity and report on impact. Planet Labs. Available at: https://www.planet.com/pulse/the-nature-tech-alliance-planet-erm-salesforce-and-naturemetric-launch-initiative-to-measure-biodiversity-and-report-on-impact/ [Accessed: 8 May 2025].
Pouyan, N. (2016). A voyage to beyond the human eye by microscope, Leeuwenhoek's invention. OALib Journal, 3: 1–9.
Prigogine, I., and Stengers, I. (1984). Order Out of Chaos: Man's New Dialogue with Nature. Internet Archive. Available at: https://archive.org/details/orderoutofchaosm00prig/page/n5/mode/2up [Accessed: 8 May 2025].
Rejeb, A., Rejeb, K., Keogh, J. G., and others. (2025). When Industry 5.0 meets the circular economy: A systematic literature review. Circular Economy and Sustainability.
Richardson, L. (2023). Green Thinking. The ACU Review. Available at: https://www.acu.ac.uk/the-acu-review/green-thinking/ [Accessed: 7 May 2025].
Saracco, R. (2019, June). Augmented machines and augmented humans: Converging on transhumanism (ID: 195293602). IEEE Digital Reality Initiative.
Scan2CAD. (2024, January 12). How CAD Has Evolved Since 1982. Available at: https://www.scan2cad.com/blog/cad/cad-evolved-since-1982/ [Accessed: 8 May 2025].
Soma Bioworks. (2024). Soma Bioworks – Sustainable biotech solutions. Available at: https://www.soma.eco/ [Accessed: 8 May 2025].
Steffen, W., Broadgate, W., Deutsch, L., Gaffney, O., and Ludwig, C. (2015). The trajectory of the Anthropocene: The Great Acceleration. The Anthropocene Review, 2(1): 81–98.
Tazzi, F. (2017). Biomimicry in organizations: Business management inspired by nature. CreateSpace Independent Publishing Platform.
Team EMB. (2024, February 9). Exploring the Limits of Human Enhancement Technology. EMB Global Blog. Last updated: March 23, 2024. Available at: https://blog.emb.global/limits-of-human-enhancement-technology/ [Accessed: 8 May 2025].
Upadhyayula, G. (n.d.). Gokul Upadhyayula. Biosciences Area – Lawrence Berkeley National Laboratory. Available at: https://biosciences.lbl.gov/profiles/gokul-upadhyayula/ [Accessed: 11 May 2025].
Upadhyayula, G. (2020, October). The life unfolding inside your cells, revealed in 3D. TEDMED. Available at: https://www.ted.com/talks/gokul_upadhyayula_the_life_unfolding_inside_your_cells_revealed_in_3d [Accessed: 8 May 2025].
Upadhyayula, G. (2024). Cell scientist to watch – Gokul Upadhyayula. Journal of Cell Science, 137(20), jcs263549.
Vaiani, B. (2022, December 7). Wunderkammer: Il luogo della meraviglia. Barnebys Magazine. Available at: https://www.barnebys.it/blog/wunderkammer-il-luogo-della-meraviglia [Accessed: 15 September 2023].
Villanueva-Hernández, P. J. (2023). Quantum vacuum: Fundamentals and frontiers. TechRxiv.
Vuylsteke, B., Detand, J., and Ostuzzi, F. (2023). Exploring the intersection of Design and Biology in today's industry. IOP Conference Series: Earth and Environmental Science, 1256, 012004.
Watts, E., Hossfeld, U., and Levit, G. (2019). Ecology and evolution: Haeckel's Darwinian paradigm. Trends in Ecology & Evolution, 34(9): 681–683.

APPENDIX

These two tables offer a concise ***CADdendum***, a clear overview of the core functions, inputs, outputs, software, and user roles for CAD, CAE, CAO, and CAM, along with a matrix illustrating the data flows and feedback loops among these four disciplines. Since current CAD systems have undergone an evolutionary leap—from simple electronic drafting tools to powerful parametric, predictive, and generative design platforms that form interconnected families of computational tools now driving advances in structural biology, biomedicine, genetic engineering, and bio-inspired materials science—they should be seen as a viral family in transit poised for further extraordinary transformations through AI, deep learning, and quantum computing.

Table 1. Cad & Co. Key Characteristics.

Acronym	Full Name	Main Function	Typical Input	Typical Output	Example Software / Tech	Primary Users
CAD	Computer-Aided Design	2D/3D creation and modeling of parts and assemblies	Project specs, sketches	Geometric-parametric models (e.g. .dwg, .step)	AutoCAD® CATIA® Fusion 360® Rhino® SolidWorks®	Designers, mechanical engineers
CAE	Computer-Aided Engineering	Performance analysis (structural, thermal, dynamic)	CAD model, boundary conditions	Simulation results (stress maps, deformations)	ANSYS® Abaqus® Nastran®	Simulation engineers, FEM analysts
CAO	Computer-Aided Optimization	Topology and parametric optimization of the design	CAE results, constraints (weight, strength)	Optimized, parameterized geometries	Altair OptiStruct® nTopology®	Optimization engineers
CAM	Computer-Aided Manufacturing	Toolpath generation and CNC programming	CAD/CAE/CAO model, machining strategies	G-code or post-processor output	Mastercam® PowerMill® Fusion 360®	Machinists, CNC programmers
CADD	Computer-Aided Drug Design	*In silico* discovery and optimization of small-molecule leads	Target structures, ligand libraries, bio-data	Libraries of bio data, predicted binding affinities, and optimized ligand designs	Schrödinger Suite® AutoDock® MOE®	AutoDock, MOEComputational chemists, medicinal chemists
AI	Artificial Intelligence	Data-driven design, generative modeling, predictive analytics	CAD/CAE/CAO data, sensor streams, historical DBs	Generative designs, anomaly detection, performance forecasts	Autodesk Generative Design® Siemens NX AI® TensorFlow®	Data scientists, design engineers
QC	Quantum Computing	Solving large combinatorial and simulation problems	Problem Hamiltonians, optimization formulations	Quantum states encoding optimal/near-optimal solutions	IBM Qiskit® Google Cirq® D-Wave Ocean SDK®	Quantum researchers, algorithm developers

Table 2. Interrelations Matrix.

From → To	**CAD**	**CAE**	**CAO**	**CAM**	**AI**	**QC**
CAD	–	**Geometry feed:** CAD model → meshing & analysis setup	**Parametric seed:** CAD constraints define optimization initial state	**Toolpath basis:** CAD geometry → CAM strategy	**Training data:** CAD datasets fuel generative design models	**Problem encoding:** CAD geometry → quantum optimization Hamiltonians
CAE	–	–	**Objective metrics:** Stress/thermal results drive topological layout	**Process checks:** Simulated loads/ temps validate machining approaches	**Model refinement:** Simulation outputs refine AI predictive models	**Quantum simulation:** CAE problems (e.g., material behavior) mapped onto QC solvers
CAO	**Design feedback:** Optimized shapes → CAD refinement	**Re-analysis:** Optimized geometries → new CAE runs	–	**Optimized toolpaths:** CAO output → CNC code generation	**Algorithm input:** Optimization runs guide AI-driven multi-objective strategies	**Hybrid loops:** QC-enhanced solvers accelerate CAO convergence
CAM	**Manufacturability feedback:** Tool constraints → CAD	**Manufacture simulation:** CAM processes → CAE-style validation	**Workflow tuning:** Machining data → further CAO cycles	–	**Process optimization:** AI analyzes machining logs to improve CAM parameters	**Quantum-aided scheduling:** QC algorithms optimize toolpath sequencing and resource allocation
AI	**Design suggestions:** Generative AI proposes CAD variants	**Predictive analytics:** AI forecasts CAE outcomes pre-simulation	**Surrogate models:** AI replaces some CAO runs for faster exploration	**Adaptive machining:** AI adjusts CAM parameters in real time	–	**Hybrid intelligence:** AI steers QC problem formulation and interprets quantum outputs

Table 2 contd. ...

...Table 2 contd.

From → To	**CAD**	**CAE**	**CAO**	**CAM**	**AI**	**QC**
QC	Quantum-driven design: QC yields novel CAD patterns	Accelerated simulation: QC tackles certain CAE computations	Quantum optimization: QC solves complex CAO formulations	**Resource optimization:** QC optimizes CAM job scheduling	**Quantum-ML synergy:** QC boosts AI training on design/ manufacturing datasets	–
CADD	**Structural motifs:** Protein/drug scaffolds guide CAD-like fragment assembly	**Molecular validation:** Proposed ligands → MD and FEA of binding sites	**Lead optimization:** Binding scores & SAR feed CAO-style algorithms	**Synthesis planning:** Cheminformatics routes inform CAM-style processes	**AI augmentation:** ML-driven ligand generation & virtual screening	**Quantum chemistry:** QC computes high-accuracy binding energies & transition states

1

Nova Hybridalia

Beyond Biomimesis, Towards Co-Evolution

"Vis unita fortior."
Latin proverb

"Given the causes, nature produces effects
in the shortest possible way."

Leonardo da Vinci

1.1 The Return of Nature's Lesson: Bionics, Biomimetics, Biomimicry, Bioinspired Design, Next Nature, Hypernature and Material Ecology®

It can be hypothesized that the history of human technology may be interpreted as a relentless attempt to understand and replicate the mechanisms and functions of Nature through a continuous dialogue between inspiration, imitation, and innovation. Today, in the reimagined contemporary WunderCAD/CAMmers, within this evolutionary trajectory of human technoscience, we are able to model, simulate, and produce biological systems using the same logic as computational design software. Researchers can create an unprecedented design convergence between: Synthetic Biology (programming living organisms for specific functions), the CAD/CAM-Bio digital ecosystem (assisted design systems for the creation of biological components and bioprinting), and AI & Machine Learning (advanced simulation of biological processes). This current technoscientific perspective enables the design of new Bioinspired Materials—self-healing, adaptive, and even self-replicating—opening unexplored scenarios in structural engineering, architecture, design, regenerative medicine, soft robotics, and the manufacturing industry. Nature Research (2025) defines bioinspired materials as synthetic substances engineered to emulate both the architecture and performance of natural tissues or living organisms. Examples

DOI: 10.1201/9781003438946-1

include light-harvesting photonic systems that replicate photosynthetic processes, composite materials that reproduce nacre's layered microarchitecture, and metal actuators modeled on jellyfish locomotion. The very term mimesis, from ancient Greek (μίμησις), from which the modern term imitation is derived, was once used with the meaning of the essence and origin of art—a creative act born from human intellect and craftsmanship. In Platonic and Aristotelian philosophy, as well as in classical culture in general, it was conceptually regarded as a faithful reproduction of the forms of the surrounding nature (Blok 2022). In reality, ancient Greek did not have a word equivalent to our modern aesthetic concept of art. Instead, the term téchne (τέχνη) was used, meaning skill, expertise, and the ability to act and operate according to a set of applied rules within an activity, whether purely intellectual or also manual. From this fundamental archaic Western root word, Latin—and later the Anglo-Saxon languages—derived all the terms related to technique and technology (Tulley 2008). In light of the advancements in the various branches of Natural Sciences, Geology, and other Earth-related disciplines, Nature is now regarded as a vast and complex molecular technological laboratory. Over billions of years of biological evolution, it has given rise to incredibly sophisticated solutions. The design exhibited by all living organisms can be compared to a true Academy of Advanced Technology, where humanity has developed its unique ability to invent increasingly advanced tools, methods, and artificial materials—artifacts that have ensured and sustained the birth and development of our civilization on Earth for hundreds of thousands of years (Hodgkin 1990). Since the dawn of civilization, the first tools and weapons were crafted using materials from the natural habitat—such as flint, obsidian, and animal bones—mimicking the shape and function of teeth and claws. This pursuit has continued through to contemporary nanotechnologies that simulate molecules and cells. The human technical spirit has consistently endeavored to decipher nature's strategies of conflict, preserve their insights, and transform them into practical applications (Emlen 2014). However, the evolution of the human artificial universe has been marked by intrinsic limitations. The dominant approach has often been exponentially linear, extractive, and unsustainable, prioritizing immediate resource exploitation over long-term regeneration. Unlike nature—where processes unfold in a gradual, circular, and self-sustaining manner—human technology has historically forced production models that generate enormous amounts of toxic, non-degradable waste, contributing to the destruction of ecosystems. Nature's technologies, in contrast, operate on circular principles. Waste from one process becomes a resource for another, creating a self-sustaining cycle that reduces dependence on non-renewable materials and fosters ecosystem regeneration. This model also enhances resilience and adaptability, as natural systems continuously adjust to environmental changes. Such flexibility is crucial for addressing crises and ensuring long-term sustainability. Human technologies, however, often lack this adaptability, being designed with short-term efficiency and profit maximization in mind. To overcome this, science, technology, and design are now looking to Nature as a model. This challenge towards necessity, according to the Material Ecology® Manifesto (Table 1), was proposed by Neri Oxman, an international pioneer in biodesign philosophy of Co-Evolution (Antonelli and Burckhardt 2020).

Table 1. The Neri Oxman Material Ecology®'s Manifesto.*

Design Principle	Overview
Nature as co-client	We call for a shift in how we perceive clientele and commercial viability, turning Nature into a co-client within the design practice.
Nature-centric design	Nature-centric design invites the creation of design solutions at the service of Nature while advancing humanity.
Activist design	Activist design utilizes technology stakes in urging commissioning corporations to implement Nature-centric values across the design, construction, and manufacturing industries.
Growth over assembly	Computational design through self-organization enables the generation of vast, diverse forms similar to those that emerge through the biological growth processes found in Nature.
System over object	System-centric design considers any design construct—wearable, product, or building—a system of quantified interrelations across natural, biological, and digital domains.
Technology over typology	Platform technologies position the scientific method and technological innovation as the kernel of the architectural practice, revisiting timeworn typologies.
Difference over repetition	Perceived repetition in the natural world exists only through underlying difference and variation in response to environmental constraints, pointing toward design for behavior over form.
Integration over segregation	Functionality is not embedded in design constructs via discrete material components but rather distributed inseparably across a single intelligent system which is greater than the sum of its parts.
Decay over disposal	Programmed decay may be designed such that matter is consumed by the same ecosystem which sprouted it to fuel new growth, thereby empowering resource cycles rather than depleting them.

* **N.B.** The following table is an original reworking of the Material Ecology® Manifesto, which can be consulted on the website of OXMAN (https://oxman.com./)

In his reserches, Oxman integrates biology, technology, and innovative materials, rethinks the relationship between Humans, Nature, and Artifacts (Ecologies of Knowing 2023). It is a global reconsideration of how to think about the design of artifacts, recontextualizing it within a modified system of Human-Nature and Design-Economy relations, also with the aim of surpassing mere biomimesis (Materially 2025). Lawrence Halprin's 1982 Sea Ranch project established an architecture in close dialogue with its coastal landscape, rooted in untreated wood, local stone and earth-toned hues, and organized through a perceptual sequence of interconnected indoor and outdoor spaces (The Sea Ranch Association 2020). By embracing the natural contours of the terrain and inviting residents into an ongoing process of co-creation, Halprin framed design as a living collaboration between people and place. Today, Neri Oxman's Material Ecology® extends these ideas into the very substance of architecture, shifting the focus from site-specific materials to living, programmable matter (Fig. 1).

In Oxman's framework, cells, bacteria, and polymers co-evolve with form, while computational algorithms replace traditional spatial sequencing to generate dynamic, immersive environments. Where Sea Ranch celebrated the tactile warmth

Fig. 1. Lawrence Halprin's 1982 Sea Ranch design principles as a backdrop to OXMAN's Material Ecology design principles.

of timber, Material Ecology employs "living" resins and biosynthetic composites that self-organize and respond to external stimuli at the molecular level. This advanced co-design process—uniting human designers, machines, and microorganisms—reimagines architecture not as built artifacts but as evolving organisms, echoing Halprin's ethos of collective creativity and extending it into the microscopic realm of material life.

There is a growing interest from industry, research, and cultural debates in learning from nature's optimization strategies and sustainable design principles. As previously mentioned , this renewed focus on Biomimicry has been greatly facilitated by advancements in Digital Bioimaging (BioImaging n.d.), 3D Molecular Simulation (POV-Ray n.d.), Computational Design (Trimble Constructible n.d.), CAD/CAM/CAE-/AO biodigital protocols (Michigan State University 2023), Machine Learning, and Deep Learning algorithms (Levity 2025), which are accelerating the integration of nature-inspired solutions across industries, encouraging professionals and theorists to master these emerging tools. The advent of Quantum Computing (Gamble 2019) further expands the possibilities, paving the way for a new era of technological innovation rooted in Nature's wisdom. In the immediate future, we are entering a new design phase—the phase of hybridization, a Nova Hybridalia, in which artificial and natural technologies are no longer separate and conflicting worlds but are instead converging toward a shared horizon offered by the emergence of the so-called Next Nature (Next Nature Network 2025). This perspective of techno-scientific innovation is expected to materialize in the near future through the adoption of experimental Technoethics protocols (Coutinho 2024). In a few seconds, when considered within the scope of Earth's geological eras, Science, Technology, and Human Economy have profoundly shaped the planet's fate and physiognomy, leading it to its current environmental break-point. This state of affairs, in the conflict between Man and Nature, is unprecedented in Earth's evolutionary history. During the environmental consequences of the Industrial Revolutions since the 18th century, a turning point emerged with the rise of practices like biomimetics and biomimicry. These disciplines are among the most innovative scientific and design methodologies, focused on finding sustainable technological and production solutions in the artificial world. Researchers in these fields aim not only to emulate natural forms in utilitarian terms but also to contribute to systemic design, understanding the technological solutions of natural organisms in relation to the deep energy and ecological principles that govern planetary life, and analogically transferring them to new models of technological development and circular economy. A significant example of how biomimicry is transforming sustainable design is provided by the article from UGREEN (2023), which describes how this approach is revolutionizing design methods for architects, interior designers, and construction companies. According to the author, biomimicry, inspired by natural principles and processes, offers innovative solutions that improve material efficiency and reduce environmental impact, representing a true game-changer in the field of sustainable design. Over the course of the present two decades, there has been a progressive, exponential, and unprecedented explosion of studies, research, publications, and industrial applications in the field of biomimicry. Another relevant example of the application of biomimicry in sustainable design is provided by the article from Perch Energy (2024), which explores how natural solutions can be adopted in the field of design. The article highlights and confirms the growing interest in adopting nature-inspired approaches that optimize materials and processes, leading to results that reduce ecological impact and promote the responsible use of natural resources. Many market analyses and projections estimate that, in the coming decades, significant impacts will be achieved by products directly related to biomimetic research and applications, which will shape the main trends in the global market. An important mapping of the state of the art of biomimicry was proposed by Aman et al. (2024), who analyzed how biomimicry research intersects with the United Nations Sustainable Development Goals (SDGs). The study provides a detailed examination of the areas where biomimicry can make significant contributions, highlighting how nature-inspired approaches can address global challenges related to environmental sustainability, resource management, and technological innovation. This mapping offers a comprehensive overview of the current landscape of biomimicry research, identifying the main directions in which science and design are evolving to support the achievement of SDGs. The research also demonstrates how biomimicry can integrate with other global strategies to promote innovative and sustainable solutions in industrial, construction, and technological sectors. Ciulla et al. (2023) provide a comprehensive review of recent advances in the development of biomimetic materials, highlighting the growing importance of this field in material science. The authors discuss how biomimetic approaches, inspired by natural systems and processes, are being applied to create innovative materials with improved properties and functionalities. This study not only underscores the rapid progress in biomimetic material development but also emphasizes the potential of these materials to address complex engineering and environmental challenges. The review provides valuable insights into the current trends and future prospects of biomimetic materials, which are increasingly being adopted in various industries, including architecture, healthcare, and sustainability. This research aligns with the broader theme of integrating nature-inspired solutions into the design and production of sustainable materials. Sharma and Sarkar (2019) explore the current landscape of biomimicry research, identifying key challenges, gaps, and tools in the field. Their work highlights the need for further exploration of the principles underlying biomimicry, as well as the integration of more advanced tools and methodologies to bridge the existing gaps. The authors provide a critical analysis of how biomimicry is applied in various industries, and the potential it holds for addressing complex environmental and technological challenges. This research is particularly valuable for understanding the broader framework of biomimicry, as it sheds light on the theoretical and practical obstacles that need to be overcome to fully leverage nature-inspired solutions in design and innovation. The paper also emphasizes the importance of interdisciplinary approaches and collaboration in advancing biomimicry research. This integration, which in its most advanced research aspects takes on the characteristics of Hypernature, is made possible by advancements in synthetic biology, which allows for the modification and creation of biological systems for human purposes, nanotechnology, which mimics natural processes at the molecular level, and artificial intelligence, which learns cognitive algorithms and problem-solving routines from natural models capable of surpassing human intelligence in both power and speed when handling vast amounts of data and solving highly complex structural problems. Hypernature represents a critical reflection on the evolution of nature in the technological age. This term describes a man-made nature that surpasses the natural one in aesthetics, functionality, and safety. It manifests in forms such as genetically modified organisms, fluorescent animals, and botanical gardens containing species from every corner of the world. Essentially, Hypernature is an exaggerated simulation of a nature that never existed but is more beautiful, shinier, and safer than the original. In this context, Hypernature is not seen as a distortion of nature but as its evolution, where human intervention is celebrated as an integral part of our natural environment. After examining the concept and its features, it can be stated that Hypernature represents a potential modern iteration of the ancient Wunderkammer, where the distinction between the natural and the artificial dissolves. Just like the Wunderkammern of the Renaissance, which collected objects and natural and artificial curiosities to astonish and educate the viewer, Hypernature collects and amplifies nature through technology, creating a reality that goes beyond the original. However, in this case, the Wunderkammer is contemporary, and instead of being physically built with exotic objects, it is the result of design assisted by computation, where traditional CAD/CAM has evolved and specialized into new forms of computational design. Hypernature, thus, becomes a designed world where technology not only replicates nature but transforms it, enhances it, and reproduces it in ways never seen before, generating new aesthetic and functional experiences that amplify the creative and sustainable possibilities of design. The Artificial Remnants project, created by the artistic duo Entangled Others (Sofia Crespo and Feileacan McCormick), explores the intersection between artificial intelligence and biodiversity through the generation of digital insects (Entangled Others 2021). Using deep learning techniques, the artists create new digital organic species that, while inspired by real biological forms, exist solely in the virtual realm. These native digital species do not aim to faithfully replicate existing insects but rather to evoke an insect-like essence through algorithmic interpretation, offering an alternative perspective on natural diversity and prompting reflection on our interactions with the non-human world. The project fits within the concept of Hypernature, a fascination with life forms that, although not found in nature, are artificially created to inspire wonder and ecological awareness. Through Artificial Remnants, Entangled Others invites us to consider how technology can enhance our understanding of biodiversity, fostering a deeper connection with both the digital and natural ecosystems. The aesthetic and conceptual dimension in which biodiversity is reinvented through technology represents one of the most intriguing frontiers of contemporary generative art. In projects like Artificial Remnants, nature is not mimetically imitated, but rewritten through algorithmic logic that generates new life forms—plausible yet never existing. This synthetic autrenature does not merely astonish with exotic and captivating morphologies: it challenges our ability to recognize the living, redefining what we mean by natural, organic, or ecologically authentic. Technology, in this context, acts as both an epistemological and creative lens: it makes visible alternative evolutionary possibilities—simulations of biodiversity that have no counterpart in the known biosphere but evoke structures, symmetries, and complexities consistent with the aesthetics of life. This process invites a new gaze on ecology, as it does not simply document nature but expands it theoretically and visually, suggesting an ecology of computational imagination. Ultimately, this is a work on the imaginary: these artificial remnants populate a digital ecosystem that invites the viewer to reflect not only on what is natural, but also on what forms of life we might learn to see, embrace, or design in a world where nature and artificiality are increasingly entangled. Ecology thus becomes not only a science of the living but also an aesthetic and critical practice of interconnection.

Among the many factors to consider in this convergence-hybridization between Nature & Artifice, the main ones are:

- The development of advanced analytical techniques that utilize the specific algorithms of *Computational Ecology*, which aim to create innovative models and data processing techniques to better understand complex ecological systems, and to develop computational models that allow for rapid testing of ecological ideas through simulation, while also providing the means to simulate projections and *what-if* scenarios that would otherwise be difficult or impossible to predict (University of Maine 2025);

- The acceleration that the algorithms of computational biology, bioinformatics, and computational chemistry bring to the world of scientific experimentation, the pharmaceutical business, and the science and engineering of advanced materials. CD ComputaBio offers, for example, advanced computational biology solutions for research and pharmaceutical development, using molecular simulations and bioinformatics to optimize drug design and the analysis of complex biological systems (CD ComputaBio n.d.);
- The continuous refinement of advanced instrumental microscopy techniques and bioimaging technologies, capable of providing us with the spatial configurations of the nanometric domains of the phenomenological reality, and through which we are able to analyze with increasing precision and digitally visualize the stunning 3D geometric matrices of molecular hierarchies nested in cells, organelles, DNA, and functional components such as cellular motors and other organic ultrastructures (National Institute of Standards and Technology n.d.). The continuous development and refinement of these Quantum-Based Bioimaging Systems promise to accelerate the creation of more advanced operational tools for studying everything from cellular processes to new materials with potential uses in medicine, biotechnology, and manufacturing;
- The impressive evolution and widespread use in research, design, and production of artificial artifacts through increasingly advanced and integrated CAD/CAM/CAE/CAO protocols, now more than ever interfaced with AI and AGI systems. Specifically, CAD, since its inception in the post-World War II era, has undergone an extraordinary transformation. In just a few decades, it has evolved from a simple digital drafting tool used to assist designers, engineers, and architects in the laborious task of 2D representation, to a sophisticated 3D digital technology and functional big data management system. CAD has become indispensable for addressing all phases, from conception to simulated representation, leading to the creation of artificial artifacts. The infographic created by CADENAS GmbH retraces 60 years of evolution in CAD, highlighting the key milestones that have transformed technical design. Starting in 1957 with the creation of PRONTO by Patrick Hanratty, considered the father of CAD, a series of significant developments followed (CADENAS 2017-2025).

From observing complex biological processes at the molecular level to applying discoveries in biology, nanotechnology, and materials science, BioDesign has gained momentum as a crucial paradigm in scientific and technological innovation. Jens Martin Skibsted stresses the importance of integrating biodesign into technology and industry to tackle environmental challenges. By drawing inspiration from natural processes and organisms, biodesign enables the creation of sustainable, resilient solutions that reduce environmental impact and promote innovation. He argues that biodesign is essential for a sustainable future, not merely a passing trend (Skibsted 2023).

Here is a concise summary of the key historical steps:

Historical Pillars of Bio Design

Ancient Times

Mimesis: The observation of Nature for inspiration in the forms and functions of natural organisms dates back to ancient civilizations. Egyptians and Greeks studied animals' and plants' forms and proportions to inspire the design of buildings, artifacts, and weapons.

Biomathematics: Even early attempts to apply mathematics to Nature go back to ancient times, with Pythagoras (between 580 BC and 570 BC – around 495 BC) and Plato (between 428/427 BC – 347 BC) studying natural proportions and forms, such as the Golden Ratio and Sacred Geometry.

Renaissence and Modern Science born

Leonardo da Vinci (1452–1519)

Bio-inspired Design and Bionics: Leonardo's drawings of flying machines based on the dynamic study of bird flight and his interest in anatomy represent an early example of a bio-inspired and bionic approach.

Biomathematics: Leonardo analyzed and codified proportions and geometries of the human body and nature, such as observing the growth of tree branches, laying the groundwork for a mathematical approach to biology.

Galileo Galilei (1564–1642)

Biomechanics: Galileo can be considered the founder of biomechanics for analyzing and examining both the balance of the human body at rest and the dynamics of human jumping. He also applied physical laws to understand tree growth.

Giovanni Alfonso Borelli (1608–1679)

Biomechanics: In his book De motu animalium published in 1680, Borelli was the first scholar to graphically represent the body of animals and humans, considering them as machines, where movements can be explained through the application of schematic physical laws describing levers and pulleys. This work is available on the Internet Archive (Borelli 1680).

18th Century

Morphology/Johann Wolfgang von Goethe (1749–1832): Goethe's observations of plants and animals led him to develop his unique morphological theory, offering a perspective on the forms and functions of Nature that contrasted sharply with the materialist and scientific theories of his time. His book Metamorphose der Pflanzen (1790) presents an emotive scientific approach, which has been widely embraced and developed by the holistic and ecological cultural movements. In this pioneering work, Goethe explores plant morphology, proposing that various plant organs—such as cotyledons, leaves, sepals, petals, stamens, and pistils—are homologous, representing different forms of a fundamental archetype. This concept laid the groundwork for comparative morphology and had a lasting influence on later developments in botany and evolutionary biology. Goethe's approach emphasized the dynamic and transformative nature of plant structures, challenging static classifications and highlighting the continuity and variation within plant forms. The original German edition of this work is available on the Internet Archive, providing access to this seminal text in botanical science (Goethe 1790).

19th Century

Evolutionism/Biotechnics:

In Europe, within the evolutionary theories of Jean-Baptiste Lamarck (1744–1829) and Charles R. Darwin (1809–1882), original research emerged, such as that of Ernst Haeckel (1834–1919), who coined the term ecology, considering the aesthetic-geometric-morphological aspects of unicellular organisms like radiolarians and other marine organisms, and D'Arcy Wentworth Thompson (1860–1948), who is best known for his book *On Growth and Form* (1917), in which he applied mathematical and physical models to the analysis of living organisms. His work explored the relationship between biological forms and mathematical principles, laying the groundwork for understanding biological morphology through the lens of geometry and physics, anticipating a technical and technological view of natural design (Thompson 1945). In his original work *Plants as Inventors*, Raoul Heinrich Francé (1874–1943) examines the innovative mechanisms and structures developed by plants, emphasizing their role as natural engineers. This work aligns with the biotechnical thread, where the natural processes of plants are seen as a source of inspiration for human technological advancement. By exploring the complex forms and functions in plant biology, Francé highlights the potential for nature to inform sustainable design and innovation, making his work an important contribution to the field of bio-inspired technology (Francé 1943).

20th Century: The Birth of Modern Bio-inspired Disciplines

Early years:

The Bauhaus, founded by Walter Gropius in 1919, revolutionized art and design by merging aesthetics and functionality. Paul Klee (1879–1940) and László Moholy-Nagy (1895–1946), two of its key figures, explored structural morphology and visual perception. Klee, in his famous work Theory of Form and Figuration, particularly in the second volume Infinite Natural History, focused on geometry and organic forms as visual languages. Moholy-Nagy applied scientific principles, such as light and movement, to the study of perception. Both profoundly influenced modern design by integrating art, science, and technology. Marianna Charitonidou (2020) examines the shared biocentric design philosophies of László Moholy-Nagy and Alvar Aalto, focusing on their admiration for natural forms and their innovative approaches to standardization. Both figures integrated biological principles into their design processes, challenging conventional modernist rigidity. Charitonidou highlights their efforts to reveal the "invisible structures" of biological functions and relate them to the socio-economic and psychological dimensions of the creative process. Their work laid the groundwork for contemporary paradigms in industrial, architectural, and urban design, emphasizing mechanisms that imitate nature. The article underscores the importance of considering emotional and human factors in design, advocating for a more organic and flexible approach to standardization.

1940s:

Biomimesis: The concept of applying biological principles to technology emerged during World War II when scientists began studying natural systems like bats and dolphins to develop radar and sonar.

Cybernetics: Introduced by Norbert Wiener (1894–1964) in 1948, it focuses on studying control and communication systems in natural organisms, drawing parallels with machines and artificial control systems.

1950s:

Bionics: Officially coined in 1958 by engineer Jack E. Steele (1924–2009), the term refers to the design of devices inspired by biological structures. This historical phase coincides with the rise of cybernetics. In *Les inventions de la nature et la bionique (1987)*, Yves Coineau and Biruta Kresling explore the field of bionics, focusing on how natural structures and processes have inspired technological innovations. The book examines various examples of how nature's designs—such as propulsion mechanisms in water, architectural forms, and locomotion strategies—have influenced human engineering and design. This work contributes to the understanding of biomimicry and its applications in modern technology.

Biomathematics: With the spread of computers, the mathematical modeling of biological systems became increasingly relevant, with pioneers such as Alan Turing (1912–1954) applying algorithms to explain morphogenesis and the molecular dynamics underlying natural forms—such as the decorative patterns found in animal skins—laying the groundwork for a mathematical foundation of Developmental Biology (Herrero 2014).

Cellular Automata: Mathematical models introduced by John von Neumann (1903–1957) and Stanislaw Ulam (1909–1984) to study the evolution of complex systems. A cellular automaton consists of a grid of cells, each of which can assume a finite set of states. The system evolves in discrete steps, where each cell's state depends on the states of neighboring cells, according to precise rules. Cellular automata aim to model processes that simulate organic growth, replication, and interaction between simple components, reflecting principles of self-organization and emerging complexity typical of biological systems.

1960s:

Bionics and Bioengineering: The development of the first bionic prostheses and artificial organ research begins. NASA uses bionic technology for space exploration systems. In Germany, architect Frei Otto (1925–2015) establishes in early '60s the Biologie und Bauen movement, which first seeks to integrate biology, architecture, and structural engineering. Frei Otto: Thinking by Modeling (2017) is a landmark monograph that investigates his groundbreaking architectural philosophy and practice, renowned for his integration of scientific methods, natural principles, and architectural design. Edited by Georg Vrachliotis and published in conjunction with the ZKM retrospective in Karlsruhe and the Wüstenrot Foundation, the volume presents a richly illustrated exploration of Otto's work through photographs, drawings, archival documents, and analytical essays. Central to the book is Otto's radical use of analog physical models—from stretched membranes to suspended nets and, most famously, soap films and soap bubbles. These experimental techniques enabled Otto to simulate minimal surfaces, geometries that naturally minimize area while enclosing volume, governed by the distribution of tension. Long before the advent of digital modeling environments, Otto used these models as dynamic physical simulations of architectural form and structural logic. What today we know as parametric mesh or NURBS geometries in CAD software—surfaces defined by complex mathematical parameters—were, in Otto's hands, realized through simple materials and analog processes. In this sense, Otto's design approach can be described as a form of MAD – Modeling-Aided Design, a conceptual precursor to CAD – Computer-Aided Design. He anticipated the generative and computational thinking that would later become central to digital generative and parametric design and architecture. His analog modeling methods resonate with those of Antoni Gaudí, who employed hanging chains and funicular models to study equilibrium structures, revealing a lineage of architectural exploration based on gravity, form-finding, and material intelligence. The book contextualizes Otto's work within the larger trajectory of 20th-century architecture, bio-inspiration, and systems thinking. It also reflects on his influence in contemporary computational design, biomimicry, and sustainable construction. Through essays, interviews, and Otto's own writings, Frei Otto: Thinking by Modeling reaffirms the profound relevance of his design philosophy in an era increasingly dominated by digital simulation and material experimentation. This is not only a retrospective of a great pioneer of biodesign, but a critical lens on how making models—especially analog ones—can still offer powerful ways of thinking, designing, and projecting architecture, rooted in natural laws and transformed by technological evolution.

Biomathematics: Early use of mathematics to model population dynamics and study complex systems, such as Lotka-Volterra equations for predation dynamics (Shim and Fishwick 2008).

1970s–1990s: Expansion and Specialization

1970s:

Bioengineering: Institutionalized as a discipline with academic programs worldwide. New medical technologies like pacemakers and cardiac assist devices emerge.

Biomimetics: The term starts to describe technological applications inspired by biology, although it is not as popular as today.

Biomimetic Design: Primarily applied in architecture, notably in the Organic Architecture movement (e.g., works by Frank Lloyd Wright 1867–1959).

Cellular Automata: British mathematician John Horton Conway (1937–2020) creates The Game of Life (1970), simulating the evolution of a population of cells on a two-dimensional grid. It demonstrates how simple local interactions can result in emergent complexity, similar to living cellular systems (Beltoforion 2025).

1990s:

Biomimicry: Janine Benyus publishes *Biomimicry: Innovation Inspired by Nature* (1997), popularizing biomimicry with a focus on sustainability.

Bionics and Bioengineering: The first bionic hand implant occurs, along with advances in tissue engineering and artificial organs.

Biomathematics: The field expands with the rise of bioinformatics, applying mathematics to genomics and DNA sequence analysis.

2000s: Intersection of Biology, Technology, and Sustainability

2000s:

Bioengineering: Advances in stem cell research and genetic engineering contribute to synthetic biology.

Biomimetics and Biomimicry: Nature-inspired materials and surfaces are adopted in automotive, aerospace, and material industries.

Bionics: Development of advanced human-machine interfaces, such as neural control systems for bionic prostheses.

Biomathematics: Grows in modeling complex biological systems, including virus dynamics (e.g., SARS and influenza spread).

2010s: Technological Developments and New Frontiers

2010s:

Biomimicry: Becomes a key approach in sustainable design, with projects that emulate natural ecosystems to reduce environmental impact.

Biomimetic Design: Architects and designers like Neri Oxman and his *Material Ecology*® field of study, combine biology, art, and technology to create new materials and structures. Advances in CAD tools lead to explorations of nature-inspired designs in the work of designers such as Zaha Hadid and Achim Menges.

Bionics: Significant advancements in soft robotics and the development of exoskeletons for medical and industrial applications.

Bioengineering: CRISPR-Cas9 technology enables more precise genetic editing, offering new possibilities for genetic engineering and synthetic biology.

Biomathematics: Plays an increasing role in computational biology, particularly in protein folding research.

2020s: Convergence of Bioscience and Global Impact

2020-2024:

Bionics and Bioengineering: Brain-Computer Interfaces (BCIs) and 3D printing of organs advance rapidly. Synthetic biology developments lead to the creation of artificial organisms.

Biomimicry and Biomimetics: Integral to corporate sustainability strategies, with new bio-inspired materials developed for CO_2 capture to tackle climate change.

Biomimetic Design: Expands into living, self-healing materials, blending biology and architecture.

Biomathematics: Becomes crucial in predictive modeling, especially during the COVID-19 pandemic to analyze and predict virus spread.

Biomimetics, bioengineering, and synthetic biology: An increasingly deeper integration between advanced software design (e.g., AlphaFold® or Rosetta® for protein folding), deep learning algorithms, and artificial intelligence accelerates the development of personalized drugs, CADD (Computer-Aided-Drug-Design), biological treatments, and biosynthetic prostheses. Synthetic biology, continuing the evolution of CAD/CAM systems in molecular and biological applications, is transforming the way we address diseases and personalized medicine, using engineering principles to create more precise and targeted solutions.

Next Nature: The convergence of the technosphere and biosphere reaches an epochal moment, pushing bio design beyond imitation into the creation of self-adapting, growing, and self-repairing products. Biological processes replace industrial systems, blurring the boundaries between organisms and objects.

Hypernature: Nature, through human intervention, becomes more "artificial"—genetically modified organisms like fluorescent fish and engineered tomatoes become more perfect than their natural counterparts.

Mediated Matter® at MIT (2010-2021): A project by architect Neri Oxman at MIT exploring the mediation between objects and environments, enhancing the relationship between natural and artificial settings (Mediated Matter n.d.).

Material Ecology®: Neri Oxman's company Oxman.com develops new design processes integrating computational design, biology, robotics, and ecological engineering, emphasizing collaboration with Nature for co-production, cohabitation, and communication.The diverse range of experiments developed by different teams at OXMAN.com converge in defining the characteristics of a new Material Ecology®. These research efforts share the goal of establishing new design forms and material practices at the intersection of computing, materials engineering, design, and ecology, with broad applications on multiple scales:

We are a design company advancing the unification of top-down design engineering with bottom-up biological growth toward the mutual empowerment of Nature and Humanity. We operate at the intersection of computational design, biology, robotics, green chemistry, and ecological engineering. We envision a world

Table 2. Comparison between Human Technology and Natural Technology.

Negative Characteristics of Human Technology and Industrialization (EXTRACTIVE)	**Systemic and Sustainable Qualities of Living Organism Technology (REGENERATIVE)**
Excessive exploitation of natural resources and the environment, with dependence on non-renewable energy sources and a consumerist approach (oil/hydrocarbons). Human technology tends to maximize economic goals without considering the long-term impacts on the environment and society.	Responsible use of resources and promotion of renewable energy sources (solar), with an approach based on regeneration and recycling. On the other hand, the technology of living organisms optimizes resource use, integrating efficient processes and minimizing waste.
Air, water, and soil pollution due to the extraction and combustion of fossil fuels and industrial production. This type of technology often causes irreversible damage to ecosystems and human health.	Reduction of environmental impacts caused by the metabolic waste of living organisms, through continuous cycles of material reuse. Natural technology, on the other hand, integrates processes that promote the cleaning and organic preservation of the physical-chemical characteristics of soil, water, air, and the broader environment, as well as the regeneration of ecosystems.
Production of toxic and non-biodegradable waste resulting from the production and consumption of energy and material goods. Human technology tends to generate large quantities of waste that contribute to environmental pollution and the accumulation of toxic waste.	Minimization of waste through the efficient use of resources and the recycling of organic materials, contributing to the creation of a closed-loop material cycle. The technology of living organisms is characterized by processes that reduce waste and promote the reuse of resources.
Negative impact on biodiversity and plant and animal species due to the destruction of natural habitats and pollution. Human technology often leads to habitat loss and the decline of species populations.	Promotion of biodiversity and ecosystem well-being, as well as the conservation of natural habitats. Natural technology supports biodiversity and the maintenance of balanced ecosystems.
Greenhouse gas emissions and contribution to climate change caused by the burning of fossil fuels and heavy industry. Human technology is often responsible for greenhouse gas emissions that contribute to global climate change.	Natural technology, both animal and plant-based, promotes processes that absorb atmospheric carbon and reduce climate impact.
Concentration of power and social inequalities in the energy and industrial sectors. Human technology tends to concentrate economic power in the hands of a few actors, creating social inequalities and negative impacts on communities.	Natural technology promotes regeneration, interconnection, and cooperation among communities, fostering highly organized and hierarchical social structures that aim for self-sufficiency.
Linear Approach: Human technology often follows a linear model of production and consumption, leading to the exploitation and depletion of resources without effective recycling of materials.	Circular Approach: The technology of living organisms adopts a circular approach, where materials and resources are recycled and reused within the system in a continuous cycle, minimizing waste and maximizing efficiency. This circular approach enables a more sustainable use of resources and reduces the overall environmental impact.

where Nature and humans coproduce, cohabitate, and communicate as one. Through the creation of products and platforms, we explore three dimensions of collaborating with Nature: coproduction, cohabitation, and communication (Oxman 2025a).

Aware that this topic is particularly complex and specialized, for the purposes of our discussion, it is possible to consider the synthetic terms of a comparison between human technologies and natural ones. A concise comparison of the main characteristics that distinguish them is proposed, based on the principles outlined by the Biomimicry Institute (2025). This table is the result of a reworking of these core principles, which emphasize how Nature's solutions to various challenges have inspired the design of human technologies. By examining these fundamental attributes, the similarities and differences that shape the operation of each system are highlighted, taking into account the sustainability, efficiency, and adaptability found in natural processes compared to the engineered solutions developed by humans:

This agenda allows us, in any case, to consider the essential objectives for creating, in the near future, a human artificial system that can truly be defined as sustainable and in harmony with Nature. The generic indicators used in this chart can be applied to allow a systematic comparison of the technic performance of two similar materials (polymers) - one created by human technology and the other by organic design: Kevlar® and chitin, that represent two paradigms in material science.

The first material, Kevlar®, accidentally invented and patented in 1973 by Stephanie Kwolek (1923–2014), is a high-performance polymer engineered through human technology, while the second is chitin, an organic biopolymer evolved through natural selection (Van Stolk 2025). Both materials exhibit exceptional mechanical properties, especially in terms of tensile strength and impact resistance, making them ideal for protective applications. Kevlar®, synthesized through a complex chemical process, is renowned for its lightweight yet robust structure and is widely used in bulletproof vests, aerospace, and industrial applications. In contrast, chitin forms the exoskeletons of arthropods and the structural components of fungal cell walls, demonstrating remarkable adaptability, self-repair mechanisms, and biodegradability. From a design perspective, Kevlar® showcases the efficiency of synthetic engineering, optimizing molecular alignment for maximum durability, while chitin exemplifies nature's evolutionary strategy, integrating hierarchical structures that enhance resilience and multifunctionality. However, the production of Kevlar® requires high energy input and involves hazardous chemicals, such as concentrated sulfuric acid, which present significant environmental risks. Additionally, Kevlar® is not biodegradable and is difficult to recycle, contributing to long-term waste accumulation and environmental toxicity. Kevlar® excels in high-performance applications due to its hardness, resistance, and ability to withstand extreme conditions, whereas chitin, being a natural material, offers advantages in terms of sustainability, biodegradability, and adaptability. Understanding the interplay between these materials highlights the convergence of biomimetic inspiration and advanced engineering, paving the way for hybrid innovations that merge the best attributes of both realms. According to the United Nations Environment Programme (UNEP), over 300 million tons of plastic are produced globally every year, causing devastating environmental impacts. Less than 10% of this polymer production is properly recycled, while the rest ends up in landfills, rivers, and oceans. Furthermore, plastic materials are derived from

Fig. 2. Detail of the Aguahoja module assembly.

resources extracted at unsustainable rates and processed using environmentally harmful methods. Biomimetic research is developing alternatives, as natural organic structures have engineered materials similar to plastics but with greater efficiency, adaptability, and no negative environmental consequences. Using basic molecular components like cellulose, chitin, and pectin, which are abundant in biomass from trees, crustaceans, and fruit peels, natural systems create a surprising variety of functional biomaterials that surpass the capabilities of artificial synthesis. Chitin, for example, is the second most abundant organic material on Earth after cellulose and forms the basis of transparent dragonfly wings, soft fungal tissues, and the incredibly tough exoskeletons of crustaceans. These materials and the living systems that produce them not only offer unparalleled functional diversity but are also highly resilient, sustainable, and environmentally adaptable (as highlighted in Table 2 and in Table3). In line with the interest in chitin's properties, architect Neri Oxman's team developed the Aguahoja® project from 2014 to 2020 (Oxman 2019a).

This collection of biopolymeric materials presents an alternative to traditional plastics (Fig. 2).

It was introduced to the international public in the form of a pavilion designed for World Water Day, established by the United Nations in 1992 to raise awareness about the planet's most essential yet at-risk resource. The research group carried out an ambitious study and design project for biocompatible materials capable of decomposing in a programmed manner, offering a new disposal method that does not harm the ecosystem. Displayed in the MIT Media Lab lobby and already acquired by the MoMA in San Francisco, Aguahoja® consists of two pavilions that feature structures and objects made from shrimp shells, insect exoskeletons, and dead leaves. Designed using CAD/CAM software, it is built by a robot using a special water-based 3D printing technology and colored with natural pigments (Oxman 2018). The result is a material-organism composed of biopolymeric compounds that capture carbon dioxide, enhance pollination, increase soil microorganisms, and provide nutrients. Additional characteristics—such as patterns, shapes, surface

Table 3. Technological Comparison Between Kevlar® and chitin.

Negative Characteristics of Human Technology and Industrialization (EXTRACTIVE)	**Systemic and Sustainable Qualities of Living Organism Technology (REGENERATIVE)**
Kevlar®	Chitin e.g., Lobster claws/shell Nanometric Hierarchical Ultrastructures
Intensive extraction of raw materials, such as oil, for the production of the polymer with high environmental impact.	Chitin is a biological material produced by lobsters through metabolic processes, not requiring intensive extraction of non-renewable natural resources.
High energy consumption and use of non-renewable resources during the production process, with negative environmental impacts.	The formation of chitin occurs within the lobster's body without the use of external energy, minimizing the environmental impact of its production process.
Release of toxic substances during the polymer synthesis process, contributing to air, water, and soil pollution.	Lobsters produce chitin as a natural component of their body structures, without the need for synthetic chemical processes that may generate harmful substances.
Poor recyclability of material, leading to the production of non-biodegradable plastic waste with a negative environmental impact.	Lobsters can recycle and reuse chitin in their body structures, promoting a sustainable material cycle within marine ecosystems (chitosan).[1]
Dependence on chemical and petrochemical industries for the production and processing of Kevlar®, leading to negative social and economic impacts.	The use of chitin promotes the self-sufficiency of lobsters and marine communities, reducing dependence on external resources and fostering ecologic independence.
Linear Approach: the production cycle of Kevlar® follows a linear model of extraction, production, use, and disposal.	Circular Approach: the chitin formation cycle is part of a circular biological process, where chitin is continuously recycled and reused within marine ecosystems.
Lower resistance and durability compared to the biological structures of lobsters, resulting in the need for continuous replacements and repairs.	The nanometric hierarchical ultrastructures of chitin that make up the exoskeletons and claws of shrimp and lobsters are highly resistant and durable, providing effective protection against predators and injuries.

[1] Chitosan is a polysaccharide primarily composed of β-(1→4)-linked D-glucosamine and N-acetyl-D-glucosamine units. In nature, it can be formed through the partial enzymatic deacetylation of chitin during microbial degradation processes. Industrially, chitosan is obtained via the alkaline deacetylation of chitin, a reaction that removes acetyl groups (–$COCH_3$) from the polymer backbone, thereby converting chitin into chitosan. This structural modification makes chitosan soluble in acidic aqueous environments ($pH < 6$, such as those found in the gastrointestinal tract), unlike chitin, which remains insoluble. As a natural biopolymer, chitosan is biocompatible and biodegradable, making it suitable for numerous applications, particularly in biomedical and pharmaceutical fields. It also exhibits antimicrobial activity, which is advantageous for food preservation and in medical devices to prevent infections. Additionally, chitosan shows gel-forming and viscosity-enhancing properties, which are valuable in pharmaceutical and cosmetic formulations. It is also employed as an anti-cholesterol agent, in weight management treatments, and in skincare products.

colors, and various mechanical, optical, olfactory, and gustatory properties—can be created using organic waste streams, thus without interfering with ecological niches. Aguahoja® is one of many projects conceived in the innovation hub of Mediated Matter, which Oxman developed at MIT from 2010 to 2020. Updates on the Material Ecology® program and Neri Oxman's design process can currently be found on her official website. Aguahoja®, in addition to its intrinsic aesthetic and functional beauty, is a project that aims for a future where the industrial cycle of overproduction and obsolescence is overturned through the use of abundant natural materials. Through this project, the researchers envisioned the ability to temporarily divert materials from healthy ecosystems, integrate them into human designs, and allow for natural decomposition in the environment to foster new growth. By enabling digital design and fabrication with biopolymers, they sought to create systems that promote the protection and strengthening of ecosystems, while offering humans a new frontier in design and production. An extremely interesting technological feature of this experiment is the use of innovative CAD/CAM suites that involve the use of 3D printers and robots capable of additive digital manufacturing of water-based biomaterials (Mogas-Soldevila et al. 2014). It has been demonstrated that the shape and composition of the material can be directly influenced by physical properties (such as stiffness and opacity), environmental conditions (such as load, temperature, and relative humidity), and fabrication constraints (such as degrees of freedom, arm speed, and nozzle pressure). Each structure in the collection contains a unique combination of organic materials, whose allocation, consistency, and distribution within the final object are computationally guided and produced with high-resolution additive manufacturing. This approach allows for control over specific physical properties and adaptation to changing environmental conditions. Unlike most synthetic materials, the structures in this collection react to their environment throughout their lifespan, adjusting their geometry, mechanical behavior, and color in response to variations in heat, humidity, and sunlight. This temporal behavior, based on time-dependent changes, is used as a design feature, enabling the material to sense, respond, and adapt to fluctuating environmental conditions. The robotic fabrication platform is designed to convert cellulose, chitosan, pectin, and other abundant biopolymers into high-performance, sustainable hydrogels that can be 3D printed into objects for applications ranging from millimeter to meter scale. In the field of advanced and systemic research proposals within Synthetic Biology and Biosynthetic Materials Design, it is worth revisiting the work that Neri Oxman has been conducting for several years, both through MIT and now with her own company. Her projects integrate results from diverse but interconnected areas of expertise, such as computational design, robotics, materials science, green chemistry, biology, and ecological engineering. This interdisciplinarity enables her to develop innovative solutions that merge technology and Nature, promoting a hybrid symbiosis between the built environment and the biological world. Oxman and her team have generated a series of research projects that bring together all the design principles and operational philosophies of the Mediated Matter project, developed during its decade-long tenure at MIT. These projects represent an exemplary case of some of the most advanced and forward-thinking visions of bio-design mediated by digital modeling technologies such as CAD/CAM. At its core, her design philosophy transcends mere biomimicry, embracing co-evolution and symbiosis—harnessing the adaptive intelligence of living organisms to fuel human creativity. Her method integrates data-driven workflows with biosynthetic bacteria embedded in custom resins enriched with molecular inducers. This framework enables the programmable fabrication of objects with intricate geometric, physical, and biochemical complexity. The objects created through this framework demonstrate an iterative approach for computational design of localized cell behavior through material fabrication and also leverage advanced digital fabrication tools that uniquely incorporate techniques from synthetic biology. This framework enables the creation of complex free-standing objects or scaffolds for a variety of applications, including point-of-use synthesizers, bioremediation devices, and whole-cell sensing matrices:

"to date, the fields of biomaterials and tissue engineering have made great advances towards the creation of material gradients in microenvironments and scaffolds that direct cell behavior. However, existing fabrication methods for creating concretely heterogeneous substrates often focus on a single property through which to influence cells – such as surface stiffness, porosity, patterning, roughness, hydrophilicity, or chemical concentration. We reasoned that by incorporating multiple material properties, we can build more dynamic and complex microenvironments and substrates, leveraging more functionalities of living cells within a single integrated object" (Oxman 2019b).

Interdisciplinary expertise and advanced computational skills are fundamental to pioneering sustainable design—where avant-garde material and production systems hybridize Nature and Artifice—as exemplified by the Computational Design Engineer/Computer Scientist profile on the Oxman website (Oxman 2025b). To operate in a bioinspired eco-design environment of excellence like OXMAN's, a computational design engineer must be able to translate project requirements into advanced software solutions that interface seamlessly with industrial machinery, bespoke prototypes, and both parametric and generative CAD/CAM systems. Mastery of simulation techniques to predict mechanical motions and material behaviors is vital, as is the ability to convert algorithms in computational geometry, digital fabrication, and artificial intelligence into ready-for-production 3D models, scripts, and process descriptions. 3D visualization, virtual prototyping, and parametric workflow adjustments become key tools for steering design decisions and conveying complex ideas instantly. A solid track record of at least four years in design-focused software development ensures the ability to integrate robotics, CNC, and unconventional 3D-printing technologies alongside parametric and generative design processes. Deep familiarity with tools like Houdini, Rhino, or Grasshopper supports efficient workflows, while fluency in languages such as Python or C++23—paired with specialized libraries (e.g., libigl, fenics, PyTorch)—provides the flexibility needed to build custom tools. Additional expertise in real-time platforms (TouchDesigner, Unreal Engine, Unity) expands the horizon for novel interaction paradigms. Finally, a strong mathematical foundation (linear algebra and multivariable calculus), a discerning eye for form and composition, and a willingness to collaborate across disciplines are essential for anyone aiming to drive innovation in contemporary design.

1.2 WunderCAD/CAMmer's Relevance in BioDesign Approach: The ELM's Manifesto

Contemporary design and engineering are undergoing a paradigm shift, as practitioners move beyond the industrial sourcing of materials and conventional notions of application. Instead, it's now possible to develop customized materials, possibly from waste, and create personalized product catalogs. We're seeing an increasing number of bioengineered materials, such as leather made from pineapples (Watson and Wolfe 2020) or plastic made from algae (Jaynes 2023). This trend also applies to design students: many of the mycelium-based materials entering the market come from students dissatisfied with existing materials, opting to develop their own (Camilleri et al. 2025). As the digital world becomes more prominent, it will be fascinating to see how materials adapt to this new space, much like the technologies of the natural world, where living creatures are made of materials programmed by genetic information. Research in Biodesign is inspired by technological solutions that Nature has refined over billions of years, optimizing them in every aspect. This approach competes with Nature itself, aiming to engineer synthetic materials with superior functional properties to organic materials in various fields. At the intersection of *Synthetic Biology* and *Materials Science, Engineered Living Materials (ELMs)* have emerged as a new discipline (Wang et al. 2022). In an article by Cell Guidance Systems (2022), the development of engineered living materials (*ELMs*) is explored, highlighting their ability to adapt to environmental changes. This emerging field combines synthetic biology and materials science, offering innovative solutions in medicine, biotechnology, and sustainability. Combining bioengineering with traditional materials science practices has led to extraordinary innovations in creating biologically active materials for applications in therapeutics, electronics, and construction. However, the rapid development of this field has made it challenging to gain a comprehensive view of the tools, techniques, and applications of ELMs across disciplines (Horizon Europe 2021).

ELMs represent the most significant revolution in materials science today, aiming to outperform current examples of smart, active, or multifunctional materials. *Living materials* offer unique properties like autonomy, intelligent responses, self-repair, and self-replication. ELMs are mainly classified into biological ELMs (bio-ELMs), made entirely of cells, and Hybrid Living Materials (HLMs), which combine abiotic scaffolds with living cells. A 2022 study proposed a classification system for ELMs, dividing them into two superkingdoms: bio-ELMs (Wakle and Khuntia 2023), and HLMs (Smith et al. 2020). ELMs are not just biomaterials, but hybrid systems, true metaphors of *Nova Hybridalia*, as they are capable of:

- **Self-assemble**, as they grow and develop autonomously (Wang et al. 2022);
- **Respond to stimuli**, thus showing the ability to adapt to the environment (Molinari et al. 2021);
- **Self-repair**, mimicking the regenerative abilities after structural damage found in many living organisms like planarians or salamanders (Imperial College London 2021);

- **Naturally degrade**, embracing the Zero Waste ideology (Altin-Yavuzarslan et al. 2023).

To create these materials, researchers combine:

- **Engineered bacteria and fungi:** *E. coli*, Bacillus, mycelium (Li et al. 2023);
- **Bioactive polymer matrix:** Proteins, chitin, alginates (Axpe et al. 2018, Jabeen and Atif 2023);
- **Biological modeling software:** Leal et al. (2017) describe the application of the BioCAD methodology in simulating complex anatomical structures, using finite element models to analyze stress and deformation. A practical example involves the biomechanical analysis of a human molar, highlighting the importance of collaboration between healthcare professionals and engineers to validate the simulation models.

The study also outlines criteria for defining phyla, classes, families, and species inspired by material classifications (Lantada et al. 2022). The ELMs' taxonomy generally reflects the types of living organisms involved in the design and implementation of these materials. These two main phyla are:

Bacteria-based ELMs

- **Characteristics**:

These engineered living materials use bacteria as the central component. Bacteria can be designed to perform a variety of functions, such as producing organic compounds, mineralizing, or synthesizing biomaterials.

Examples and applications:

- **Self-repair**: Bacteria that produce calcite to repair cracks in concrete or other construction materials.
- **Biomineralization**: Using bacteria to produce minerals that aggregate into a solid matrix, creating materials that appear organic but have robust properties.
- **Sustainable materials**: Using bacteria to create biodegradable materials, for example, through the production of natural polymers.

Fungi-based ELMs

- **Characteristics**: Fungi, particularly their mycelium, are used to create living structures. Fungi are highly versatile organisms and can be used to create lightweight, durable, and biodegradable materials.

Examples and applications:

- **Creation of bio-composites**: Fungi can be used to create materials that serve as an eco-friendly alternative to plastics or synthetic foam. These materials are often lightweight and durable but also completely biodegradable.
- **Growth on substrates**: Mycelium can grow on organic substrates, forming structures that can be used for construction or as packaging materials.
- **Adaptability and self-repair**: Fungi can also adapt to various environments and self-repair if damaged, due to their ability to expand and regenerate.

Bacteria are much smaller and capable of performing complex biological functions such as biomaterial production and self-repair. Fungi are more complex organisms, primarily used to create composite structures, with the ability to adapt and grow continuously. This division into two phyla reflects the biological approach used to obtain living materials with specific functions, taking advantage of the unique characteristics of each organism. The main applications currently being experimented with are in the field of:

- Bioactive architecture: To create buildings and biomaterials capable of replacing cement, such as mycelium-based biocomposites (Proksch 2024);
- Biodegradable packaging: As an alternative to traditional plastics (SynBioBeta 2024).[2]
- Regenerative medicine: In the creation of synthetic tissues and bioprinted skin (Yu et al. 2021, VoxelMatters 2025);
- Soft robotics: Enabling the creation of flexible, adaptive materials and structures (Liu and Xu 2020, Appiah 2019).[3]

Proksch (2024), in particular, discusses the emerging field of Engineered Living Materials (ELMs), which combine synthetic biology and material science to mimic the properties of natural living materials. ELM research aims to produce novel materials with tailored functions using genetically engineered organisms coupled with synthetic or biological polymers to create desirable features like self-repair and enhanced mechanical properties. Proksch is part of a National Science Foundation (NSF)-funded project, where ELMs are defined as composite materials of engineered living cells encapsulated within a polymeric matrix. The interdisciplinary research team comprises chemists, biochemists, bioengineers, mechanical engineers, and architects who develop ELMs for the built environment, including CAD/CAM 3D-printable resins with engineered living cells that exhibit different functionalities. The focus of the lab-based research addresses three main issues around innovative ELMs: (1) the ability to thrive in changing hydration levels (outdoor environment) and survive periods of low hydration levels, (2) the integration of photosynthetically active, productive cells in high-tech building membranes, and (3) the creation of ELM

[2] This article describes an innovative research conducted at the Shenzhen Institute of Advanced Technology (SIAT), Chinese Academy of Sciences, where Bacillus subtilis spores were engineered to produce lipases capable of degrading plastics such as polycaprolactone (PCL). These spores are incorporated into plastic materials during production and remain dormant until environmental factors such as surface erosion or composting activate them, initiating the complete degradation of the plastic material.

[3] This author explores the early emerging field of biohybrid soft robotics, which combines living materials with robotic devices to create systems capable of self-healing, sensory response, and dynamic adaptation. The author discusses the challenges and opportunities associated with integrating living cells into soft robotic structures, highlighting the potential of these materials for applications in medicine, robotics, and human-machine interfaces.

clusters as resilient bioreactors for bioproduction. The study highlights the need for disciplinary collaboration on ELM research between chemistry, molecular biology, bioengineering, material science, architectural design, and beyond, juxtaposing the science, built environment, and design perspectives on ELMs. It also touches on essential questions in this emerging field.

Through the continuous advancement of bioevolutive CAD/CAM algorithms, state-of-the-art parametric and generative algorithms are now capable of replicating the intricate morpho-functional characteristics of living organisms. This enables the creation of bio-inspired designs that seamlessly integrate with natural processes. This groundbreaking approach opens new possibilities for innovation across multiple fields, including architecture and bioengineering. For example, living cells convert substrates into raw materials such as proteins, sugars, or lipids through metabolism, and then coordinate the assembly of these materials to control the resulting structure's morphology. Evolution has endowed natural systems with functions like photosynthesis and underwater adhesion, properties challenging to replicate with conventional materials. Significant efforts in bioinspired technologies focus on developing hybrid materials that control the functionality of biological agents. Additive manufacturing platforms have emerged as a key hybrid approach, enabling the deposition of bio-inks, such as whole-cell suspensions, into controlled forms (scaffolds). These platforms also model exogenous chemical or environmental signals. A team of researchers, including Neri Oxman, from prestigious centers such as the Media Lab (MIT), Wyss Institute (Harvard University), and Dana-Farber Cancer Institute (Harvard Medical School), has recently developed an innovative digital fabrication platform called Hybrid Living Material (HLM). In a 2020 article, Vialva (2020) describes this method, which involves 3D printing objects capable of controlling organisms. Using the HLM platform, the team employed a multi-material inkjet 3D printer and customized recipes to combine resins and chemical signals. The signals activate responses in genetically engineered microbes, opening the door to medical devices with therapeutic agents. This approach integrates CAD/CAM design, additive manufacturing, and Synthetic Biology, achieving the replicable fabrication and control of hybrid materials. The Objet Connex500® multi-material inkjet-based 3D printer is used in the CAM phase for HLM fabrication, chosen for its high resolution, unique multi-material control, and the ability to create complex self-supporting structures (Stratasys 2025). The process modifies 3D print descriptions to ensure proper distribution of materials, allowing the fabrication of structured photopolymers that maintain precise spatial distributions of regulatory chemical signals within polymerized architectures. Hydrogel is applied to immobilize engineered E. coli colonies on the printed objects' surfaces, enabling material-guided chemical signaling, which produces programmed responses regulated by genes in a reproducible and predictable way. With this platform, researchers were able to print self-supporting HLM constructions up to half a meter long, generating freeform shapes and incorporating specific mechanical and optical properties. The engineered bacterial response was defined through digital printing descriptions, enabling programmable interactions between 3D-printed materials and genetic regulation circuits. This experiment integrated the workflow from design to output, merging signal diffusion models and bacterial responses in a CAD environment. The HLM platform's ability to guide 3D structures, bacterial functions, and material property distributions simultaneously provides unprecedented control over functional outputs. For a complete technical description of the platform, refer to the patent document US20190276793A1 (Bader et al. 2019). In recent decades, integrated research in science and technology has advanced our understanding of the chemical and physical mechanisms regulating molecular processes in organic life. We have identified various elements and genetic modules within these networks, leading to the design of synthetic genetic circuits like switches, oscillators, feedback loops, and logic gates. By leveraging these circuits, it has been possible to create advanced genetic machines capable of making programmable decisions. As a result, these discoveries have led to new applications, such as the dynamic and autonomous regulation of metabolic networks, the targeted evolution of biological units, remote and specific diagnostics and therapies, as well as biological containment solutions to prevent the spread of microorganisms and modified genetic material (Xia et al. 2019). The goal is to create simple genetic circuits that function similarly to electronic circuits, with their dynamics predicted by mathematical models. In the early 2000s, pioneering studies successfully engineered genetic circuits to program cells with desired functions (Luo et al. 2023). *Genetic Circuit Engineering* (GCE) shows strong analogies with the design of electronic circuits through CAD and is a fundamental discipline of *Synthetic Biology*. It represents an unprecedented point of intersection between Molecular Biology, Computer Science, and Engineering (Nature Portfolio 2025). GCE deals with the manipulation of entire genomes or large portions of them within a living organism, in order to radically modify its chemical characteristics at a systemic level, such as altering its metabolism, resistance to external factors, or other complex functions. Researchers are now capable of designing entirely synthetic genomes that are completely artificial or rewriting natural genomes. Genomic science has significantly advanced our comprehension of the biological realm. Researchers are now able to *decode* the genetic information of organisms across various life forms by sequencing the four nucleotide bases that form DNA. The process of sequencing genomes has become commonplace, resulting in thousands of genomes being stored in public databases. In effect, scientists are transforming biology into a digital format, converting the chemical components of DNA—A, C, T, and G—into binary code (1s and 0s) for computers. However, the possibility of reversing this process and starting with 1s and 0s in a computer to define the characteristics of a living cell remains an open question. This is the issue that was explored (J. Craig Venter Institute 2010). As early as 2010, a scientific team headed by Drs. Craig Venter, Hamilton Smith, and Clyde Hutchison of the J. Craig Venter Institute® (JCVI), achieved the final step in their quest to create the first synthetic bacterial cell. The synthetic cell was named *Mycoplasma mycoides JCVI-syn1.0*, and it represented the first proof of principle that genomes can be computer-designed, chemically synthesized in the lab, and transplanted into a recipient cell to produce a new self-replicating cell controlled solely by the synthetic genome. The creation of this first synthetic bacterial cell, the result of over 15 years of research that saw the development of new laboratory protocols, advanced tools, and computing and digital technologies capable of synthesizing unnatural DNA segments that encode proteins or RNA molecules controlling each other's levels, marked a highly significant achievement in the development of the GCE technological sector at the time. In the GCE field, one of the most emblematic examples of how biological circuits can be designed and manipulated to control cellular processes is that of *Genetic Toggle Switches* (GTS). Genetic circuits (GTC) consist of interconnected genes that collaborate to carry out specific tasks. They serve as the fundamental components of biological systems and are crucial for sustaining life. The field of computing has significantly impacted genetic circuits, providing tools and methods to design and modify them. For example, computational algorithms now enable the creation of genetic circuits that can sense and react to particular environmental cues. These circuits have diverse uses in biotechnology and medicine, such as developing synthetic biological systems like biosensors or drug delivery mechanisms. They can also be applied to reprogram cells for specific roles, such as generating insulin for diabetes treatment. Furthermore, genetic circuits are useful for investigating complex biological phenomena like growth and disease mechanisms. With the reduction in costs and the increased accessibility of DNA synthesis, scientists can more easily design and assemble synthetic circuits. Recent advancements in DNA synthesis have facilitated the development of more intricate circuits, incorporating multiple inputs and outputs. As a result, new technologies, such as biosensors capable of detecting particular molecules, have emerged. The future of synthetic circuits will be shaped by modular design, a principle in synthetic biology that breaks biological systems into smaller, interchangeable components. This approach allows scientists to mix and match parts to create new circuits with various functionalities. Modular design has already been used to construct genetic circuits that detect and respond to different environmental signals, and its significance is expected to increase as synthetic biology continues to evolve (FasterCapital 2025). The creation of the first GTS was proposed in 2000. It consisted of two genes that inhibited each other, with external inputs (inducers such as IPTG (*isopropyl-β-D-thiogalactoside*) or tetracyclines) used to switch the system (Gardner 2000). Specifically, IPTG is a chemical molecule capable of activating the genes involved in the metabolism of the Lac system, which stands for lactose. This genetic system is found in bacteria such as E. coli and controls the metabolism and utilization of lactose. The *Lac system* includes the *Lac operon*, a group of genes essential for lactose metabolism, and LacI, a repressor that turns off the *Lac operon* unless lactose or *IPTG* is present to deactivate it. Just as in the design of electronic circuits, *GCE* designs components can be combined into complex systems such as genetic oscillators (repressilators), biological sensors, or genetic memory systems. A synthetic biologist often relies on computational tools very similar to CAD systems used in the fields of mechanics or electronics.[4]

Starting from pioneering applications of CAD of biological circuits using TinkerCell (Chandran et al. 2010), synthetic biology has developed a well-

[4] The *repressilator* is an artificial genetic circuit created to produce regular oscillations in gene activity within a cell. It consists of three genes that inhibit each other in sequence: the first blocks the second, the second blocks the third, and the third blocks the first. This cyclic configuration generates a predictable rhythm of gene activation and deactivation, similar to a biological oscillator. The repressilator was first introduced by Michael Elowitz and Stanislas Leibler in 2000.

characterized library of parts, genetic regulatory circuits, and design principles, which serve as the foundation for programming cells to perform various tasks. The CAD approach offers several advantages, including the ability to test different designs and predict how the circuits will behave under different conditions. CAD suites are used to design circuits capable of controlling gene expression, detecting toxins, and even producing biofuels. Two notable examples of this software in synthetic biology are Cello® and SynBioCAD®, which are directly comparable to traditional CAD protocols in engineering. Cello® is a computer-aided-design program created by CIDAR Labs (Boston University), the Synthetic Biology Center at MIT, and the Biosystems and Biomaterials Division at the NIST, which applies the concept of *Circuit-Design Automation* (CDA) from computer circuit design to genetic parts in living organisms. CDA refers to the automation of the process of designing electronic circuits using software tools that help engineers create, simulate, and verify circuits. It is a specific component within the broader field of Electronic Design Automation (EDA), which includes the software, hardware, and services used to design electronic systems, such as integrated circuits (ICs), printed circuit boards (PCBs), and complex electronic systems. Cello® 2.0 is a free software that enables the automatic design of programmable genetic circuits. It uses a high-level logical description and a library of characterized DNA parts, assigning each biological component its intended role within the network. The software constructs the corresponding DNA sequence and provides annotated representations ready for experimentation and fabrication. Thanks to new formal rules and integration with the SynBioHub® repository, Cello® 2.0 extends the application of genetic circuits beyond *Escherichia coli* to other organisms and more complex genetic contexts, including complete genomes. Processing time varies from a few minutes to several hours depending on the complexity of the project (Jones et al. 2022).

As synthetic biology advances, researchers are envisioning a future where designing biological circuits becomes as straightforward as designing electronic integrated circuits. Cello® is a platform that functions like a programming language, enabling the creation of computational circuits within living cells. These circuits, when placed on plasmids and expressed in *Escherichia coli*, need to be isolated from their genetic environment, yet they generally performed as intended. For instance, they could control cellular activities in response to various environmental stimuli. This approach has the potential to simplify the creation of increasingly intricate circuits through genetic modification. Cello® translates electronic design specifications for combinational logic into complete DNA sequences, encoding transcriptional logic circuits that can be executed in bacterial cells. A repository of transcriptional repressors, developed in the Voigt Lab, provides genetic gates such as NOT and NOR, which can be combined to form any desired logical function. The use of an automated workflow and custom genetic gates will enhance the reproducibility and accessibility of circuit design in biological engineering laboratories (CIDAR Lab 2023).

From the Nona Research Foundation® Inc. website, based in Boston, it is possible to download for free Cello® along with a series of other software, CAD, and open-source tools dedicated to design operations for *Synthetic Biology*, including (Nona Research Foundation 2024):

- **3DμF®**: A visual CAD tool for the design of microfluidic devices; outputs STL for CNC milling and SVG for lithography, as well as a textual JSON representation;
- **DAFD®**: The first microfluidic design automation software that can deliver a user-specified desired performance using machine learning;
- **Fluigi®**: A software that automates the design of microfluidic devices used in synthetic biology by optimizing their layout; specifications are in the text-based MINT language;
- **GeneDesign®**: A web-based synthetic gene designer including various modules such as reverse translator, codon juggler, etc., for sequence manipulation;
- **Puppeteer®**: A web-based planning and instruction generation tool which automates the task of DNA assembly by tracking, managing, and controlling laboratory tasks, reagents, and equipment. It is designed to translate protocols into sets of human- and machine-readable instructions, reducing the labor cost of building genetic devices;
- **Raven®**: A sophisticated dynamic programming algorithm to design high-quality DNA assembly plans. Raven produces SBOL-compliant images that visualize your assembly plan as well as human-readable instructions;
- **SBOL Designer®**: A simple, biologist-friendly CAD software tool for creating and manipulating the sequences of genetic constructs using the Synthetic Biology Open Language (SBOL) 2.2 data model;
- **SynBioHub®**: A design repository for people designing biological constructs. It enables DNA and protein designs to be uploaded and then provides a shareable link to allow others to view them.

Specifically, Cello® applies the *Electronic Design Automation* (EDA) approach to genetic circuits and involves several interconnected steps, which require standardized input/output formats. Among these formats are Verilog® for representing a logical function and Eugene® for encoding a set of parts and constraints between them (Bilitchenko et al. 2011, Nielsen et al. 2016). Recently, the EDA industry has seen significant developments, including the integration of artificial intelligence to enhance efficiency and automate the design process (Li 2024). The system operates with libraries of known genetic components, such as promoters, RBS (Ribosome Binding Sites), coding sequences for regulatory proteins, and terminators. By mapping digital logic functions like AND, OR, and NOT into biological sequences, it enables researchers to design and construct genetic circuits, translating a desired logical specification into a DNA sequence (e.g., Boolean logic functions where a gene is activated only when two inputs are present) that can be implemented in a cell. The software converts this logic into a precise circuit diagram, selects the most suitable genetic components based on the host organism, and simulates and optimizes circuit behavior to ensure proper function, making adjustments if needed. Using a modular approach inspired by electronic design, synthetic biologists can define how input signals (e.g., external or internal cellular conditions such as molecule concentrations or biochemical signals) interact with outputs (e.g., controlled gene expression like fluorescence or protein production). Cello® employs mathematical models to predict circuit behavior within the cell, assess stability, ensure compatibility with the host organism (e.g., bacteria or yeast), and optimize component interactions to minimize biological background noise.[5] At the end of the process, the program is capable of generating an optimal design of the desired genetic circuit, producing a DNA sequence ready to be synthesized and implemented in the laboratory. The main applications of Cello® are:

- **Custom Biosensors**: Creating circuits capable of detecting specific environmental or biological molecules, such as bacteria designed to detect contaminants in a liquid (Jones et al. 2022).
- **Therapeutic Cells**: Designing cells that respond to selective biological signals, for example, to release drugs only in the presence of tumors (Chladt et al. 2021).
- **Industrial Production**: Optimizing microorganisms to produce biofuels or chemicals (Consensus n.d.).
- **Basic Research**: Studying interactions between genes and regulators through synthetic circuits (Kubaczka et al. 2024).

The quality of Cello®'s design depends on the completeness and reliability of the genetic libraries available for the target organism. Operationally, compatibility issues often arise, and not all designed circuits function correctly when inserted into complex organisms. Ultimately, fluctuations in gene expression can affect the stability of the designed circuit. In the field of *Synthetic Biology*, we are currently at a stage where design is not far from being fully guided by specific CAD systems and computer software products. First, these tools are often difficult to locate, as they are scattered across different platforms and repositories, with no standardized keywords for efficient online searches. Second, accessibility remains an issue—some tools require registration, purchase, or access fees, creating barriers for users. Third, interoperability is a major limitation; most tools do not seamlessly integrate with one another, making it difficult to establish computational workflows that ensure clear communication and reproducibility of results. Finally, and perhaps most critically for widespread adoption, many of these tools have a steep learning curve, requiring specialized expertise that limits their usability for a broader

[5] *Biological background* noise refers to random fluctuations that affect cellular processes. These variations can come from various sources, such as gene expression differences (leading to variable protein levels), environmental changes (like temperature or nutrient availability), and stochastic molecular interactions (due to probabilistic chemical reactions within the cell). In digital renderings of the molecular environment, this noise is represented by chaotic movements simulating Brownian motion around the cell or organelle. Background noise can influence cellular behavior, causing differences in gene activity even among genetically identical cells. While some cells use this variability to adapt or differentiate, excessive noise can disrupt biological processes. In brain activity, background noise was once seen as a mere byproduct of biochemical activity, but recent studies suggest it could offer insights into sleep, aging, and conditions like coma or deep anesthesia.

scientific community. Scientific workflows offer a solution to these challenges by providing open, web-based platforms that facilitate findable and accessible data analysis while linking computational processes to experimental protocols. These workflows promote interoperability and reproducibility across different platforms, enabling researchers—regardless of their computational expertise—to integrate diverse tools into a cohesive analysis pipeline. For scientists without programming skills, computational approaches can be daunting, from installing tools to selecting appropriate parameters and efficiently combining multiple applications. Scientific workflows address these issues by structuring data processing into configurable steps, often featuring user-friendly graphical interfaces that enhance usability and efficiency. Moreover, workflow platforms create opportunities for tool developers to reach a wider audience, fostering broader integration and accelerating advancements in their respective fields. Among existing workflow systems, *Galaxy* stands out as a platform originally developed for genome analysis. Today, it hosts over 8,500 tools available in the public ToolShed, providing a powerful ecosystem for reproducible computational research. One of the most advanced applications interfacing with Galaxy is SynBioCAD®; this integration led to the creation of the Galaxy-SynBioCAD® portal, an open-source workflow environment designed for the design and visualization of genetic circuits within the fields of synthetic biology, metabolic engineering, and industrial biotechnology, which derives from CAD engineering design environments (Hérisson et al. 2022). Its architecture is based on the *Synthetic Biology Open Language* (*SBOL*) and the *Systems Biology Markup Language* (*SBML*). The former is a free and open-source standard for representing biological designs, developed by the synthetic biology community to create a standardized format for the electronic exchange of information about the structural and functional aspects of biological projects. The latter is a standardized XML-based format designed to facilitate collaboration between researchers and the integration of models across various software platforms for representing and exchanging computational models of biological processes, including metabolic pathways, cellular signaling pathways, and gene networks (SBML.org n.d.). Galaxy-SynBioCAD® "combines the proven efficiency of RetroPath2.0 and automated script generator to drive robotized workstations for vector assembly and strain transformation" (MICALIS 2024).

Galaxy-SynBioCAD® is an open-source computational platform that integrates all stages of synthetic biology design into a single environment. It combines retrosynthesis software based on generalized reaction rules that generate networks linking the target compounds to the metabolites of the chassis strain, downstream analysis methods using flux analysis, and experimental design tools that identify the DNA parts to be assembled for building pathway libraries. The platform also includes an automated script generator that controls robotized workstations for vector assembly and strain transformation, offering a versatile one-stop destination that connects multiple tools in one place. It enables users to construct customized workflows for pathway generation and analysis, pathway ranking, genetic part design, and the programming of robotic equipment, all of which can be seamlessly integrated with existing data files. A user-friendly graphical interface allows for rapid testing and prototyping, even for users with little or no programming experience (JFAulon.com 2022).

In conclusion, the ELM approach and its application in genetic circuit design represent a significant step toward the synthesis of increasingly sophisticated biological systems. The integration of artificial intelligence and automated design not only facilitates the realization of complex genetic circuits but also fosters a new vision of biology as a continuously evolving discipline. In this context, the role of bio CAD/CAM systems emerges as a crucial element: computer-assisted design applied to biology allows for the modeling and simulation of genetic circuits with unprecedented precision, similar to what happens in electronic engineering. The *Co-evolution* between nature and artifact is becoming increasingly evident in emerging *Techno-Biological Symbiosis* (Brangier and Adelé 2011), as demonstrated by pioneering projects such as Neri Oxman's *Material Ecology® Manifesto*, where biomimicry goes beyond mere emulation of nature to become a fusion process between biology and engineering, giving rise to materials and structures that evolve in symbiosis with their environment. This paradigm finds a striking analogy with the concept of the *Wunderkammer*, where the collection of *naturalia* and *artificialia* was not merely an aesthetic exercise but an attempt to understand and organize the mysteries of the world through objects that embodied the intersection of science, art, and technology. Similarly, bio CAD/CAM platforms such as Cello® (*Cello Design*) and SynBioCAD® (*Galaxy-SynBioCAD*) are transforming synthetic biology into a new form of digital *Wunderkammer*, where the ability to design and simulate genetic circuits helps reduce biological uncertainty and enhance the robustness of engineered biological systems. This approach not only redefines the relationship between design and biology but also paves the way for a new era of intelligent materiality, where the boundary between what is natural and what is artificial becomes increasingly blurred, leading us toward a future in which life itself may be co-designed with the support of advanced computational tools.

References

Altin-Yavuzarslan, G., Brooks, S. M., Yuan, S. F., Park, J. O., Alper, H. S., and Nelson, A. (2023). Additive manufacturing of engineered living materials with bio-augmented mechanical properties and resistance to degradation. Advanced Functional Materials, 33(24): 2300332.

Raman, R., Sreenivasan, A., Suresh, M., and Nedungadi, P. (2024). Mapping biomimicry research to sustainable development goals. Scientific Reports, 14(1): 18613.

Antonelli, P., and Burckhardt, A. (2020). Neri Oxman: Material Ecology [Exhibition preview]. Museum of Modern Art. Available at: https://www.moma.org/d/pdfs/W1siZiIsIjIwMjAvMDMvMTgvN3h0aTB0ZDIwZV9Nb01BX094bWFuX1BSRVZJRVcucGRmIl1d/MoMA%20Oxman%20PREVIEW.pdf?sha=b9de8a989ee0332c [Accessed: 9 May 2025].

Appiah, J. K. (2019). Living materials herald a new era in soft robotics. Advanced Materials, 31(44): 1807747. Available at: https://doi.org/10.1002/adma.201807747 [Accessed: 10 May 2025].

Axpe, E., Duraj-Thatte, A., Chang, Y., Kaimaki, D.-M., Sanchez-Sanchez, A., Caliskan, H. et al. (2018). Fabrication of Amyloid Curli Fibers–Alginate Nanocomposite Hydrogels with Enhanced Stiffness. ACS Biomaterials Science & Engineering, 4.

Bader, C., Smith, R., Sharma, S., Kolb, D., and Oxman, N. (2019). Methods and apparatus for regulation of gene expression across a large-scale solid structure. U.S. Patent Application US20190276793A1. Available at:https://patents.google.com/patent/US20190276793A1/en [Accessed: 10 May 2025].

Beltoforion (2025). John Conway's Game of Life – An Introduction to Cellular Automata. Beltoforion. Available at: https://beltoforion.de/en/game_of_life/ [Accessed: 10 May 2025].
Bilitchenko, L., Liu, A., Cheung, S., Weeding, E., Xia, B., Leguia, M. et al. (2011). Eugene – A domain-specific language for specifying and constraining synthetic biological parts, devices, and systems. PLOS ONE, 6(4): e18882.
Blok, V. (2022). Technology as mimesis: Biomimicry as regenerative sustainable design, engineering, and technology. Techné: Research in Philosophy and Technology, 26(3): 426–446.
Borelli, G. A. (1680). De motu animalium Io. Alphonsi Borelli ... opus posthumum. Romae: Ex typographia Angeli Bernabò. Available at: https://archive.org/details/McGillLibrary-osl_motu-animalium_B731m1680_81_v1-20170 [Accessed: 9 May 2025].
CADENAS GmbH (2025). 60 Years of CAD Infographic: The History of CAD since 1957. Available at: https://www.3dfindit.com/en/engiclopedia/60-years-of-cad-infographic-the-history-of-cad-since-1957 [Accessed: 9 May 2025].
Camilleri, M., Narayan, S., Lingam, D., and Blundell, R. (2025). Mycelium-based composites: An updated comprehensive overview. Biotechnology Advances, 79: 108517.
CD ComputaBio (n.d.) CD ComputaBio. Available at:https://www.computabio.com gad_source=1&gclid=Cj0KCQiA1Km7BhC9ARIsAFZfEIvLjrkPAgq3tFn_7pcq0LfK8ul_lW9hErralRe_imi74ttJpznoqwIaAmQyEALw_wcB [Accessed: 9 May 2025].
Cell Guidance Systems (2022). ELMs: Engineered living materials adapting to life. Available at: https://www.cellgs.com/blog/elms-engineered-living-materials-adapting-to-life.html [Accessed: 10 May 2025].
Chandran, D., Bergmann, F. T., and Sauro, H. M. (2010). Computer-aided design of biological circuits using TinkerCell. Bioengineering Bugs, 1(4): 274–281.
Charitonidou, M. (2020). László Moholy-Nagy and Alvar Aalto's Biocentric Vision of Design and their Admiration for Invisible Structures: Elastic Standardization as Biosemiotics. CARTHA Magazine. Available at: https://www.carthamagazine.com/wp-content/uploads/2021/02/CARTHA_2020_INVISIBLE-STRUCTURES_Marianna_Charitonidou.pdf [Accessed: 9 May 2025].
Chladt, T., Engelmann, N., Kubaczka, E., Hochberger, C., and Koeppl, H. (2021). Automated design of robust genetic circuits: Structural variants and parameter uncertainty. ACS Synth. Biol., 10(12): 3316–3329.
CIDAR Lab. (2023). Cello. Available at: https://www.cidarlab.org/cello [Accessed: 9 May 2025].
Ciulla, M. G., Massironi, A., Sugni, M., Ensign, M. A., Marzorati, S., and Forouharshad, M. (2023). Recent advances in the development of biomimetic materials. Gels, 9(10): 833.
Coineau, Y., Kresling, B. (1987). Les inventions de la nature et la bionique. Hachette.
Consensus. (n.d.) What are the applications of synthetic biology in biofuel production and bioremediation? Available at: https://consensus.app/questions/what-applications-synthetic-biology-biofuel-production/ [Accessed: 9 May 2025].
Coutinho, A. R. (2024). When Nature Meets AI: Reimagining Our Relationship with the Natural World for the Common Good. IIPP Student Ideas. Available at: https://medium.com/iipp-mpa-blog/when-nature-meets-ai-reimagining-our-relationship-with-the-natural-world-for-the-common-good-b63016444eeb [Accessed: 9 May 2025].
Ecologies of Knowing (2023). Neri Oxman. Ecologies of Knowing. Available at: https://ecologiesofknowing.com/tag/neri-oxman/ [Accessed: 9 May 2025].
Elowitz, M. B., and Leibler, S. (2000). A synthetic oscillatory network of transcriptional regulators. Nature, 403(6767): 335–338.
Emlen, D. J. (2014). Animal weapons: The evolution of battle. Henry Holt and Company.
Entangled Others (2021). Artificial Remnants. Available at: https://entangledothers.studio/artificial-remnants/ [Accessed: 9 May 2025].
FasterCapital (2025, April 4). Biology of Wiring: The Influence of CSB on Genetic Circuits. Available at: https://fastercapital.com/it/contenuto/Biologia-del-cablaggio--l-influenza-di-CSB-sui-circuiti-genetici.html [Accessed: 10 May 2025].
Francé, R. H. (1943). Plants as Inventors. Available at: https://archive.org/details/b29819301/page/n5/mode/2up [Accessed: 10 May 2025].
Gamble, S. (2019). Quantum Computing: What It Is, Why We Want It, and How We're Trying to Get It. In: National Academy of Engineering. Frontiers of Engineering: Reports on Leading-Edge Engineering

from the 2018 Symposium. Washington (DC): National Academies Press (US). Available at: https://www.ncbi.nlm.nih.gov/books/NBK538701/ [Accessed: 9 May 2025].

Gardner, T. S., Cantor, C. R., and Collins, J. J. (2000, January 20). Construction of a genetic toggle switch in *Escherichia coli*. Nature, 403(6767): 339–342.

Global BioImaging (n.d.) Imaging Technologies and Global Goals. Global BioImaging. Available at: https://globalbioimaging.org/articles/imaging-technologies-and-global-goals [Accessed: 9 May 2025].

Goethe, J. W. Von. (1790). Attempt to Explain the Metamorphosis of Plants. Gotha: Carl Wilhelm Ettinger. Available at: https://archive.org/details/metamorphoseder00goetWikipedia+3 [Accessed: 9 May 2025].

Herrero, M. (2014). Alan Turing´s work on morphogenesis. Arbor, 189, a081.

Hodgkin, R. A. (1990). Techne, technology and inventiveness. Oxford Review of Education, 16(2): 207–217.

Horizon Europe (2021). Engineered Living Materials. Available at: https://www.horizon-europe.gouv.fr/engineered-living-materials-25028 [Accessed: 10 May 2025].

Imperial College London (2021). Self-healing living materials used as 3D building blocks. Available at: https://www.imperial.ac.uk/news/228633/self-healing-living-materials-used-3d-building/ [Accessed: 10 May 2025].

J. Craig Venter Institute (2010). First self-replicating synthetic bacterial cell. Available at: https://www.jcvi.org/research/first-self-replicating-synthetic-bacterial-cell [Accessed: 10 May 2025].

Jabeen, N., and Atif, M. (2023). Polysaccharides based biopolymers for biomedical applications: A review. Pharmaceutical Technology, 6203.

Jaynes, C. H. (2023). These biodegradable plastics made from algae are a new hope for the environment. World Economic Forum. Available at: https://www.weforum.org/stories/2023/08/biodegradable-plastics-spirulina-algae/ [Accessed: 10 May 2025].

JFAulon.com (2022) The automated Galaxy-SynBioCAD pipeline for synthetic biology design and engineering, Nature Communications. JFAulon.com. Available at: https://jfaulon.com/galaxy-synbiocad-portal/ [Accessed: 21 May 2025].

Jones, T., Oliveira, S., Myers, C., Voigt, C., and Densmore, D. (2022). Genetic circuit design automation with Cello 2.0. Nature Protocols, 17: 1–17.

Kubaczka, E., Gehri, M., Marlhens, J. J. M., Schwarz, T., Molderings, M., Engelmann, N. et al. (2024). Energy aware technology mapping of genetic logic circuits. ACS Synthetic Biology, 13(10): 3295–3311.

Lantada, A. D., J Korvink, Islam, M. (2022). Taxonomy for engineered living materials. Cell Reports Physical Science, 3(4): 100807.

Leal, I., Noritomi, P., Idogava, H., and Kemmoku, D. (2017). BioCAD applications in anatomical structures. XXV Congresso de Iniciação Científica da UNICAMP.

Levity (2025). Deep Learning vs. Machine Learning – What's The Difference? Available at: https://levity.ai/blog/difference-machine-learning-deep-learning [Accessed: 9 May 2025].

Li, J. (2024). A study of advances in machine learning-based electronic design automation. Applied and Computational Engineering, 102: 96–101.

Li, K., Wei, Z., Jia, J., Xu, Q., Liu, H., Zhong, C. et al. (2023). Engineered living materials grown from programmable Aspergillus niger mycelial pellets. Materials Today Bio., 19: 100545.

Liu, S., and Xu, W. (2020). Engineered living materials-based sensing and actuation. Frontiers in Sensors, 1, Article 586300. Available at: https://doi.org/10.3389/fsens.2020.586300 [Accessed: 10 May 2025].

Luo, J., Chen, J., Huang, Y., You, L., and Dai, Z. (2023). Engineering living materials by synthetic biology. Biophysical Reviews, 4(1): 011305.

Materially. (2025). Nature X Humanity: Neri Oxman's guidelines to ensure that design regenerates nature instead of exploiting it. Available at: https://www.materially.eu/en/nature-x-humanity-neri-oxmans-guidelines-to-ensure-that-design-regenerates-nature-instead-of-exploiting-it/ [Accessed: 10 May 2025].

Mediated Matter (n.d.) Overview. MIT Media Lab. Available at: https://www.media.mit.edu/groups/mediated-matter/overview/ [Accessed: 10 May 2025].

MICALIS (2024). Galaxy SynBioCAD. MICALIS. Available at: https://www.micalis.fr/equipe/galaxy-synbiocad/ [Accessed: 21 May 2025].

Michigan State University (2023). What Is the Difference Between CAD, CAE and CAM? MSU Online Engineering. Available at: https://online.egr.msu.edu/articles/cad-vs-cae-vs-cam-what-is-the-difference/ [Accessed: 9 May 2025].

Mogas-Soldevila, L., Duro-Royo, J., and Oxman, N. (2014). Water-based robotic fabrication: Large-scale additive manufacturing of functionally graded hydrogel composites via multichamber extrusion. 3D Printing and Additive Manufacturing, 1(3): 141–151.

Molinari, S., Tesoriero, R. F., and Ajo-Franklin, C. M. (2021). Bottom-up approaches to engineered living materials: Challenges and future directions. Matter, 4(10): 3095–3120.

National Institute of Standards and Technology (n.d.) Quantum Bioimaging. U.S. Department of Commerce. Available at: https://www.nist.gov/programs-projects/quantum-bioimaging [Accessed: 9 May 2025].

Nature Portfolio. (2025). Genetic circuit engineering. Available at: https://www.nature.com/subjects/genetic-circuit-engineering#:~:text=Genetic%20circuit%20engineering%20is%20the,%2Ddown%2C%20global%2Dgenomic%20engineering. [Accessed: 10 May 2025].

Nature Research (2025). Bioinspired materials. Available at: https://www.nature.com/subjects/bioinspired-materials#:~:text=Bioinspired%20materials%20are%20synthetic%20materials,natural%20materials%20or%20living%20matter [Accessed: 8 May 2025].

Next Nature Network (2025). Next Nature Magazine. Next Nature Network. Available at: https://nextnature.org/en/magazine [Accessed: 9 May 2025].

Nielsen, A. A., Der, B. S., Shin, J., Vaidyanathan, P., Paralanov, V., Strychalski, E. A. et al. (2016). Genetic circuit design automation. Science, 352(6281): aac7341.

Nona Research Foundation (2024). Creating a Synthetic Biology Software Commons. Nona Research Foundation. Available at: https://www.nonasoftware.com/ [Accessed: 21 May 2025].

Otto, F. (2017). Thinking by Modeling (G. Vrachliotis, Ed.; S. Cowper & A. Kossack, Trans.). Spector Books.

Oxman (2018). Water-Based Fabrication. Mediated Matter Group, MIT Media Lab. Available at: https://oxman.com/projects/water-based-fabrication [Accessed: 10 May 2025].

Oxman (2019a). Aguahoja. Mediated Matter Group, MIT Media Lab. Available at: https://oxman.com/projects/aguahoja [Accessed: 10 May 2025].

Oxman (2019b). Hybrid Living Materials. Mediated Matter Group, MIT Media Lab. Available at: https://oxman.com/projects/hybrid-living-materials [Accessed: 10 May 2025].

Oxman (2025a). Computational Design Engineer. Available at: https://jobs.lever.co/OXMAN/7e016ab6-c0ab-4437-9f75-bb12e8194da9 [Accessed: 10 May 2025].

Oxman (2025b). Work. OXMAN. Available at: https://oxman.com/work [Accessed: 10 May 2025].

Perch Energy (2024). Biomimicry: Sustainable Design Solutions. Perch Energy. Available at: https://www.perchenergy.com/blog/innovation/biomimicry-sustainable-design-solutions [Accessed: 9 May 2025].

POV-Ray (n.d.) Chemical/Molecular Modelling Programs. POV-Ray. Available at: https://www.povray.org/resources/links/3D_Programs/Chemical_Molecular_Modelling_Programs/[Accessed: 9 May 2025].

Proksch, G. (2024). Engineered living materials (ELMs) for the built environment. (pp. 35–36). *In:* Barnes, G., and Satterfield, B. (Eds.), 112th ACSA Annual Meeting Proceedings, Disruptors on the Edge.

SBML.org (n.d.) SBML: Systems Biology Markup Language. SBML.org. Available at: https://sbml.org/ [Accessed: 21 May 2025].

Sharma, S., and Sarkar, P. (2019). Biomimicry: Exploring research, challenges, gaps, and tools. In Proceedings of ICoRD 2019 Volume 1 (pp. 87–97). Springer.

Shim, H., and Fishwick, P. A. (2008). Lotka-Volterra Model. Encyclopedia of Ecology. ScienceDirect. Available at: https://www.sciencedirect.com/topics/earth-and-planetary-sciences/lotka-volterra-model [Accessed: 10 May 2025].

Skibsted, J. M. (2023, August 18). Why Embracing Biodesign Is Essential For The Planet's Future. Forbes Technology Council. Available at: https://www.forbes.com/councils/forbestechcouncil/2023/08/18/why-embracing-biodesign-is-essential-for-the-planets-future/ [Accessed: 10 May 2025].

Smith, R. S. H., Bader, C., Sharma, S., Kolb, D., Tang, T.-C., Hosny, A. et al. (2020). Hybrid living materials: Digital design and fabrication of 3D multimaterial structures with programmable biohybrid surfaces. Advanced Functional Materials, 30(19): 1907401.

Stratasys. (2025). Objet350/500 Connex 1-2-3 3D Printers. Stratasys Support Center. Available at: https://support.stratasys.com/en/Printers/PolyJet-Legacy/Objet350-500-Connex-1-2-3 [Accessed: 10 May 2025].

SynBioBeta. (2024). Engineered spores create living plastics and the future of biodegradable materials. SynBioBeta. Available at: https://www.synbiobeta.com/read/engineered-spores-create-living-plastics-and-the-future-of-biodegradable-materials [Accessed: 10 May 2025].

The Sea Ranch Association (2020). The Sea Ranch Design Manual and Rules. Available at: https://www.tsra.org/wp-content/uploads/2020/06/DM_v7.pdf [Accessed: 28 May 2025].

Thompson, D. W. (1945). On Growth and Form. Cambridge University Press. Available at: https://archive.org/details/ongrowthform0000thom [Accessed: 10 May 2025].

Trimble Constructible (n.d.) What Is Computational Design? Trimble Constructible. Available at: https://constructible.trimble.com/construction-industry/what-is-computational-design [Accessed: 9 May 2025].

Tulley, R. (2008). Is there techne in my logos? On the origins and evolution of the ideographic term—technology. International Journal of Technology, Knowledge and Society, 4(1): 93–104.

UGREEN (2025). Interior Designers, and Construction Companies. Available at: https://ugreen.io/biomimicry-for-sustainable-design-a-game-changer-for-architects-interior-designers-and-construction-companies/ [Accessed: 9 May 2025].

University of Maine (2025). Computational Ecology. University of Maine. Available at: https://umaine.edu/cs/research/features/computational-ecology/ [Accessed: 9 May 2025].

Van Stolk, C. (2025). How One Failure Changed the World: The Story of Kevlar. The Edge of Yesterday. https://www.edgeofyesterday.com/time-travelers/how-one-failure-changed-the-world-the-story-of-kevlar.

Vialva, T. (2020). Researchers develop Hybrid Living Materials using inkjet 3D printing. 3D Printing Industry. Available at: https://3dprintingindustry.com/news/researchers-develop-hybrid-living-materials-using-inkjet-3d-printing-167762/ [Accessed: 10 May 2025].

VoxelMatters (2025). Medical additive manufacturing. VoxelMatters. Available at: https://www.voxelmatters.com/category/medical/ [Accessed: 10 May 2025].

Wakle, A., and Khuntia, S. (2023). Developments and studies in bio-emulsion liquid membranes (Bio-ELM) synthesis, parameters studies, characterizations, and applications. Journal of Water Process Engineering, 56: 104300.

Wang, A., Liu, Y., Li, J., Chen, Y., Liu, S., and Zhong, C. (2022). Engineered living materials (ELMs) design: From function allocation to dynamic behavior modulation. Current Opinion in Chemical Biology, 70: 102188.

Watson and Wolfe (2020). What is Pineapple Leather? Available at: https://www.watsonwolfe.com/2020/08/29/what-is-pineapple-leather/ [Accessed: 10 May 2025].

Xia, P., Peng-Fei, X., Ling, H., Foo, J. L., and Chang, M. (2019). Synthetic genetic circuits for programmable biological functionalities. Biotechnology Advances, 37(7): 107398.

Yu, Y., Wang, Q., Wang, C., and Shang, L. (2021). Living materials for regenerative medicine. Engineered Regeneration, 2: 96–104.

2

Making New High Tech Biomaterials in the Anthropocene Era

"There is a fundamental error in treating the Earth as if it were a business in liquidation."

Herman E. Daly

"Nature has strange laws,
but at least she abides by them."

Leo Longanesi

2.1 The *Great Acceleration* and the *Biomimetic Brake*

To understand the current remarkable development of studies, research, and concrete proposals aimed at improving the sustainability of human impact on the natural ecosystem, it is helpful to briefly outline the historical and social factors that led to the concept of *Environmental Sensitivity* (*ES*) (Tang et al. 2024). *ES* is the ability to design and develop products, processes, or services that assess and minimize their ecosystem impact from the very beginning, choosing materials and construction methods that respect natural resources and local habitats. It manifests as the promotion of resource circularity—encouraging the reuse, recycling, or biodegradation of components to extend their useful life and reduce waste. It also involves integrating energy efficiency and raw-material conservation criteria throughout the entire life cycle—from production to disposal—and actively engaging users, communities, and stakeholders, educating and empowering them so that technological innovation always operates in harmony with environmental well-being. Emerging as a significant theme in the 1960s and 1970s, this eco-concept was fueled by the growing awareness of human impact on the environment, spurred by key events like the 1972 Limits to Growth report by the Club of Rome (Meadows et al. 1972). While some economists rejected its warnings about resource depletion, the report ignited global debates and highlighted environmental challenges. The roots

DOI: 10.1201/9781003438946-2

of the environmental movement trace back to the 19th century, with pioneers like John Muir and Beatrix Potter advocating for nature conservation. In the 1960s and 1970s, grassroots movements emerged, catalyzed by events like the Cuyahoga River disaster and Rachel Carson's Silent Spring (Natural Resources Defense Council 2015). These movements gained momentum with organizations like Greenpeace and Friends of the Earth. Major conferences, such as the 1972 Stockholm Conference and the 1992 Earth Summit in Rio de Janeiro, laid the foundation for international environmental action. The 2015 Paris Agreement further advanced global commitments to tackle climate change. Despite progress, challenges remain. Deforestation, species extinction, pollution, and climate change continue to pose urgent threats. The environmental movement has sparked global efforts to protect the planet, but it faces significant opposition. The concept of ecological transition, first introduced at the Earth Summit, has gained prominence, calling for a structural shift towards more renewable economic and social models. The 2030 Agenda for Sustainable Development, adopted by all 193 UN member states, outlines 17 Sustainable Development Goals (SDGs) that tackle poverty, hunger, inequality, and environmental sustainability. Among these, SDG 12 (Responsible Consumption and Production), SDG 13 (Climate Action), SDG 14 (Life Below Water), SDG 15 (Life on Land), and SDG 9 (Industry, Innovation, and Infrastructure) directly relate to bioinspired design innovations, aiming to reduce environmental impact through eco-friendly materials and practices. Despite these efforts, political and economic tensions threaten to slow progress. National and economic interests often undermine the effectiveness of environmental policies, and the divide between developed and developing nations complicates global cooperation. However, ongoing debates continue to push the environmental agenda forward, with some material progress being made. The ecological transition remains essential for ensuring a sustainable future for both people and the planet. In recent decades, humanity and Planet Earth have witnessed an unprecedented phenomenon known as the *Great Acceleration*, a term first used in 2004 by a group of scientists led by Will Steffen, Paul Crutzen, and John McNeill, who introduced it in an article published in the scientific journal Nature (Steffen et al. 2015). These scholars were examining global changes in the Earth's environment and in biological and geophysical systems starting from the mid-20th century. Since 2004, the initial concept of the Great Acceleration, which was verified from the 1950s onwards, has been continuously updated, evolving into a monitoring model of the parameters that signal the exponential increase of human activity and its impacts on the environment. The authors had established a series of key indicators that showed a significant acceleration in many human activities, such as global population growth, the expansion of the global economy, the use of natural resources, greenhouse gas emissions, deforestation, urbanization, and other phenomena linked to global industrialization. Essentially, this analysis provides evidence—from the perspective of the Earth system—that the onset of the Great Acceleration marks the beginning of the Anthropocene, the Age of *Humans*, a period of rapid change and growth in human activity with significant consequences for the Earth's environment (Ellis 2021).

This term *Anthropocene*, coined in 2000 by Nobel laureate Paul Crutzen and scientist Eugene Stoermer (Crutzen and Stoermer 2017), refers to a geological epoch where human activity has a significant, lasting impact on the Earth's environment. While many of these trends are associated with increased well-being and economic development, they are leading to severe environmental consequences that threaten the stability of our global ecosystem—impacting the biosphere, climate, biodiversity, and other fundamental aspects of the Earth system. The concept of the Great Acceleration has become a key part of our understanding of global environmental changes and their impacts on the planet's sustainability. It has helped to catalyze attention on the urgent need to address environmental challenges and to adopt concrete measures to mitigate the negative impacts on both the environment and society. One of the most alarming indicators associated with the Great Acceleration is the exponential rise in greenhouse gas emissions, which poses serious risks for future generations. Carbon dioxide (CO_2) emissions have more than doubled since 1950, reaching unprecedented levels in human history (Ritchie and Roser 2020). This has significantly contributed to the rise in global temperatures and climate change, with devastating consequences such as extreme heatwaves, extreme weather events, and rising sea levels. In addition to CO_2 emissions, other concerning parameters reflect the Great Acceleration. For instance, deforestation, species extinction, and marine pollution have increased at alarming rates. Deforestation, primarily caused by agricultural expansion and habitat loss, has led to the destruction of valuable forest ecosystems, with severe consequences for biodiversity and the water cycle. Similarly, marine pollution, caused by plastic waste, chemicals, and industrial discharges, threatens marine life and food security for millions of people who rely on the oceans for their livelihood. Another alarming parameter is the increase in the use of non-renewable natural resources. The Great Acceleration has led to a rapid expansion in the extraction of minerals, fossil fuels, and other materials—finite resources of our planet. This has not only led to unsustainable exploitation of resources but also severe environmental impacts such as soil and water pollution, biodiversity loss, and ecosystem destruction. Finally, there is the issue of accelerated population growth. The global population reached 7 billion people in 2011 and continues to grow at a rapid pace. This population increase further strains natural resources and global ecosystems, contributing to the acceleration of negative environmental trends. In summary, this blacklist of alarming indicators associated with the Great Acceleration further highlights the negative impact of human activity on the environment (CBC Radio 2023). These include (Fig. 1):

- **Water resource consumption**: Excessive and unsustainable use of water has become a critical issue in many parts of the world. The Great Acceleration has led to increased demand for freshwater, causing water reserves to shrink, desertification, and competition for resources.
- **Soil loss**: Soil erosion, driven by unsustainable farming practices, urbanization, and deforestation, threatens food security and ecosystem balance by reducing the land's ability to support plant and animal life.
- **Air pollution**: Emissions from industrial sources, vehicles, and other human activities have raised air pollution levels, leading to serious health problems and contributing to global warming.

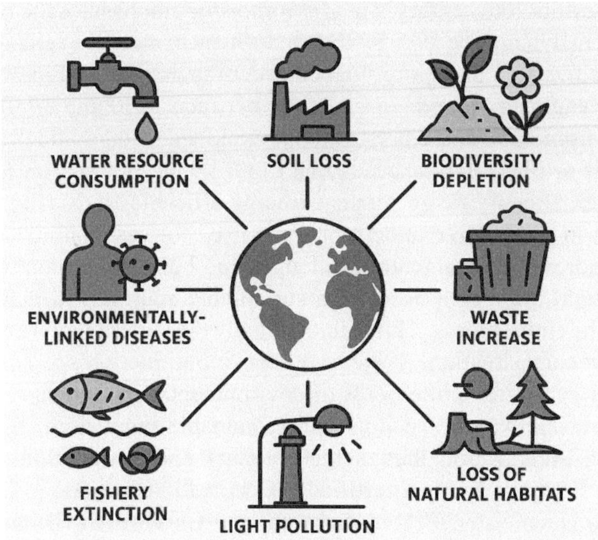

Fig. 1. Negative Impacts of the Great Acceleration.

- **Biodiversity depletion**: The Great Acceleration has caused an unprecedented loss of biodiversity, with species extinctions occurring much faster than natural rates, severely impacting ecosystems.
- **Waste increase**: Waste production, including plastics and e-waste, has surged with population and consumption growth, causing environmental damage and threatening wildlife.
- **Ocean acidification**: Increased CO_2 emissions are absorbed by oceans, leading to acidification, which harms marine ecosystems, especially coral reefs and mollusks.
- **Loss of natural habitats**: Urbanization, intensive agriculture, and resource exploitation have led to the fragmentation and destruction of natural habitats, putting many species at risk.
- **Light pollution**: Excessive artificial lighting disrupts human health and wildlife, interfering with sleep patterns and migration, and reproduction cycles.
- **Fishery extinction**: Overfishing threatens food security for millions and destabilizes marine ecosystems, risking fish population collapse.
- **Environmentally-linked diseases**: Pollution and climate change contribute to rising incidences of respiratory diseases and diseases transmitted by vectors, such as malaria and dengue.

The alarming parameters associated with this phenomenon highlight the urgent need for a radical change in the way we live and operate on Planet Earth. Only through real and collective actions can we hope to reverse these trends and protect our planet for future generations. These additional parameters further illustrate the urgency and complexity of the environmental crisis. This concept is closely linked to this, as both reflect the exponential increase in human activity and its environmental effects since the mid-20th century. The *Anthropocene* emphasizes the broader change in Earth's system dynamics driven by human influence, including changes to climate, landscapes, biodiversity, and more. Just consider the exponential increase in cement production for engineering and construction: between 1930 and 1950 it saw only a slight rise, but from 1950 to the present it has exploded so dramatically that the mass of human-made artifacts now exceeds that of all living biomass on Earth (Syvitski et al. 2022, International Energy Agency. Energy efficiency n.d.) (Fig. 2).

This shift in perspective underscores the need for sustainable and responsible strategies to address environmental challenges and build a sustainable future. Both concepts highlight the urgent need for a sustainable approach to managing natural resources and the environment. This is undoubtedly the current picture of the alarming global environmental situation. As we have seen, biomimicry represents a substantial opportunity for evolution in the fields of environmentalism and sustainable design. Founded to translate Nature's genius into sustainable innovation, Biomimicry 3.8 is a pioneering organization that merges biology and design thinking to inspire transformative solutions and is a certified B Corporation founded in 2010 by Janine Benyus, Dayna Baumeister, Bryony Schwan, and Chris Allen (Biomimicry 2023). The "3.8" in its name refers to the 3.8 billion years of evolution on Earth—a legacy of natural intelligence that the organization seeks to translate into sustainable innovation. It is a consultancy and training firm that promotes nature-inspired innovation through professional development and corporate services, working with companies such as Nike, Kohler, and Seventh Generation. AskNature.org is an open-access digital platform launched in 2008 by the Biomimicry Institute, a nonprofit organization founded in 2006 by Janine Benyus and Dayna Baumeister. The site functions as a biological knowledge base, offering a rich collection of natural strategies to support sustainable design and innovation (Biomimicry Institute 2025). The Biomimicry 3.8 design movement aims to inspire sustainable innovation by offering access to biological strategies developed over millions of years of evolution. Metaphorically,

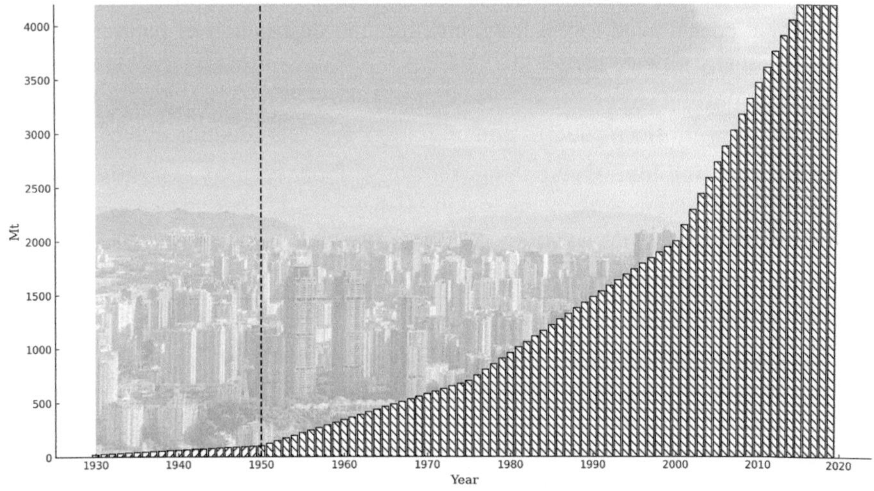

Fig. 2. 1930–2025 – The Great Acceleration of Cement's Mass on Earth.

it acts not only as a brake on the reckless momentum of the Great Acceleration, but also as a counter-acceleration—driving us toward a co-evolved, life-centered future where design aligns with the principles of regeneration and resilience. Biomimicry 3.8, also thanks to its widespread global diffusion since its founding—through a network of certified educators and practitioners—has progressively colonized the design and technology innovation landscape. In ecological terms, it spreads like a beneficial symbiosis, seeding a life-centered philosophy that concretely positions itself on the side of nature, regeneration, and long-term planetary viability. The platform's mission is to facilitate the adoption of these natural solutions in addressing today's environmental, social, and economic challenges. Biomimicry 3.8 collaborates with businesses, academic institutions, governments, and other stakeholders to apply biomimetic strategies across a range of sectors—including architecture, industrial design, engineering, agriculture, and healthcare. A central achievement of the movement has been the promotion of biomimicry in educational curricula and industrial design practices. As a result, a growing number of companies are integrating biomimicry into their innovation and sustainability strategies, developing products that are more efficient, less polluting, and better adapted to their operational environments. Through its publications, Janine Benyus and her team clarify the role of nature in biomimicry, breaking it down into three fundamental concepts: Nature as model, measure, and mentor—a framework that continues to guide the development of sustainable technologies and design.

Nature as Model: Biomimicry is described as a new science that studies Nature's models and seeks to emulate or take inspiration from them to solve human problems. In this perspective, natural organisms, systems, and processes become sources of inspiration for the development of innovative technologies and solutions. For example, the structure of leaves can inspire the development of water-repellent materials, while animal locomotion may suggest new designs for robots and vehicles (Roh et al. 2024).

Nature as Measure: Biomimicry adopts an ecological standard to evaluate the appropriateness of human innovations. After 3.8 billion years of evolution, Nature has accumulated a vast repository of knowledge about what works, what is suitable, and what endures. As such, humanity can look to it as a benchmark for assessing the effectiveness and sustainability of its solutions and innovations (Sowińska-Świerkosz and García 2022).

Nature as Mentor: Biomimicry introduces a new way of perceiving and valuing life, shifting the focus from what we can extract from Nature to what we can learn from it. In this paradigm, nature becomes a mentor—offering profound lessons in design, adaptation, and sustainability that can guide human innovation toward more effective and ecologically integrated solutions (Fiorentino and Montana-Hoyos 2014).

This approach—integrating biology and its research methods as an essential part of industrial design—can open new pathways for sustainable innovation, transforming the way we conceive and develop technologies and solutions for the future. Moreover, the development of biomimetic analysis is indirectly fostering a deeper and more comprehensive understanding of natural design, significantly contributing to the systematic knowledge of the dynamics and technologies employed by living organisms, including humans. From its early theorization and international dissemination, biomimicry and biomimetics have profoundly influenced the evolution of advanced methodologies and frameworks within CAD (Computer-Aided Design), CAM (Computer-Aided Manufacturing), and CAE (Computer-Aided Engineering) systems. This influence extends beyond formal or aesthetic inspiration, finding deep resonance in the emerging fields of parametric and generative design. Here, bioinspired algorithms, adaptive geometries, and material strategies are not merely decorative, but function as systemic tools that reconnect design processes with the dynamic flows of Earth's living systems—offering an ecologically attuned approach to innovation that aligns with the principles of life itself (Diaz Lantada et al. 2012). The unprecedented exploration of Nature's domain, stimulating new morphological and functional perspectives in design culture, plays a key role in the evolution of tools and methodologies that emulate natural forms and bio-processes to address design and manufacturing challenges in a more sustainable, efficient, and innovative manner. For example, starting from the early 2000s, research and study streams have developed that connected the *NIST Functional Basis*, developed by the National Institute of Standards and Technology in the United States, with the *Biomimicry Taxonomy*, providing designers with access to databases of natural phenomena to inspire new ideas in biomimetic design (Biomimicry Institute 2024). In the field of design engineering, the need for a shared and formalized language to describe the functions of technical systems has led to the development of taxonomic structures capable of standardizing functional analysis processes. In particular, the Biomimicry Taxonomy represents a paradigm shift in engineering design, moving the focus away from form or technology toward a deeper understanding of the functions that nature has refined over billions of years of evolution. This convergence of engineering design methodologies and bioinspired frameworks has led to the development of standardized vocabularies aimed at enhancing conceptual modeling and innovation. Among these, the NIST Functional Basis (Hirtz et al. 2002) provides a structured system for representing technical functions through a modular syntax based on two core components: flows (matter, energy, signal) and functional operators (such as convert, transport, control, store, or sense). Functions are thus described as simple yet expressive pairings—e.g., *Convert Energy, Transport Material, Control Signal*—which facilitate abstraction, comparison, and computational manipulation. This design approach resonates strongly with the Biomimicry Taxonomy developed by the Biomimicry Institute, which classifies biological strategies according to the functions organisms perform to survive and thrive (e.g., capture, protect, communicate, move, transform energy). While the NIST Functional Basis arises from an engineering lineage, and the Biomimicry Taxonomy from a biological one, both converge on the recognition that function is a universal language bridging nature and technology. Their integration enables designers to translate natural strategies into actionable engineering functions, fostering a systematic exploration of Nature's time-tested solutions within generative design environments. The integrated conceptual structure—from problem-setting to problem-solving—of this framework enables designers to explore and apply the hierarchical functional logic of natural strategies in a systematic and innovative way. By leveraging advanced tools such as generative design, CAD, CAE, and computational biology protocols, biomimicry facilitates a functional translation of Nature's evolved solutions into engineering contexts. In doing so, the biomimicry methodology not only fosters sustainable innovation, aiming to counteract the momentum of the Great Acceleration, but also opens new pathways for the rational design of solutions that are more efficient, adaptable, and ecologically aligned (Alexandridis et al. 2016). This approach proves particularly useful not only for the analysis of existing products but also for the generation of new design concepts. In particular, within the biomimetic design paradigm, the NIST Functional Basis proves to be a powerful tool for establishing functional correspondences between biological solutions and technical solutions, facilitating the mapping of natural functions into engineering terms. Moreover, in advanced and generative design environments, where algorithms and artificial intelligence play a role in decision-making processes, the standardization of functions according to the NIST Functional Basis enables effective semantic communication between functional inputs and design outputs. This approach facilitates the implementation of CAD systems that incorporate biological knowledge, enhancing innovation in computer-aided design. Biomimicry methodology is inspired by the models and solutions found in living organisms, introducing concepts that range from the geometry of natural structures to the optimization of production processes. It encourages the use of forms and materials that minimize waste and enhance structural performance. Parametric design, in particular, has greatly benefited from biomimicry through the adoption of algorithms capable of generating shapes and structures inspired by nature's ingenuity. Biological forms—such as the honeycomb cells of bees, plant growth patterns, and the geometry of seashells—have been integrated into the digital design process to optimize material usage and improve structural strength. These natural geometries are translated into digital models within CAD/CAM and 3D printing workflows, enabling the fabrication of lighter, more efficient, and high-performing solutions—often beyond the reach of conventional design methods (Siddique et al. 2022). The adoption of these models has led to the development of design techniques that emulate natural behavior to improve product sustainability and functionality. In parallel, CAM systems have incorporated concepts of self-organization and optimization drawn from natural behaviors. The use of algorithms inspired by natural phenomena, such as the collective behavior of ants or the organization of insect swarms, has optimized manufacturing processes, improving production flow efficiency and resource management. These algorithms are applied, for example, in optimizing CNC machine movements and reducing production times, allowing CAM technologies to dynamically adapt to production needs and material characteristics. The influence of biomimicry also extends to the design and simulation of advanced materials. CAD simulations today allow the modeling of materials that replicate the behavior of biological tissues, such as skin or plant tissues, and self-healing materials that mimic the repair capabilities of biological systems. Software like SolidWorks®, Autodesk®, and COMSOL® provide models that replicate the material response to external damage, reacting similarly to how human skin or a plant would, enhancing product durability and efficiency through the use of biomimetic materials. Additionally, biomimicry has strongly influenced Additive

Manufacturing (3D printing), where natural structures are emulated to design lightweight and durable objects. CAD/CAM technologies such as Rhino® and Grasshopper® are used to generate designs inspired by complex biological models, such as the vascular networks of plants or the reticulated structure of bones, to create products that optimize material use and production cost. These developments are supported by an increasing body of academic literature examining the intersection of biomimicry and digital design. For instance, the work of Michael Pawlyn (2011) explores how Nature-inspired solutions can be integrated into design and manufacturing processes, including CAD software for architecture and industrial design. Scientific articles address the impact of biomimicry in manufacturing, highlighting how Nature-inspired methodologies can be used to optimize production processes and improve sustainability (Raman et al. 2024). Researchers explore how natural algorithms, such as evolutionary or genetic algorithms, are applied in parametric design processes and CAM simulations (Zang et al. 2010). In addition, in advanced and generative design environments, where algorithms and artificial intelligence play a role in decision-making processes, the standardization of functions according to the NIST Functional Basis enables effective semantic communication between functional inputs and design outputs. This approach facilitates the implementation of CAD systems that incorporate biological knowledge, enhancing innovation in computer-aided design (Howells 2015). These examples demonstrate how the biomimicry approach is transforming design and manufacturing by integrating natural principles into advanced technological processes, fostering innovation and sustainability in the field of digital design and industrial production. The biomimetic design approach involves a series of phases and processes that integrate biological principles and evolutionary strategies into industrial, architectural, and engineering design (Mohammed 2023).

Outlined below are the primary steps in his usual workflow. :

1. **Research and Observation:** The first phase consists of researching and observing models, processes, and structures found in nature that are relevant to the design problem. This includes studying living organisms, ecosystems, and natural phenomena to identify the solutions that nature has developed to address challenges similar to those being tackled in the design process.
2. **Analysis of Biological Principles:** Once the relevant biological models have been identified, the next step is to analyze the underlying biological principles and the evolutionary strategies used by organisms to adapt and thrive in their environments. This may include examining the structure, function, behavior, and interactions of organisms with their surroundings.
3. **Transfer of Biological Principles:** In this phase, the identified biological principles are transferred and applied to the specific design problem. This may involve adapting biological strategies to the design context, identifying analogies between biological mechanisms and design requirements, and integrating biological insights into the design process.
4. **Idea and Solution Generation:** Using biological principles as a source of inspiration, innovative ideas and solutions are generated to address the design

challenge. This may involve developing new design concepts, combining existing solutions with biomimetic approaches, and exploring creative, nature-based alternatives.

5. **Performance and Sustainability Evaluation:** Once the ideas and solutions are developed, their performance and environmental, social, and economic impacts are assessed. This may include evaluating functionality, efficiency, durability, eco-compatibility, and the feasibility of the proposed solutions.
6. **Iteration and Refinement:** Finally, solutions are iterated and refined based on evaluation results and feedback. This ongoing iteration process helps optimize the solutions to ensure they effectively and efficiently meet the design requirements.

In the context of the potential contribution to the design and development of patentable and practically implementable solutions, biomimetic basic research can be divided into two main approaches: bottom-up research and top-down research. These two approaches differ in their methodology and the way they address design and development problems. The AskNature.org biomimetic design approach is structured as an online searchable digital archive, which collects and catalogs a wide range of strategies and adaptations found in nature, providing inspiration for innovation and sustainable design. The database structure is organized to facilitate the search and exploration of information according to:

- **Categories and subcategories:** The database is divided into various categories and subcategories that reflect the main areas of inspiration, such as biodiversity, energy, materials, mobility, and more. This division allows users to navigate through different areas of interest and access the specific information they are seeking.
- **Adaptation strategies:** Within each category and subcategory, the information is organized based on the adaptation strategies observed in nature. For example, under the *Materials* category, there may be subcategories such as *structural strength* or *thermal insulation*, each containing various adaptation strategies developed by organisms to address specific challenges.
- **Examples and case studies:** Each adaptation strategy is accompanied by examples from nature and case studies that illustrate how organisms use them to solve problems and adapt to their environments. These examples are supported by images, detailed descriptions, and scientific references to provide an in-depth understanding of the strategies.
- **Related research and cross-connections:** The database allows users to explore related research and cross-connections between different strategies and categories. For example, a user exploring adaptation strategies for water conservation might find connections with strategies for heat resistance or predator protection, demonstrating how organisms address multiple challenges in an integrated way.
- **Advanced search tools:** AskNature.org also offers advanced search tools that allow users to filter information based on specific parameters, such as the type of environment, geographic region, organism type, and more. This helps users narrow down their search and find relevant information more efficiently.

The following comparative table contrasts Bottom-Up vs Top-Down approaches in biomimetic design:

Table 1. Comparative Characteristics of Bottom-Up vs Top-Down Approaches in Biomimetic Design.

Characteristics	Bottom-Up Research	Top-Down Research
Basic Approach	Starts with the analysis of details and basic components to arrive at a broader understanding of the problem.	Starts with a general overview and focuses on global solutions to address the problem.
Exploration Method	Explores solutions from the bottom, observing individual phenomena or components and integrating them.	Defines initial requirements and objectives, then designs solutions to meet those parameters.
Initial Focus	Analysis of materials, basic system characteristics, or component functionalities.	Definition of global project requirements, objectives, and specifications.
Context of Use	Suitable when the problem is not well-defined or when new creative possibilities are to be explored.	Suitable for well-defined problems with specific requirements to meet.
Biomimetic Design Research	Focuses on detailed analysis of biological models and natural mechanisms developed through evolution.	Starts with a global view of the design problem, identifying requirements, objectives, and challenges.
Application Method	Identifies analogous biological solutions to apply them in the industrial design process.	Guides design toward targeted solutions that meet specific project requirements.
Outcome	Discovery of innovative biological principles that can be transferred to industrial design.	Targeted design solutions that meet technical, economic, and market constraints.

From a methodological and strategic point of view, both approaches have their advantages and limitations, and they can be combined depending on the needs and characteristics of the design problem. Bottom-up research can lead to innovative and unexpected solutions, while top-down research provides a structured and targeted way to solve specific problems. Ultimately, the choice between these two design-research approaches depends on the nature of the design problem, the available resources, and the project objectives. By integrating both in biomimetic design, engineers and designers can develop solutions that draw on biological principles and evolutionary strategies while remaining focused, functional, and capable of addressing market needs and industrial challenges. A hybrid approach in biomimetic design means drawing freely on nature's ingenious solutions—its shapes, materials, and processes—while simultaneously grounding those ideas in real-world constraints like manufacturability, cost, regulatory standards, and performance targets. In practice, this looks like:

- **Harvesting inspiration** from biological models (for example, the microstructure of lotus leaves for self-cleaning coatings or the geometry of bird wings for efficient aerodynamics).

- **Evaluating feasibility** against technical criteria: Can we reproduce that microstructure at scale? Will the materials hold up under industrial conditions?
- **Iterating between the two**: Testing prototypes, feeding back lessons learned into further biological research, and refining designs to balance novelty with reliability.

By continually checking each creative insight against manufacturing processes, budgetary limits, safety regulations, and end-user needs, the hybrid approach ensures that breakthroughs inspired by nature don't remain lab curiosities but evolve into durable, market-ready products. This synergy both accelerates the innovation cycle—by avoiding dead ends—and steers development toward solutions that are ecologically responsible, economically viable, and industrially robust. Below in Table 2, a schematic representation of three practical examples of biomimetic material design, contextualized using both *bottom-up* and *top-down* different approaches.

In recent years, the rapid evolution of CAD/CAM/CAE suites has significantly expanded the potential for integrating biomimicry into digital design and manufacturing workflows. These systems now support a seamless continuum of tools. They range from initial concept modeling to structural analysis and fabrication. Not only are these tools more powerful and interoperable, but they are also increasingly aligned with ecological thinking and nature-based strategies. The following table presents a comparative overview of selected software platforms that are particularly well suited to biomimetic design. It highlights their main features and typical applications within a framework that embraces parametric modeling, generative logic, performance simulation, and advanced manufacturing—enabling designers to explore and implement biological principles with unprecedented precision and flexibility.

2.2 *Euplectella aspergillum*: Modeling Fluid Dynamics in CAD/CAE Using Nature-Inspired High-Tech Principles

Inspired by biomimetics, the sciences of Nature are showing that, even in the invisible depths of the bodies of living organisms that seem simple and primitive, such as marine sponges, shells, or radiolarians, functional biotechnologies are created and act, producing molecular ultrastructures capable of creating materials and morphological structures with astonishing control of energy performance (Blasiak et al. 2022). The literature on this subject has now become vast, encompassing thousands of publications, magazines, reports, and scientific writings that present research and proposals for technological transfers and functional patents, largely inspired by biomimetic analyses (Jatsch et al. 2023). An emblematic example is the case of *Euplectella aspergillum*, commonly known as the glass sponge or Venus' basket, due to the intricate primordial beauty of its elongated body, woven with thin optical fibers of pure silica, 0.2 mm thick (Porfiri 2021). The body of this sponge, which belongs to the class *Hyalospongiae* and is predominantly found in the Pacific Ocean, consists of a highly flexible white cylindrical lattice—and has been modeled using one of the world's most powerful supercomputers. It is assembled from a structure of six-rayed, amorphous hydrated silica spicules, which are freely interconnected in a

Table 2. Examples of Bottom-Up and Top-Down Design Approaches.

Approach	Example	Description	Development Phase	Objective	CAD-CAM-CAE Integration
Bottom-up	Self-healing material inspired by human skin	Scientists study the self-healing capability of human skin to develop synthetic materials that can repair themselves. By analyzing the structure and function of skin tissues, they identified mechanisms that allow for efficient healing.	Analysis of biological mechanisms and design of synthetic materials.	Create materials that can self-repair.	CAD is used to design the material's microstructure; CAM for prototyping; CAE to simulate the material's response to damage and healing processes.
Bottom-up	Super-hydrophobic leaf surface	By analyzing micro and nanostructural features of leaf surfaces, researchers identified mechanisms that allow water rejection and self-cleaning. Development of super-hydrophobic surfaces inspired by plant leaves, using nanotechnology and material engineering.	Study of natural structures to create coatings and surface treatments.	Develop water-repellent and self-cleaning surfaces.	CAD to design the nanostructure of the surfaces; CAM to produce large-scale surface treatments; CAE to simulate hydrophobic behavior and long-term durability.
Bottom-up	Lightweight and strong materials inspired by bone structures	Engineers study animal bone structure to design lightweight, strong materials that mimic the composition and trabecular architecture of bones.	Exploration of bone structure to develop composite and engineered materials.	Design lightweight, strong materials for industrial applications.	CAD to model bone structures and analyze optimal geometries; CAM for producing components with light yet strong structures; CAE to simulate mechanical performance and resistance.
Top-down	Self-healing material inspired by human skin	Engineers define the need for more durable materials for sectors like construction, automotive, and electronics, setting self-healing as a key requirement. They then explore biomimetic solutions inspired by the skin's self-healing ability.	Definition of performance requirements and design of biomimetic solutions.	Meet the need for durable materials with self-healing capabilities.	CAD to design material geometries and self-healing mechanisms; CAM to prototype and test healing capacity; CAE to analyze material resistance and repair processes.
Top-down	Super-hydrophobic leaf surface	Engineers define the need for self-cleaning and hydrophobic materials for environments with high humidity or contamination, selecting appropriate biomimetic solutions.	Design of materials for specific environmental applications and selection of suitable technologies.	Meet the need for hydrophobic and self-cleaning surfaces in contaminated environments.	CAD to design and optimize nanostructured surfaces; CAM for producing functional coatings; CAE to simulate hydrophobic performance and test the treatment's durability.
Top-down	Lightweight and strong materials inspired by bone structures	Engineers define the need for lightweight and strong materials for high-performance applications, based on bone design.	Design and evaluation of materials to optimize lightweight and strength properties.	Create lightweight, strong materials for high-performance applications.	CAD to design and optimize material structures; CAM for producing lightweight and strong prototypes; CAE to simulate mechanical performance and optimize strength and lightness.

network structure organized into a highly regular and hierarchical cylindrical lattice, with a vitreous appearance of great elegance, reminiscent of the weaving of *Vienna straw*. Its unique basket-like structure ensures exceptional flexibility and resistance to damage caused by the forces of currents and underwater movements. The weave of the *basket* is itself made up of two layers that enclose a mesolayer of thin collagen and various cellular elements (archeocytes, sclerocytes, and choanoblasts). For the most part, they exhibit radial symmetry, with lengths varying from 7–10 cm to over 1 meter. The resulting design appears paradoxical as it overcomes the fragility of the material it is made of—glass—showing exceptional strength and mechanical stability. A research team, supported in part by the COPMAT project (Full-scale COmputational design of Porous mesoscale MATerials), funded by the EU, has focused on a largely unexplored aspect of deep-sea sponges (CORDIS 2021). Through biomimetic studies, they investigated the hydrodynamic fields surrounding and penetrating these organisms (Falcucci et al. 2021). In this original study, the application of software tools akin to the CAD/CAE/CAO family is particularly prominent. The findings revealed a strong correlation between the sponge's structure and its function, offering valuable insights into how its reticular skeleton reduces overall hydrodynamic stress from the surrounding water mass. The research operates within the domain of computational physics, with a specific focus on fluid dynamics, and more precisely, Computational Fluid Dynamics (CFD). This methodology leverages numerical analysis and advanced algorithms to solve and interpret fluid-related phenomena through computer-based simulations, using platforms such as COMSOL Multiphysics® (COMSOL 2025).[1] The research team—composed of scientists from Australia, Italy, and the United States—investigated whether the intricate woven patterns of silica filaments that form the sponge's tubular skeleton not only enhance its mechanical strength but also contribute to optimizing hydrodynamic flow inside and outside its body cavity. The skeleton's morphology appears capable of generating a nutrient-rich vortex within the sponge at low flow velocities, thus supporting its biological functions. For their simulations and experiments, the researchers employed the MARCONI100 supercomputer (21.6 petaflops), located at the Cineca high-performance computing center in Italy, along with specialized software developed by Giorgio Amati, Senior Technology Officer at Cineca and co-author of the study. Through the CFD software module, it is possible to perform computational fluid dynamics analyses and simulations, including:

- Internal and external flows
- Incompressible and compressible flows
- Laminar and turbulent flows
- Single-phase and multiphase flows

[1] The COMSOL CFD Module is an advanced simulation tool used to model fluid flow phenomena, including laminar, turbulent, multiphase, and porous media flows. Thanks to its ability to couple multiple physics, it is particularly suitable for studying complex systems such as biomimetic structures interacting with fluid environments, where structural, thermal, and flow dynamics need to be analyzed simultaneously.

Table 3. Comparative Overview of Some CAD/CAM/CAE and Computational Tools Used in Bioinspired Design Workflow.

Software	Type	Main Features	Use
ANSYS®	CAE	Engineering simulations for structural, thermal, fluid dynamics, and more	Performance verification of biomimetic materials and physical simulations
Abaqus®	CAE	Software for nonlinear and complex deformations	Simulating structures and biomimetic materials under extreme stress
Altair HyperWorks®	CAE	Suite for structural simulations and advanced optimization	Structural optimization and simulations for high-performance solutions
AutoCAD®	CAD	Widely used for 2D and 3D design, suitable for engineers and architects	Designing technical details and sketches
Autodesk PowerMill®	CAM	Advanced CNC machining software, particularly suitable for complex shapes	Producing complex components with high precision, ideal for biomimetics
Autodesk Revit®	BIM/CAD	Building Information Modeling (BIM) platform integrating architecture, structure, and MEP; supports parametric components and scripting via Dynamo®	Designing and coordinating architectural and engineering systems with parametric and performance-driven features; useful for integrating biomimetic design strategies in the early planning phase
Blender®	CAD	Powerful open-source software for generative and parametric forms	Modeling complex biomimetic surfaces, rendering, and visualizations
CATIA®	CAD	Advanced software used in aerospace and automotive sectors	Designing complex components, advanced simulations, ideal for bioengineering
COMSOL Multiphysics®	CAE	Physical, mechanical, thermal, and chemical simulations, highly versatile	Advanced analysis for new materials and biomimetic solutions
Cura® (Ultimaker)	CAM	Open-source slicing software for 3D printing, supports multi-material setups and bio-compatible materials	Preparing biomimetic components for FDM (Fused Deposition Modeling) printing, including materials with mechanical gradients
Dynamo®	CAD/Parametric	Visual scripting environment for Autodesk Revit®; facilitates parametric control and data-driven design	Automating architectural workflows, generating responsive structural systems
Freeform® (3D Systems)	CAD	Organic modeling with voxel-based sculpting and direct interface with haptic devices	Designing anatomically accurate, biomimetic forms for medical, industrial, or wearable applications
Fusion 360®	CAD/CAM/ CAE	Cloud-based software integrating CAD, CAM, and CAE, useful for collaborative design	Prototype design and CNC milling optimization, 3D printing
Galapagos ®	Plug-in (Optimization)	Evolutionary solver for multi-variable optimization within Grasshopper®	Optimizing form-finding processes based on fitness criteria (e.g., structural efficiency, daylight)

Grasshopper®	CAD/Parametric	Visual algorithm editor integrated with Rhino®; supports custom scripting, parametric modeling	Creating biomimetic geometries, adaptive façades, and complex surface manipulations
Kangaroo ®	Plug-in (Physics)	Physics engine for interactive simulation of forces, constraints, and relaxation (Grassopper®)	Simulating material behavior, tension-based structures, and minimal surfaces
Ladybug® / Honeybee®	Plug-in (Environmental)	Toolset for environmental analysis and building performance simulations	Simulating solar exposure, energy use, thermal comfort, and climate-driven design
LS-DYNA®	CAE	Nonlinear dynamic simulations to analyze structural behavior in extreme conditions	Impact, collision, and stress testing for biomimetic materials and complex structures
Mastercam®	CAM	CAM software for CNC machining, suitable for a wide range of machines and processes	Optimizing tool paths for the production of mechanical components
Materialise Magics®	CAM	Professional STL (STereoLithography) editing, part nesting, and data prep for additive manufacturing	Optimizing 3D-printed parts derived from biomimetic geometry with precision and industrial standards
Netfabb® (Autodesk)	CAD/CAM	Tools for 3D printing support structures, mesh repair, and topology optimization	Preparing complex parts for additive manufacturing with biomimetic detail and structural validation
nTopology®	CAD/CAE	Advanced implicit modeling; ideal for lattice structures, topology optimization, and simulation-driven design	Creating lightweight, biomimetic structures optimized for 3D printing and functional performance
Pufferfish ®	Plug-in (Morphology)	Geometry manipulation tools focused on morphogenesis and topological transformations	Creating smooth transitions, morphing biomorphic forms, and gradient-based structures
PTC Creo®	CADCAMCAE	Integrated software for design, simulation, and production, ideal for complex sectors	Design, simulation, and production of industrial and biomedical components
RhinoCAM®	CAM	Integration of Rhino with CAM module, useful for creating CNC tool paths directly	Prototyping biomimetic materials and generative design
Rhinoceros (Rhino)®	CAD	Flexibility in 3D design, ideal for parametric and generative design with Grasshopper	Designing complex geometries, optimizing structures inspired by nature
Siemens NX®	CAM	Advanced software for design, manufacturing, and engineering, ideal for large industries	Designing and producing high-performance components, aerospace, and biomedical sectors
Siemens NX®	CAD/CAM/ CAE	Integrated platform combining design, engineering, and production in one environment	Designing and producing biomimetic solutions with real-time simulations

...Table 3 contd.

Table 3 contd. ...

Software	Type	Main Features	Use
SolidWorks®	CAD	Mechanical design with a user-friendly interface, ideal for parametric components	Modeling mechanical components, including biomimetic materials
Slicer for Fusion 360®	CAM	Tool for converting 3D models into layered 2D slices for laser cutting, CNC, or lamination	Rapid prototyping of organics forms through planar assembly techniques
SolidWorks Simulation®	CAE	Integrated with SolidWorks for mechanical simulations and analysis during the design phase	Mechanical analysis of biomimetic components and design op
ZBrush®	CAD	High-resolution sculpting of complex, organic surfaces; used in concept modeling and detailed biomorphic design	Shaping naturalistic geometries, textures, and surfaces in early or artistic stages of design

- Flows in free and porous media

The researchers used this computational arsenal to create the first simulations of the *Venus Basket*, capturing its response to and influence on the surrounding water flow at the deep-sea floor, where it lives. The simulations were based on Lattice Boltzmann Methods (LBM) used to simulate the dynamic behavior of liquid flows. In computational fluid dynamics, the LBM methods are a set of techniques used for fluid simulation. Instead of solving the Navier-Stokes equations, the Boltzmann equation is solved to simulate the flow of a Newtonian fluid using collision models, such as the Bhatnagar-Gross-Krook (BGK) method. By simulating the interaction of a limited number of particles, the behavior of the viscous flow emerges automatically from the intrinsic motion of the particles themselves and the ensuing collision processes. The researchers, including Professor Maurizio Porfiri from New York University, Dr. Giacomo Falcucci from the University of Tor Vergata in Rome, and Dr. Sauro Succi from the Center for Life Nanosciences at the Italian Institute of Technology at La Sapienza University in Rome, as well as Research Affiliate of the Physics Department and Recurrent Visiting Professor at Harvard University, focused on analyzing the flow of liquids inside and outside the sponge's body cavity. They demonstrated how the organism has built its unique morphological structure to optimize its specific environmental adaptation. The sponge's morphology reduces resistance while also facilitating the creation of low-speed vortices inside the body cavity, which are used for feeding and reproduction. From these observations, analyses, and sophisticated CAD/CAE simulations and computer graphics, the team derived valuable technological and physical-functional insights useful for the design of skyscrapers or, in fact, any mechanical structure, from skyscrapers to new low-resistance structures for ships or airplane fuselages. The particular structures created by *Euplectella* could inspire biomimetic structural engineering solutions that apply a similar lattice of crescents and holes to reduce aerodynamic resistance and optimize the distribution of dynamic forces. The accuracy of the method, combined with access to one of the world's best supercomputers, allowed for calculations that had never been attempted before, shedding light on the role of fluid flows in the adaptation of organisms living in the depths. This study, focusing for the first time on analyzing marine water flows inside the sponge's structure, where all elements that contribute to its formation reduce water thrust, enabling the sponge to remain anchored to the seabed effortlessly, could inspire innovative solutions for the structural engineering of advanced materials. It could also help understand how to use this extraordinary *natural technology* to build lighter airplanes, ships, and skyscrapers that are highly resistant to turbulent air and water motions. As is typical of natural multifunctional design, where adopted technological solutions address more than one functional-structural aspect, the same structures that reduce water resistance create vortices inside the body cavity, allowing the sponge to feed and reproduce effectively. The involvement of a biologist in the research team proved essential to interpret the function of the vortices—phenomena that remained unclear to engineers and physicists. These internal flows serve to prolong the fluid's residence time within the sponge, thereby enhancing its ability to filter suspended particles from the surrounding water. This case exemplifies the profound value of interdisciplinary collaboration in the biomimetic field, where biology, physics, engineering, and computational design intersect to uncover mechanisms evolved over hundreds of millions of years. It also highlights the critical importance of research that transcends disciplinary boundaries, revealing how the forms and processes of living systems can inform the technologies of the future (Anter Italia 2022). In this context, the *WunderCAD-CAMmmer* metaphor emerges as a symbolic and operative space where *Artifice* and *Nature* hybridize—where digital simulation tools meet the ancestral intelligence of biological structures. It is through this convergence, this sense of engineered wonder, that we can access the deep-time logic of natural design: a form of technology not invented but discovered—rooted in the geological memory of life itself.

2.3 A Virtuous Ring

In the context of the Anthropocene, human impact on the environment and the growing demand for resources have prompted a radical rethink of our economic and production model. The concept of the *circular economy*, based on the principles of reuse, regeneration, and waste reduction, emerges as a crucial response to these global challenges. The *Virtuous Ring (VR)*, therefore, becomes a powerful metaphor to describe a technological system that self-regulates in a continuous and endless cycle, where materials are no longer seen as resources to be exhausted, but as links in a process that constantly regenerates itself. In a seminal study on the intersection between economic models and fiscal policy, Walter R. Stahel explores how taxation systems can be restructured to support circular economy principles. By highlighting the need to shift fiscal pressure away from renewable human labor and toward the consumption of non-renewable resources, his work provides a framework for aligning economic incentives with sustainability goals (Stahel 2022). It represents a key concept in systems thinking and the theory of positive change. It is characterized as a continuous, self-reinforcing cycle in which positive actions or behaviors feed back and amplify each other's beneficial effects. Each element within the cycle helps sustain and enhance the others, forming a self-regulating system over time. At the core of the VR lies the idea of transforming development dynamics into a positive flow—one that drives sustainable growth and ongoing improvement. Frequently used as a metaphor, the concept describes processes that, once triggered, not only maintain momentum but also exponentially amplify benefits through a cascading synergy. Early progress leads not just to immediate success but also to a chain reaction of advantages that reinforce the entire system in the long run. Such circular growth dynamics apply across diverse domains—from technological innovation and environmental sustainability to economic development and social advancement—making this virtuous-cycle framework a versatile tool for addressing today's challenges.

A Conceptual Framework (Chartered Governance Institute UK & Ireland 2022):

This powerful conceptual framework is often used to describe cyclical processes in which positive actions continuously reinforce each other, generating sustainable growth and improvement. It can be applied in various domains, including economics, technology, social systems, and environmental sustainability.

1. **Definition and Metaphor**
   - **Definition**: *Virtuous Ring* refers to a self-sustaining, positive feedback loop where initial actions create conditions that naturally lead to more beneficial actions, resulting in continuous improvement and growth. It is a cycle that perpetuates itself without requiring external intervention.
   - **Metaphor**: It is likened to a chain of interconnected actions or behaviors where each element supports and enhances the others, creating a continuously expanding cycle of positive effects.
2. **Core & Characteristics**
   - **Self-Perpetuating Nature**: Once initiated, the cycle is self-sustaining. Positive results naturally lead to more positive actions and outcomes.
   - **Exponential Growth**: As the cycle progresses, each iteration increases the overall benefit, leading to accelerated growth in terms of knowledge, productivity, or development.
   - **Feedback Loops**: Continuous feedback within the system helps refine actions, improving both the process and the results over time.
3. **Key Elements**
   - **Initial Positive Action**: The starting point of the ring, where a small positive action or change triggers the cycle (e.g., innovation, investment, education, or technology).
   - **Supporting Ecosystem**: A network or environment that amplifies and sustains the initial positive action (e.g., communities, resources, policies, or technologies).
   - **Mutual Reinforcement**: Each element in the system not only benefits from other elements but also strengthens them, creating a robust and interconnected system that encourages continuous improvement.
   - **Sustainability**: The VR is inherently sustainable, supporting long-term success without depleting resources, thus contributing to sustainable growth and innovation.
4. **Possible Applications**
   - **In Innovation and Technology**: The iterative improvement of technology creates a continuous innovation cycle, where each new development leads to further opportunities for growth and refinement.
   - **In Business**: Organizational culture, employee empowerment, and positive leadership foster a successful business environment that perpetuates growth and efficiency.
   - **In Social Systems**: Social programs, education, or healthcare can create positive feedback cycles that improve overall societal well-being over time.
   - **In Environmental Sustainability**: Ecological restoration, conservation efforts, and resource management can form a self-sustaining loop that leads to a more resilient and balanced ecosystem.

5. **Challenges and Barriers**
   - **Initial Investment**: A foundational action or change is often required to kickstart the virtuous cycle, which may require significant effort or resources.
   - **Interruptions from Negative Cycles**: External disruptions or negative feedback loops can undermine the integrity of the virtuous ring and hinder its self-perpetuation.
   - **Scalability**: Ensuring that the virtuous cycle remains effective as it expands is crucial. If the system grows too large, there is a risk that it will stagnate or become inefficient.
6. **Conclusion**
   - **The Role of the *Virtuous Ring* in Achieving Long-Term Goals**: This concept enables continuous improvement and sustainable development across diverse fields, promoting long-term success without depleting resources.
   - **Interconnectedness**: The success is dependent on the interconnectedness of the system, where each action and result strengthens the overall integrity of the process.
   - **Potential for Innovation**: By understanding and leveraging the dynamics of a virtuous ring, organizations, societies, and individuals can foster exponential growth and positive outcomes, contributing to a brighter and more sustainable future.

Current research is increasingly focused on developing highly specialized design methods for the creation of high-tech materials inspired by nature, aiming to meet the needs of an advanced circular economy. This process is based on the use of innovative solutions and biomimetic principles to optimize resource use and reduce environmental impacts, aiming for a more sustainable use of materials and natural resources. In this context, CAD/CAM/CAE (Computer-Aided Design, Computer-Aided Manufacturing, Computer-Aided Engineering) technologies play a crucial role, enabling the design, simulation, and production of highly customized materials and products optimized for sustainability and efficiency. Biotechnology, nanotechnology, and bioinspired materials science are foundational fields that facilitate the development of durable, easily disassemblable, and reusable products, thereby reducing reliance on virgin raw materials and minimizing waste generation. CAD/CAM/CAE softwares, used in the design and production of these materials, allows for the creation of highly complex and optimized models that meet specific performance and sustainability requirements and can be easily dismantled or reused. The integration of these technologies with biomimetic principles allows for the realization of solutions that mimic the efficiency and balance found in nature, adapting dynamically to different needs while minimizing waste. A specific example of this approach is represented by biocompatible materials developed through biotechnology, which can be designed to be easily biodegradable or to integrate into natural processes without leaving any contamination. Using CAD, designers can create advanced geometries and structures that optimize the mechanical and chemical properties of materials, while CAM enables precise production tailored to specifications, reducing error margins and resource consumption. Finally, CAE allows for simulating the behavior of materials under real-world conditions, ensuring that they are highly resistant and durable, with a reduced environmental impact. Current research in the design field of sustainable materials increasingly focuses on integrating advanced CAD/CAM/CAE technologies with biomimetic principles to address the challenges of a circular economy. By leveraging the capabilities of bio-digital design tools, researchers and engineers can create materials inspired by Nature that optimize resource use, reduce environmental impact, and are engineered for disassembly and reuse. Several academic institutions and industrial centers are at the forefront of this trend, developing cutting-edge materials and systems that combine Nature's efficiency with modern technology. The rising prominence of biomimicry, as both a design philosophy and a market phenomenon, was already noted by Emily B. Kennedy in 2017. In her article *Biomimicry: Design by Analogy to Biology*, she highlighted the significant growth in interest surrounding the field, along with its vast potential across industries. Citing a *Fortune* blog post from that same year, Kennedy pointed to the increasing commercial relevance of biomimetic approaches, quoting Verne Harnish's provocative question to R&D managers: "If you're not incorporating the most brilliant ideas from the natural world into what you sell, you're leaving money on the table." This framing reflects an early recognition of the tension between scientific innovation and market-driven enthusiasm, underscoring the need for strategic engagement with biomimetic ideas (Kennedy 2017). Building upon this early acknowledgment of biomimicry's ascent, the later essay titled *Revising Biomimetics: Opportunities and Ambiguities in the Bioinspired Design Approach*, authored by Langella, Vieira de Arruda, and Di Bartolo (2022), revisits the topic through a more critical and theoretically grounded lens. Among the most thought-provoking contributions to the discourse is their examination of what may be interpreted—metaphorically—as a *speculative bubble*: a dynamic in which biomimicry risks becoming more of a fashionable label than a truly rigorous or transformative design paradigm. While the authors do not explicitly use this term, their analysis points to an uncontrolled proliferation of projects and research initiatives validated primarily through affiliation with the *BIO* narrative, rather than through demonstrable methodological soundness or systemic alignment. Their work stands out for its scholarly depth and capacity to contextualize bioinspired design within broader socio-economic, political, and technological frameworks. It calls for a critical reassessment of biomimicry's role in design innovation—one that emphasizes scientific rigor, ethical integrity, and a genuine commitment to revising dominant development models shaped by globalization. Notably, the authors conclude by proposing a kind of *decalogue*—a structured set of ten principles that articulate a vision of ethical and methodological responsibility in biomimetic design. In this light, their essay offers a valuable interpretive framework through which to assess the case studies presented throughout this book, particularly in light of their ability (or inability) to transcend branding and contribute to sustainable, system-oriented innovation (Langella et al. 2022).

For a detailed and up-to-date overview of the current state of research in the field of biomimetics, one can refer to the volume *Map Research Bionics: Transdisciplinary Ecosystems in Bionics, Biodesign and Biomimicry. Scientific and Technological Aspects for a Design Culture*, edited by Amilton Arruda (2023). Published by Blucher

Open Access, the book is freely available online and brings together 31 contributions from international researchers and designers working on the convergence between natural systems and design.

The volume is organized into three main sections:

1. **Academic ecosystems and meta-design activities** – exploring how bionics and biomimicry are integrated into university-level courses in design and architecture, with case studies from institutions such as the University of Aveiro, the University of Camerino, and the Pratt Institute.
2. **Research activities and applied bionics projects** – presenting laboratories and research units from both public and private universities, including the BIODESIGN LAB at the Federal University of Pernambuco and the Hybrid Design Lab at the University of Naples Federico II.
3. **Applications of bionics in business models and design studios** – examining how biomimicry is being adopted by design studios, fablabs, and companies, with notable examples such as TÁTIL DESIGN DE IDEAS.

The book provides an in-depth panorama of methodologies and applications in biomimetics, emphasizing the importance of a transdisciplinary approach that bridges design, biology, and technology. These studies highlight the ways in which collaboration among academic institutions, industrial research centers, and private companies is driving a new wave of innovation in the development of sustainable and bioinspired material solutions. One notable example is the European educational project *BioSTEM/Biomimetic Inventions*, which brings together researchers, students, and engineers to develop advanced building materials based on biomimetic principles (European School Education Platform 2024). This initiative leverages CAD/CAM/CAE tools to simulate and optimize material behavior under real-world conditions, offering a more sustainable approach to construction and design. Academic and industrial collaborations play a pivotal role in driving the development of advanced materials that fulfill functional requirements while integrating harmoniously with the environment—thereby reducing waste and supporting a circular economy. The integration of CAD/CAM/CAE technologies with biomimetic principles is reshaping the design and manufacturing landscape, enabling the creation of high-performance, adaptable, and ecologically aligned materials. As these tools continue to evolve, they hold immense potential to transform industries and provide innovative solutions to today's most pressing environmental challenges. Among the many emerging initiatives at the intersection of biology, technology, and design, one of the most compelling examples is the *We Want To Learn – Diploma Studio 10 (DS10)* project at the University of Westminster's School of Architecture in London. This experimental academic format represents a pioneering initiative in the field of parametric and biomimetic architecture. Led by tutors Toby Burgess and Arthur Mamou-Mani, DS10 explores the integration of advanced digital technologies—such as artificial intelligence and digital fabrication—with principles of sustainability and circular design. Student projects range from the creation of eco-parametric structures to urban interventions aimed at the sustainable regeneration of public spaces. The blog *WeWantToLearn.net* serves as a platform for sharing research, projects, and educational resources, fostering an open and collaborative learning community. Tools such as artificial intelligence and parametric design are used to develop innovative solutions in the fields of architecture and urban design. Under the guidance of tutors, students are provided with the opportunity to design and build prototypes in the real world, promoting hands-on learning and a strong commitment to sustainable innovation (WeWantToLearn.net 2025).

Here are the main design issues considered in the experimental projects format:

- **Bio-Inspired and Biodegradable Materials**

  Materials derived from natural solutions, such as bioplastics, are a key example of how design can embrace the circular economy, aiming to reduce environmental impact and promote a more responsible use of resources. A notable example of this innovation is the use of bioplastics derived from marine algae or other natural sources, which are biodegradable and thus do not create permanent waste. These materials, which decompose easily without leaving harmful traces in the environment, offer a tangible response to the problem of conventional plastics, helping to close the material cycle in a harmonious and natural way. Another example is the creation of bioplastics that have characteristics similar to traditional synthetic polymers but are fully compostable, reducing the burden of plastic waste on our ecosystem. Projects from companies developing plastics derived from plant fibers or corn could be further optimized through the integration of nanomaterials that improve their strength and durability while maintaining a low ecological footprint.

- **Design for Reuse and Modularity**

  The concept of modularity, which involves designing components that are easily separable and replaceable, represents a fundamental pillar in the high-tech circular approach. An emblematic example of this philosophy is modular electronics, such as phones, computers, and electronic devices that can be easily upgraded or repaired without replacing the entire product. Companies like Fairphone, which produce modular phones, allow consumers to replace individual components like the battery, screen, or camera, drastically reducing electronic waste and extending the product's lifespan. Emerging technologies like 3D printing also enable the design and production of customized and modular components that can be easily replaced or recycled, reducing the need to produce new materials. With 3D printing, shapes can also be optimized to minimize material waste during production, ensuring that each part produced is functional and adaptable to future needs.

- **Sustainable Energy and Materials**

  Another crucial aspect of the high-tech circular economy is the creation of sustainable energy systems and materials that can be regenerated and recycled in continuous cycles. Solar panels, for example, represent an ideal solution to reduce our energy impact, but their production and disposal raise important ecological concerns. Innovations in the design of modular and easily recyclable solar panels are beginning to offer solutions to reduce their ecological footprint. Another example concerns long-lasting batteries, such as lithium batteries, which are designed to be reusable and regenerated at the end of their life cycle. Advanced energy storage systems, combined with smart charging solutions, allow energy use to be optimized, reducing reliance on non-renewable sources. Additionally, emerging technologies like solid-state batteries promise to be more efficient and safer than current batteries, with longer lifespans and a lower environmental impact throughout their life cycle.

- **Biomimicry for Material Recovery**

  Biomimicry, which studies and imitates natural processes, offers great potential in the circular economy. Self-healing materials, which regenerate autonomously, are a striking example of how technology can be inspired by nature to reduce the need for new materials and extend the life of products. For example, the concept of self-healing materials, such as those developed in nanomedicine and polymers, has been applied to the design of electronic devices that, in case of damage, can recover autonomously, avoiding the need to replace damaged parts and reducing waste. Additionally, the use of microorganisms to recover valuable materials from waste, as in bioremediation, is another example of how nature can inspire technological solutions for the recycling and regeneration of materials. Biological processes, such as using bacteria to recover rare metals, could be applied within a circular economy cycle to extract resources from end-of-life products. The Virtuous Cycle represents an ideal model of how technological innovation can align with the principles of the circular economy, creating a system where materials and resources are continually optimized, reused, and regenerated. Advances in biotechnology, nanotechnology, and materials design are essential to developing solutions that are not only efficient and high-performing but also contribute to a more sustainable future in harmony with the environment, where technology and nature can collaborate to build a virtuous cycle capable of reducing waste and optimizing resource use, in perfect symbiosis with sustainability principles. In the context of the high-tech circular economy, advanced computational and parametric design technologies like CAD/CAM/CAO/CAE play an extremely interesting role in ensuring that materials and products are not only designed efficiently but also to be easily reused, repaired, and regenerated in a continuous cycle. These digital tools allow for intelligent and optimized design, integrating sustainability into every stage of the production process, from the initial concept to the final product realization, and even its eventual disposal and reuse.

- **Optimized Design with CAD/CAM/CAE/CAO**

  At Westminster's Studio 10, students engage in cutting-edge architectural experimentation through an array of computational tools. The design process is deeply informed by parametric modeling software such as Rhinoceros 3D® combined with Grasshopper®, allowing for generative design exploration and biomimetic form-finding. For environmental analysis and dynamic simulations, tools like *Autodesk CFD®, Ladybug®, and Honeybee®* are often integrated into workflows to assess energy performance, airflow, and daylighting conditions. Visual storytelling and digital representation are enhanced through graphic and rendering software including *Adobe Creative Suite® (Photoshop®, Illustrator®,*

After Effects®), *V-Ray®*, and *Ecotec®*, *Processing®*, *Twinmotion®*, or *Enscape®* for real-time visualization. The studio also embraces emerging technologies such as AI-driven design, machine learning algorithms, and digital fabrication tools like CAD/CAM 3D printers and laser cutters, bridging the gap between conceptual speculation and material execution. By using parametric and generative software in CAD, CAM, and CAE protocols, the design process can be modeled and simulated to ensure that each part is made as efficiently as possible. Design for reuse and modularity is facilitated through advanced simulation and rapid prototyping. For example, in the case of electronic components or complex devices, CAD allows for the design of products that can be easily disassembled into individual components, while CAM and CAO optimize manufacturing and assembly processes to reduce waste and unnecessary resources. CAD/CAM/CAE/CAO-based design also enables mass customization, meaning products can be adapted to the specific needs of customers without compromising production efficiency or environmental impact. By using simulation and optimization, designers can predict and solve problems before they arise during production, ensuring a more sustainable manufacturing cycle (WeWantToLearn.net 2020.

2.4 Dynamic Product Lifecycles Managment (PLM) in the Age of Circular Design

Cycle is an extraordinarily powerful and effective term when it helps us understand the incredible and perfectly optimized ecological engineering that has supported *life* on planet Earth for billions of years. The major environmental cycles are interconnected natural processes that regulate the Earth's ecosystems. These cycles, driven by physical, chemical, and biological factors, ensure the recycling of vital resources, maintaining ecological balance and supporting life. The chemical processes govern the transformation and movement of elements like carbon, nitrogen, and oxygen, while physical factors, such as energy from the sun, shape the movement of water and the temperature of the planet. Biological factors, including the actions of plants, animals, and microorganisms, drive many of these cycles, facilitating nutrient exchange and promoting life processes that sustain ecosystems. Together, these cycles form a dynamic system that supports life and preserves the delicate equilibrium of our planet. The main ecological cycles are (Ha and Schleiger n.d.):

1. **The Water Cycle**: This cycle describes the movement of water through the environment. It includes evaporation, condensation, precipitation, and runoff of water, which moves between oceans, the atmosphere, and land. This cycle is crucial for maintaining the availability of freshwater on Earth.
2. **The Carbon Cycle**: The carbon cycle involves the movement of carbon dioxide (CO_2) between the atmosphere, living organisms, rocks, oceans, and vegetation. Carbon is a key component of life and plays a crucial role in global warming, as its accumulation in the atmosphere increases the greenhouse effect.
3. **The Nitrogen Cycle**: Nitrogen is an essential element for plant and animal growth. The nitrogen cycle describes the process by which nitrogen is fixed in

the soil by bacteria, absorbed by plants, consumed by animals, and returned to the atmosphere through processes such as decomposition and nitrification.

4. **The Phosphorus Cycle**: Phosphorus is another vital element for plant and organism growth. The phosphorus cycle involves its release from rocks, absorption by plants, passage through the food chain, and its return to the soil through the decomposition of dead organisms.
5. **The Oxygen Cycle**: This cycle involves the movement of oxygen through the atmosphere, hydrosphere, biosphere, and geosphere. Plants produce oxygen during photosynthesis, which is consumed by living organisms for respiration. Oxygen is essential for aerobic life and for maintaining metabolic processes.
6. **The Sulfur Cycle**: The sulfur cycle involves the movement of sulfur through the environment, including oceans, the atmosphere, soils, and living organisms. Sulfur is crucial for protein synthesis and other biomolecules, and its distribution affects soil acidity and atmospheric pollution.
7. **The Lithium Cycle (or Metal Cycle)**: Although less known than other cycles, the lithium cycle is becoming more relevant as it concerns the movement of metals like lithium and other minerals through the ecosystem. Lithium is essential in rechargeable batteries and modern industrial processes.

Thus, these magnificent seven are essential for life's businnes on Earth, and their balanced functioning is crucial for the health of ecosystems and environmental sustainability. Events like climate change, pollution, and deforestation can disrupt these dynamic *equilibrium*, with significant impacts on biodiversity and the well-being of living organisms, including humans. The concept of *cycle* and *cyclicality* serves as a powerful semantic bridge connecting both the Nature's ecological evolution and human industrial domains. In Nature, cycles exemplify the continuous flow and recycling of resources, maintaining ecological balance. In parallel, Industry 5.0 is seeking to adopt strategies and principles inspired by these natural, systemic models, aiming to replicate their efficiency and sustainability within industrial production (van Erp et al. 2024). By emphasizing adaptability, resource optimization, and the closed-loop principles seen in ecological systems, Industry 5.0 encourages the integration of *human-centric*, sustainable production practices that reflect the resilience and interconnectedness inherent in natural cycles. Industry 5.0 represents a bold and transformative shift in industrial production, characterized by its *holistic*, *regenerative*, and *adaptive* approach. By drawing inspiration from the seamless and interdependent processes of nature, Industry 5.0 strives to create a *resilient*, *circular*, and *human-centered* manufacturing paradigm. It is an era where technology, innovation, and sustainability are harmonized to create *synergistic* systems that optimize both human creativity and ecological balance. This model aims to build *sustainable*, *self-healing*, and *intelligent* industries that not only focus on efficiency but also integrate ethical, social, and environmental considerations, promoting a *progressive*, *inclusive*, and *future-proof* industrial ecosystem. The use of cycle in both contexts highlights the shared goal of ensuring longevity, minimizing waste, and maintaining equilibrium, whether in the environment or in industrial systems. In the same vein, the new frontier of Industry 5.0 introduces a parallel concept within industrial processes: the Eco-Industrial Cycle (Center for Strategic and International Studies 2024). Unlike traditional linear production models, which operate on a *take-make-dispose* philosophy, Industry 5.0 emphasizes the integration of human ingenuity, advanced technologies, and sustainability. This new industrial paradigm advocates for a circular production model, where materials, products, and even ideas are reused, rethought, and optimized across their entire lifecycle. The *Eco-Industrial Cycle* in Industry 5.0 draws direct analogies to the ecological cycle. Just as ecosystems thrive by recycling resources and energy, industrial processes in this new era strive to optimize the use of raw materials, minimize waste, and ensure that every product or component has a second life—whether through repurposing, remanufacturing, or adaptive reuse. This approach mirrors the ecological cycle, where every organism plays a role in maintaining the balance of nature. However, the key difference lies in the integration of human-centered design. While ecological cycles are driven by Nature's inherent processes, Industry 5.0 introduces a human touch: a conscious effort to combine the best of technology and craftsmanship. The fusion of advanced AI, robotics, and human creativity makes it possible to design systems that are not just efficient, but also adaptive, personalized, and responsible. The role of Product Lifecycle Management (PLM) in this context becomes crucial, as it ensures that the entire lifecycle of a product—right from ideation through to its end-of-life—is thoughtfully managed and optimized for sustainability. Therefore, while the ecological cycle is governed by natural laws and the flow of energy, the *Eco-Industrial Cycle* under Industry 5.0 takes a more proactive approach, seeking to harmonize human innovation with the principles of sustainability, ensuring that industry can thrive without compromising the planet's future (Hsu et al. 2024). This shift signifies a new era where industry and ecology are no longer opposing forces, but partners in creating a resilient, sustainable world. Product *Lifecycle Management (PLM)* is a system that manages and centralizes all information related to a product throughout its lifecycle, from ideation to disposal. This integrated approach enables companies to optimize the design, production, maintenance management, and end-of-life stages of products. PLM plays a crucial role in enhancing the efficiency and quality of business processes, as it facilitates collaboration among different departments and enables continuous monitoring of changes and progress (Meegle 2025). One of the key aspects of PLM is its connection with CAD and CAE softwares, which are essential during the design and engineering phases. CAD allows for the creation of detailed 3D models and the analysis of component geometries, while CAE supports the simulation and verification of the structural and functional performance of these components under different operating conditions. Together, these tools help refine the design and optimize the product's performance before physical production begins (Encycam 2025). The integration of CAD and CAE with PLM provides a powerful synergy, allowing for more efficient decision-making, better collaboration, and optimized product development (Scott 2024). The PLM system ensures that all modifications are accurately recorded and that decisions made in one phase of the process do not negatively impact later stages. Furthermore, PLM helps track the management of materials and components, optimizing production flows, reducing assembly time, and enhancing overall efficiency while minimizing errors. A practical example of early PLM application in this context can be found in the automotive industry, where companies like Ford utilize PLM to manage vehicle design and development (Raza et al. 2011). During the vehicle development process, Ford employs CAD to design components such as the engine and chassis, and CAE to simulate their behavior in cases of impact, stress, or vibration. The PLM system integrates this information and simulations, ensuring seamless collaboration and allowing for accurate tracking of design iterations, material management, and production workflows (Korba et al. 2024). In the aerospace industry, companies like Airbus use PLM to manage the design and production of complex aircraft. Airbus's PLM system coordinates work among globally distributed engineering teams, using CAD to design the various aircraft parts and CAE to simulate aerodynamic and structural performance (Dassault Systèmes 2024). This integration not only improves the efficiency and quality of the design but also allows for the monitoring and management of the aircraft's entire lifecycle, from initial design to maintenance and part renewal during its operational life. This reduces operational costs, enhances safety, and ensures the longevity and reliability of the aircraft. Thanks to the seamless integration of CAD, CAE, and PLM, companies can improve design, anticipate issues before they occur in production, and manage the entire product lifecycle more efficiently. This integrated approach supports sustainability and innovation by enabling companies to design with greater precision, reduce material waste, and develop products that are not only high-performance but also environmentally responsible. The ability to manage the full lifecycle of a product—from conceptualization to disposal—ensures that the products are designed with the future in mind, contributing to long-term sustainability and resource optimization. The use of advanced software tools such as CAD/CAE suites—including platforms like Autodesk Revit®, Rhino/Grasshopper®, and Autodesk CFD®—not only optimizes the design phases of ideation and production, but also significantly enhances the entire product Life Cycle Assessment (ResearchGate 2021). These computational tools support the creation of products and structures that are easier to repair, upgrade, and maintain, ultimately increasing durability and reducing the likelihood of components ending up as waste. Moreover, such technologies enable integrated lifecycle simulations, allowing designers and engineers to anticipate how a product or structure will respond to stress, wear, environmental conditions, or user interactions over time. This predictive modeling capability plays a key role in shaping more resilient, adaptable, and sustainable solutions from the very beginning of the design process. A key aspect of this process is the integration of material science into design workflows. The selection of materials with the right properties—such as strength, flexibility, thermal resistance, and environmental impact—is crucial. CAD and CAE tools, combined with advances in material science, allow for the simulation of material behavior under different conditions. For example, the development of advanced composites, high-entropy ceramics, and bio-inspired materials can be modeled and optimized to suit specific applications, ensuring better performance and longevity. A practical example of this application can be found in the field of sustainable architecture and infrastructure. For instance, the use of Autodesk Revit® in conjunction with Building Information Modeling® (BIM) allows for the creation of highly detailed digital twins of buildings (Gilmanova 2020). These models can simulate energy consumption, structural performance, and even maintenance needs over time, factoring in material degradation, environmental stress, and other lifecycle aspects. Similarly, Rhino/Grasshopper® enables parametric design and optimization of complex geometries, often used in green building envelopes and modular construction systems (Ajtayné Károlyfi and Szép 2023). These tools empower architects and engineers to select optimal materials and structural strategies for each phase of a building's lifecycle. By leveraging material science insights, they can predict how different materials will perform over time, ensuring the sustainability and resilience of the final design. The ability to simulate disassembly, material reuse, or adaptive reuse scenarios contributes to reducing construction waste and enhancing long-term sustainability. In this way, digital design platforms are not merely aids in visualization—they become predictive engines for ecological and circular innovation.

2.5 Integrating 3D Printing into the Circular Economy

3D printing, which can be used in combination with CAD, CAM, and CAE software, is another powerful tool to embrace the principles of the circular economy (van Erp et al. 2024). It enables the production of complex, lightweight structures with minimal material waste, facilitating localized manufacturing and the reuse of recycled feedstocks (Oladapo et al. 2024). It allows for the production of customized products with maximum material efficiency, reducing waste and optimizing the use of resources (Kassab et al. 2023). Companies like Stratasys® and 3D Systems® use 3D printing to create prototypes and final products that are not only highly customizable but also designed to be easily reused or recycled. The ability to produce only what is necessary—without the surplus waste typical of traditional manufacturing methods—and to precisely optimize materials is one of the key aspects that makes 3D printing a fundamental element of the circular economy. Integrating CAD, CAM, CAE, and CAO technologies with circular economy principles not only improves production efficiency and reduces waste, but also facilitates the design of products that are easy to disassemble, repair, and reuse. Companies adopting these advanced technologies are able to embrace the virtuous cycle of the circular economy, creating a positive impact on both the environment and the economy through material optimization, resource reuse, and waste reduction. The integration of *Additive Manufacturing (AM)* technologies into circular economy paradigms represents a critical inflection point in the evolution of sustainable production models. By enabling decentralized, on-demand fabrication with minimal waste, 3D printing offers a scalable platform for closing material loops, valorizing industrial by-products, and rethinking the lifecycle of goods through a regenerative lens. A growing number of companies and research consortia are investigating how post-consumer plastics, biogenic residues, and locally available geopolymers can serve as sustainable feedstocks for AM processes. For example, the British company Filamentive® has developed a suite of PLA-recycled filaments derived from algae, 100% recyclable cardboard spooling, and designed to universally fit most 3D printers in use worldwide, integrating rigorous traceability protocols and community-based collection systems to ensure material purity and social inclusion (Filamentive 2025). In the United States, Forust®, a venture by Desktop Metal®, utilizes binder jetting to 3D print high-resolution wooden structures from recycled sawdust and lignin, essentially reconstituting timber without cutting down trees. This process not only diverts waste from landfills but also reintegrates it into value-added applications such as furniture, architectural components, and interior design products (Forust 2025). The Italian company WASP® (World's Advanced Saving Project) has pioneered large-scale construction 3D printing using native soil and natural fibers (Wasp 2025). Their TECLA® project—developed in collaboration with Mario Cucinella Architects—demonstrates how local earthen materials can be robotically deposited to create zero-kilometer housing, addressing both the housing crisis and material scarcity (Mario Cucinella Architects 2025). TECLA® integrates parametric design, climatic adaptability, and low-carbon materiality, embodying both high-tech innovation and vernacular ecological wisdom. On a more civic-oriented scale, the Dutch-Greek initiative Print Your City, developed by The New Raw, engages citizens in transforming their household plastic waste into public urban furniture through robotic 3D printing (The New Raw 2025). This participatory model embeds user engagement, waste valorization, and circular urban design into a single process loop (Print Your City 2022). In Eastern Europe, Greenfill3D® (Poland) has engineered composite filaments incorporating agro-industrial residues such as wheat bran, expanding the functional possibilities of biowaste within thermoplastic extrusion. These bio-based filaments exhibit favorable printability and mechanical characteristics, offering an alternative to fossil-derived polymers (Greenfill3D 2025). Experimental design firms such as Emerging Objects® (USA) are pushing material boundaries further by employing cement dust, salt, coffee grounds, and crushed glass in additive systems for architectural modules, water filtration structures, and even artificial reef habitats. Such material innovation reflects a shift toward context-sensitive fabrication, where the properties of regional waste streams are algorithmically integrated into the design logic (Emerging Objects 2025). Collectively, these selected international case studies exemplify a shift from product-centric to resource-aware design frameworks. By aligning additive manufacturing with life cycle thinking, material circularity, and geo-situated ecologies, these initiatives reposition 3D printing as a core enabler of ecological intelligence and post-industrial sustainability. Alongside major technology companies, there are other emerging examples of innovative enterprises applying circular economy principles through original uses of CAD/CAM/CAE/CAO technologies:

- **Circular Ocean®**: This Portuguese startup developed a system to collect abandoned fishing nets from the sea, a serious ecological problem. Using CAD for design and CAO for optimizing material recovery and processing, they created new products from recycled materials—such as furniture and household accessories—offering practical solutions to reduce marine pollution and promote the circular economy (Circular Ocean 2025).
- **Ecovative Design®**: An example of biomaterials integrating CAD/CAM and engineering technologies to create sustainable products. The U.S.-based startup Ecovative® produces biodegradable packaging materials made from fungi and other natural substances, drastically reducing the use of plastics and improving product lifecycle sustainability. Thanks to optimized design and production processes, their environmental impact is kept to a minimum (Ecovative 2025).

- **Rimac Automobili®**: Although known for its high-performance electric vehicles, Rimac® has adopted advanced CAD and CAE to develop highly optimized components in terms of both performance and environmental impact. Rimac® uses a stunning concept car with a 3D-printed chassis that *breathes oxygen* and advanced waste management techniques during production, applying a system that facilitates part reuse and recycling, in line with circular economy principles (Yanko Design 2020).
- **Bureo®**: This company focuses on creating sustainable products using recycled fishing nets. Through CAD, CAM, and CAE technologies, Bureo® has developed skateboards and other sporting goods that utilize this recycled material. This approach reduces plastic waste and promotes sustainability, making each product a part of the virtuous cycle of resource reuse (Bureo 2025).
- **Paptic®**: A Finnish company developing biodegradable materials as alternatives to plastic. Using CAD and CAM technologies to design ecological and reusable packaging, Paptic® creates innovative solutions that naturally degrade, reducing reliance on single-use plastics and supporting the transition to a circular economy (Paptic 2025).
- **Loop Industries®**: An outstanding example of the circular economy applied to the plastics industry. The company develops and commercializes technologies for recycling PET plastics, transforming them into new products without compromising quality. Using advanced chemical processes, Loop Industries has designed a system that regenerates used plastics without virgin materials, drastically reducing waste and the need for new resource extraction (Loop Industries 2025).
- **Carbios®**: An innovative company developing advanced PET plastic recycling solutions. Using a biological process, Carbios can break down used PET and transform it into high-quality plastic ready for new products. The company has integrated advanced technologies and circular approaches to reduce the environmental impact of plastic production (Carbios 2025).
- **Fairphone®**: This company manufactures sustainable smartphones and promotes circular management of its devices. Fairphone has designed a modular phone that allows users to easily replace components like batteries or cameras, extending the device's life and reducing the need for disposal. This approach reduces electronic waste and supports the circular economy (Fairphone 2025).
- **Re:newcell®**: A company, England-based, that uses innovative technology (CIRCULOSE®) to transform textile waste into new materials, particularly for paper and fabric production. Through a process that breaks down old garments into cellulose pulp, Re:newcell® reduces the need for virgin fibers and promotes circularity in the textile sector. The company is working to change how textile waste is treated, making materials more sustainable (Renewcell 2025).

As the convergence of additive manufacturing and circular economy principles gains momentum, we are witnessing the emergence of a new design ethos—one that prioritizes regeneration over extraction, decentralization over mass production, and ecological intelligence over linear efficiency. The integration of CAD/CAM & 3D

Fig. 3. Close-up view of the layered clay structure in the 3D print.

printing into circular workflows does more than reduce waste: it redefines how, where, and why we manufacture. It enables a shift from globalized supply chains to locally responsive ecosystems of production, capable of adapting to material availability, cultural context, and environmental constraints. Looking ahead, the full realization of this potential will depend on continued interdisciplinary collaboration—between engineers, designers, biologists, urban planners, and policymakers—as well as on the development of open-source platforms and material libraries that democratize access to resource-efficient innovation. In this evolving landscape, 3D printing becomes not only a technological tool but a catalyst for reimagining our material world through the lens of circularity, resilience, and planetary stewardship. In an era defined by rising environmental awareness and the urgent need to reduce resource consumption, sustainable design has become the cornerstone of next-generation mobility. By prioritizing materials and processes that minimize waste, energy use, and carbon footprint, forward-thinking manufacturers are reshaping the way vehicles are conceived, prototyped, and produced—ensuring that performance and aesthetics go hand in hand with ecological responsibility. Honda R&D Europe, together with the Italian manufacturer WASP, is experimenting with a 3D printing process based on *industrial clay*. This material, a ceramic blend of carefully selected clays, organic binders, and additives engineered for precise extrusion, high mechanical strength, and continuous reuse, was first introduced into design studios thanks to the foresight of American designer Harley Earl, who revolutionized the making of automobile prototypes in the late 1920s by blending the sculptor's art with the modeler's skill. Clay sculpting remains essential for turning a designer's vision into a tangible form—OEMs (Original Equipment Manufacturers) in both the automotive and product design industries still rely on clay models to bring virtual 3D concepts of cars and motorcycles to life. Now, using a Delta WASP 40100 Clay printer, Honda's motorcycle design team can 3D-print industrial-grade clay prototypes of electric motorcycles that are ready for precise hand-finishing (Listek 2021) (Figs. 3–4).

Fig. 4. Honda motorcycle 3D-printing maquette.

Unlike traditional additive methods (polymer or metal printing) or subtractive techniques (CNC milling), this approach allows for the direct layering of clay onto mechanical frameworks, yielding highly complex yet robust geometries. The material's infinite recyclability addresses both economic and environmental sustainability: every scrap can be reprocessed and reused without generating waste. This not only slashes prototyping costs but also dramatically shortens development cycles by eliminating the need for specialized tooling and lengthy machining operations. Antonio Arcadu, Italian Design Modeling Coordinator at Honda R&D Europe, has orchestrated this project as a genuine automotive revolution: the new workflow marries the hands-on creativity of the designer with the pinpoint accuracy of advanced machinery, fostering a seamless human-machine dialogue (Jitchotvisut 2021). Although handcrafted clay models offer unrivaled customization when shaped by experienced artisans, achieving the perfect form can be time-consuming. By integrating 3D printing into the workflow, Honda stands to shave off significant development hours—though its real-world effectiveness has yet to be proven. Drawing on his ten years of experience as an industrial clay modeler, Arcadu reflects on his insight to merge 3D printing technology with the manual shaping of this material:

"In order to preserve the creative process, we cannot rely on technology alone, but we need modelers not to interrupt their connection with the matter in its most direct form: *the hands*. Although this is a newly introduced process, it can be considered disruptive in the automotive sector because it introduces a new flow, mostly additive, able to optimize processing times and reduce the amount of material used in the traditional and subtractive processes."

Thanks to the geometric freedom afforded by 3D clay printing, teams can rapidly explore diverse aesthetic and functional solutions, fine-tuning proportions and details in record time. Looking ahead, this methodology not only accelerates the emergence of futuristic-looking electric vehicles but also establishes a more responsible production paradigm—one that harmonizes performance, design, and respect for the planet.

References

Ajtayné Károlyfi, K., and Szép, J. (2023). A Parametric BIM Framework to Conceptual Structural Design for Assessing the Embodied Environmental Impact. Sustainability, 15(15).

Alexandridis, G., Tzetzis, D., and Kyratsis, P. (2016). Biomimicry in product design through materials selection and computer aided engineering. IOP Conference Series: Materials Science and Engineering, 161: 012046.

Arruda, A. (ed.) (2023). Map Research Bionics: Transdisciplinary Ecosystems in Bionics, Biodesign and Biomimicry. Scientific and Technological Aspects for a Design Culture. Blucher Open Access. Available at: https://openaccess.blucher.com.br/article-list/9786555502220-616/list/ [Accessed: 11 May 2025].

Biomimicry 3.8 (2023). Biomimicry 3.8 – Innovation Inspired by Nature. Available at: https://biomimicry.net/ [Accessed: 11 May 2025].

Biomimicry Institute (2024) Biomimicry Taxonomy. AskNature. Available at: https://asknature.org/resource/biomimicry-taxonomy/ [Accessed: 11 May 2025].

Biomimicry Institute (2025) Biomimicry Institute. Available at: https://biomimicry.org/ [Accessed: 11 May 2025].

Blasiak, R., Jouffray, J.-B., Amon, D. J., Moberg, F., Claudet, J., Jørgensen, P. S. et al. (2022) A forgotten element of the blue economy: marine biomimetics and inspiration from the deep sea. PNAS Nexus. Available at: https://hal.science/hal-03889645 [Accessed: 11 May 2025].

Bureo. (2025). Bureo: Makers of NetPlus® Material, 100% Recycled Fishing Net. Bureo. Available at: https://bureo.co/ [Accessed: 11 May 2025].

Carbios. (2025). Innovating for a True Circular Economy in Plastics. Carbios. Available at: https://www.carbios.com/en/ [Accessed: 11 May 2025].

CBC Radio (2023). The Great Acceleration. CBC Ideas. Available at: https://www.cbc.ca/radio/ideas/great-acceleration-1.6814569 [Accessed: 11 May 2025].

Center for Strategic and International Studies. (2024). Industrial ecology: Closing a loop in circularity. CSIS. Available at: https://www.csis.org/analysis/industrial-ecology-closing-loop-circularity [Accessed: 11 May 2025].

Circular Ocean. (2025). Circular Ocean: Turning discarded fishing nets into a new sustainable raw material. Circular Ocean. Available at: https://www.circularocean.pt/ [Accessed: 11 May 2025].

COMSOL (2025). CFD Module. COMSOL Multiphysics. Available at: https://www.comsol.com/cfd-module [Accessed: 11 May 2025].

CORDIS (2021). Is a deep-sea sponge inspiring the design of future mechanical structures? CORDIS - EU research results. Available at: https://cordis.europa.eu/article/id/430564-is-a-deep-sea-sponge-inspiring-the-design-of-future-mechanical-structures/it [Accessed: 11 May 2025].

Crutzen, P. J., and Stoermer, E. F. (2017). The Anthropocene (2000). In: Howe, J. P., and Sutter, P. S. (eds.) Making Climate Change History: Documents from Global Warming's Past. University of Washington Press. Available at: https://www.degruyterbrill.com/document/doi/10.1515/9780295741406-056/html [Accessed: 11 May 2025].

Dassault Systèmes (2024). Airbus expands commitment to PLM solutions from IBM and Dassault Systèmes. Dassault Systèmes. Available at: https://investor.3ds.com/news-releases/news-release-details/airbus-expands-commitment-plm-solutions-ibm-and-dassault/ [Accessed: 11 May 2025].

Díaz Lantada, A., Lafont Morgado, P., Echávarri Otero, J., Chacón Tanarro, E., de la Guerra Ochoa, E., Munoz-Guijosa et al. (2012). Biomimetic computer-aided design and manufacture of complex biological surfaces. In Proceedings of the International Conference on Biomedical Electronics and Devices (BIODEVICES-2012) (pp. 286–290). SCITEPRESS.

Ecovative. (2025). Ecovative: We grow better materials. Ecovative. Available at: https://ecovative.com/ [Accessed: 11 May 2025].

Ellis, M. A. (2021). From Anthropocene to Noosphere: The Great Acceleration. Earth's Future, 9(6): e2020EF001917.

Emerging Objects. (2025). Emerging Objects: 3D Printing Architecture and Materials. Emerging Objects. Available at: https://emergingobjects.com/ [Accessed: 11 May 2025].

Encycam. (2025). What are CAD, CAM, and CAE? Encycam. Available at: https://encycam.com/articles/what-are-cad-cam-and-cae/ [Accessed: 11 May 2025].

Falcucci, G., Amati, G., Fanelli, P., Krastev, V. K., Polverino, G., Porfiri, M. et al. (2021). Extreme flow simulations reveal skeletal adaptations of deep-sea sponges. Nature, 595: 537–541.

Filamentive. (2025). Filamentive: Sustainable 3D Printing Filament. Filamentive. Available at: https://www.filamentive.com/ [Accessed: 11 May 2025].

Fiorentino, C., and Montana-Hoyos, C. (2014). The emerging discipline of biomimicry as a design paradigm shift. The International Journal of Designed Objects, 8(1): 1–15.

Forust. (2025). Forust: Sustainable 3D Printed Wood Products. Forust. Available at: https://www.forust.com/ [Accessed: 11 May 2025].

Gilmanova, A. (2020). Computer-aided design systems for product life cycle stages. Journal of Physics: Conference Series, 1661, 012205.

Greenfill3D. (2025). Greenfill3D: Sustainable 3D Printing Solutions. Greenfill3D. Available at: https://greenfill3d.com/ [Accessed: 11 May 2025].

Ha, M. and Schleiger, S. (n.d.). Biogeochemical Cycles. In Environmental Science. LibreTexts. Available at: https://bio.libretexts.org/Bookshelves/Ecology/Environmental_Science_(Ha_and_Schleiger)/02%3A_Ecology/2.04%3A_Ecosystems/2.4.03%3A_Biogeochemical_Cycles [Accessed 11 May 2025].

Hirtz, J., Stone, R. B., McAdams, D. A., Szykman, S., and Wood, K. L. (2002). A functional basis for engineering design: Reconciling and evolving previous efforts. NIST Technical Note 1447. National Institute of Standards and Technology.

Howells, R (2015). Profile – Michael Pawlyn, Exploration Architects, talks about biomimicry. Attractions Management. Available at: https://www.attractionsmanagement.com/index.cfm?codeID=29664&pagetype=features [Accessed: 11 May 2025].

Hsu, C.-H., Li, Z.-H., Zhuo, H.-J., and Zhang, T.-Y. (2024). Enabling Industry 5.0-driven circular economy transformation: A strategic roadmap. Sustainability, 16: 9954.

International Energy Agency. Energy efficiency (n.d.) Available at: https://www.iea.org/energy-system/energy-efficiency-and-demand/energy-efficiency (Accessed: 29 May 2025).

Jatsch, A.-S., Jacobs, S., Wommer, K., and Wanieck, K. (2023). Biomimetics for sustainable developments. A literature overview of trends. Biomimetics, 8(3): 304.

Jitchotvisut, J. (2021). RideApart_Honda Plans To Unveil New 3D Printed Prototyping Project In 2021. Available at: https://www.rideapart.com/news/499279/honda-3d-printing-prototype-project/ [Accessed: May 30, 2025].

Kassab, A., Al Nabhani, D., Mohanty, P., Pannier, C., and Ayoub, G. Y. (2023). Advancing Plastic Recycling: Challenges and Opportunities in the Integration of 3D Printing and Distributed Recycling for a Circular Economy. Polymers, 15(19), 3881.

Kennedy, E. B. (2017). Biomimicry: Design by analogy to biology. Research-Technology Management. Available at: https://www.tandfonline.com/doi/full/10.1080/08956308.2017.1373052 [Accessed: 11 May 2025].

Korba, P., Hovanec, M., Tymofiiv, V., Al-Rabeei, S., Vencel, M., and Asnad, P. (2024). Use of Product Lifecycle Management in preparation for simulation of logistic processes. In Proceedings of the book (pp. 161–172). Springer.

Langella, C., Vieira de Arruda, A. J., and Di Bartolo, C. (2022) Revising biomimetics: Opportunities and ambiguities in the bioinspired design approach. diid – Disegno Industriale Industrial Design. Available at: https://www.diid.it/diid/index.php/diid/article/view/diid78-langella-vieira-arruda-di-bartolo [Accessed: 11 May 2025].

Listek, V. (2021) 3DPrint.com_Honda and WASP Partner for Sustainable 3D Printed Motorcycle Models. Available at: https://3dprint.com/285848/honda-and-wasp-partner-for-sustainable-3d-printed-motorcycle-models/ [Accessed: May 30, 2025].

Loop Industries. (2025) Closing the Plastic Loop. Loop Industries. Available at: https://www.loopindustries.com/en [Accessed: 11 May 2025].

Mansour, N., Callera, A., Potere, F., Micalizzi, S., Costantino, M., De Gaetano, F. et al. (2025). Circular economy and 3D printing in the healthcare sector. Frontiers in Bioengineering and Biotechnology, 13.

Mario Cucinella Architects. (2025) TECLA – Technology and Clay. Mario Cucinella Architects. Available at: https://www.mcarchitects.it/progetti/tecla-technology-and-clay [Accessed: 11 May 2025].

Meadows, D. H., Meadows, D. L., Randers, J., & Behrens, W. W. III (1972, March 2) The Limits to Growth: A Report for the Club of Rome's Project on the Predicament of Mankind. Internet Archive. Available at: https://archive.org/details/TheLimitsToGrowth [Accessed: 11 May 2025].

Meegle. (2025) Product Lifecycle Management (PLM). Meegle. Available at: https://www.meegle.com/en_us/topics/manufacturing/product-lifecycle-management-plm [Accessed: 11 May 2025].

Mohammed, A. (2023). A systematic design technique of biomimicry to correlate and integrate architecture and biology to attain green buildings. JES Journal of Engineering Sciences, 51, 239–259.

Natural Resources Defense Council (2015) The Story of Silent Spring. NRDC. Available at: https://www.nrdc.org/stories/story-silent-spring [Accessed: 11 May 2025].

Oladapo, B. I. et al. (2024). Evaluating the impact of recycling on polymer of 3D printing for sustainable product manufacturing. Resources, Conservation and Recycling, 197, 107179.

Paptic. (2025) Paptic: The Sustainable Alternative for Plastics in Packaging. Paptic. Available at: https://paptic.com/ [Accessed: 11 May 2025].

Pawlyn, M. (2011). Biomimicry in architecture (Flexible cover edition). RIBA Publishing.

Porfiri, M. (2021). Novel research revealing how a unique sponge "goes with the flow" could improve man-made structures. NYU Tandon School of Engineering. Available at: https://engineering.nyu.edu/news/novel-research-revealing-how-unique-sponge-goes-flow-could-improve-man-made-structures [Accessed: 11 May 2025].

Raman, R., Sreenivasan, A., Suresh, M., and Nedungadi, P. (2024). Mapping biomimicry research to sustainable development goals. Scientific Reports, 14: 18613.

Raza, M., Kirkham, T., Harrison, R., and Reul, Q. (2011). Knowledge based flexible and integrated PLM system at Ford. Journal of Information & Systems Management, 1.

Renewcell. (2025). Renewcell: We make fashion circular. Renewcell. Available at: https://www.renewcell.com/en/ [Accessed: 11 May 2025].

Ritchie, H., and Roser, M. (2020). CO_2 emissions. Our World in Data. Available at: https://ourworldindata.org/co2-emissions [Accessed: 11 May 2025].

Roh, Y., Lee, Y., Lim, D., Gong, D., Hwang, S., Kang, M. et al. (2024). Nature's blueprint in bioinspired materials for robotics. Advanced Functional Materials, 34(5): 2306079.

Scott, J. (2024). How PLM can improve supply chain and design synergy and move goods faster. Machine Design. Available at: https://www.machinedesign.com/3d-printing-cad/article/21283777/razorleaf-how-plm-can-improve-supply-chain-and-design-synergy-and-move-goods-faster [Accessed: 11 May 2025].

Siddique, S. H., Hazell, P. J., Wang, H., Escobedo, J. P., and Ameri, A. A. H. (2022). Lessons from nature: 3D printed bio-inspired porous structures for impact energy absorption – A review. Additive Manufacturing, 58: 103051.

Syvitski, J., Restrepo-Ángel, J., Saito, Y., Overeem, I., Vörösmarty, C., Wang, H. et al. (2022). Earth's sediment cycle during the Anthropocene. Nature Reviews Earth & Environment, 3: 1–18.

Sowińska-Świerkosz, B., and García, J. (2022). What are Nature-based Solutions (NBS)? Setting core ideas for concept clarification. Nature-Based Solutions, 2: 100009.

Steffen, W., Broadgate, W., Deutsch, L., Gaffney, O., and Ludwig, C. (2015). The trajectory of the Anthropocene: The Great Acceleration. The Anthropocene Review, 2(1): 81–98.

Tang, Y., Shu, M., Wu, Y., and Xuan, Y. (2024). Ecological sensitivity evaluation and spatial pattern analysis of Minjiang Estuary National Wetland Park based on GIS. Journal of Resources and Ecology, 15(1): 33–43.

The New Raw. (2025). Print Your City. urbanNext. Available at: https://urbannext.net/print-your-city/ [Accessed: 11 May 2025].

van Erp, T., Carvalho, N. G. P., Gerolamo, M. C., Gonçalves, R., Rytter, N. G. M., and Gladysz, B. (2024). Industry 5.0: A new strategy framework for sustainability management and beyond. Journal of Cleaner Production, 461: 142271.

WASP. (2025). WASP: 3D Printing Made in Italy. WASP. Available at: https://www.3dwasp.com/ [Accessed: 11 May 2025].

WeWantToLearn.net (2020). Ecotect – Resources. Available at: https://wewanttolearn.wordpress.com/category/resources/software/ecotect/ [Accessed 11 May 2025].

Yanko Design. (2020). The Rimac Scalatan is a stunning concept car with a 3D-printed chassis that 'breathes oxygen'. Yanko Design. Available at: https://www.yankodesign.com/2020/05/20/the-rimac-scalatan-is-a-stunning-concept-car-with-a-3d-printed-chassis-that-breathes-oxygen/ [Accessed: 11 May 2025].

Zang, H., Zhang, S., and Hapeshi, K. (2010). A review of nature-inspired algorithms. Journal of Bionic Engineering, 7(Supplement), S232-S237.

3

Make it Lighter

"To simplify also means to add lightness."

Colin Chapman

"The most amazing lesson in aerodynamics I ever had was the day
I climbed a thermal in a glider at the same time as an eagle.
I witnessed, close up, effortlessness and lightness
combined with strength, precision and determination."

Norman Foster

3.1 The Gravitational Paradox

"Compared to natural structures, *highly anisotropic* and *heterogeneous,* yet *simultaneously highly integrated* into multifunctional unitary elements, human technology appears as a set of rather rudimentary solutions" (Manzini 1989).

Natural structures display a remarkable degree of structural and functional complexity, evident not only in the materials themselves but also in their spatial arrangement. *Anisotropy*—variation of properties depending on direction—allows systems to respond effectively to different demands of strength and flexibility. For example, wood is stronger along its fibers than across them, supporting stable and efficient plant growth. *Heterogeneity,* on the other hand, refers to the use of varied materials and forms, each serving a specific function but working together cohesively. Animal bones are strong yet lightweight due to their porous structure, while bird feathers combine lightness and durability to support flight. These characteristics are not isolated in nature but are deeply interwoven into adaptive, efficient systems. A single organism can perform many different functions with the same structure (Bizzarri et al. 2020). The leaves of a plant, for instance, serve both as a means for photosynthesis and as a surface that contributes to the organism's water and temperature balance. In contrast, human technology, while evolving and improving over time, tends to have more rudimentary solutions compared to these natural structures. Each technological component is designed for a specific purpose, and its function is limited to a narrow area, lacking the ability to perform multiple

DOI: 10.1201/9781003438946-3

tasks simultaneously as happens in nature. The separation of functions, as seen in technological products, involves the use of homogeneous materials, chosen for their ability to perform a specific task but not designed to adapt as natural materials do. An example is the difference between the lightness of a bird's feathers, which serve multiple functions, and the parts of an airplane, each designed for a clearly defined purpose. The *gravitational paradox* can be seen as the contrast between the efficiency and adaptability of natural structures and the complexity and separation of functions in human technologies. While nature achieves great synthesis and efficiency through the integration of forms and functions, human technology, though advanced, tends to be less integrated and more complex, often separating functions into various components. Nature, on the other hand, has always experimented with highly bioengineered solutions for lightness and resistance, creating composite organic materials and extraordinary living structural sandwiches, such as the filamentous structures known as *actinofibrils*, discovered even in the ancient bio-design of the incredible invention of pterosaur wings (Bennet 2009). It is thanks to these types of technologies that, in the skies of the Mesozoic, some species belonging to the order *Pterosauria*, the largest living organisms ever known, were able to fly during each geological era of the Earth. These ingenious solutions have not escaped the attention of biomimetic researchers, who have studied their *less is more like structure* and functionality, considering possible applications in robotic technology (Chatterjee et al. 2013). Nature's engineering has invented and produced truly extraordinary animals, such as the family of this enigmatic *Azhdarchidae*, which in some cases, based on fossil evidence found, is estimated to have had a wingspan of over ten meters, comparable to that of a modern *Cessna 172* (Williams et al. 2021). Using keratin, muscles, and various elastic elements, lightness, flexibility, and strength are combined to reinforce the outer part of the wing, in a *patent* that consists of creating three distinct fibrous layers forming a c*risscross structure*, that is, a layering of fibers on top of each other in such a way that the fibers are crossed (Kellner et al. 2009). This approach closely mirrors the contemporary concept of weaving continuous fiber composite materials, as exemplified in the aerospace industry, where polymer matrix composites are fabricated using CAD/CAM 3D printing and integrated into the design of wings and structural components. Polymers such as PEEK (Polyether ether ketone), known for its high thermal and mechanical resistance, and Carbon Fiber-Reinforced Epoxy Resins (CFRP), are also woven into lattice configurations very similar to actinofibrils to optimize the lightness and strength of aircraft—as in the case of helicopter blades (Singh et al. 2020), or used in advanced biomedical experiments for bone implants in cases of degenerative spinal disorders (Ghermandi et al. 2023). Emerging technologies such as 3D printing with reinforced polymers, combined with biomimetic CAD design, are rapidly transforming the approach to designing materials and structures inspired by natural material solutions, as in the case of the WASP-*Trabeculae Pavilion* (Naboni et al. 2019):

"The project looks into 3D printing for answers to the emerging problem of scarcity in material resources. The design is based on a computational process that finds inspiration in nature, specifically in the materialization logics of the trabeculae, the internal cells that form the bone microstructure. From this investigation, custom algorithms have been developed to support the creation of a cellular load-responsive structure with continuous variations in sizing, topology, orientation and section, in order to maximize material efficiency" (ACTLAB 2018).

Recent discoveries by paleontologists have established that actinofibrils are anatomical support structures found in the wing membrane and the hindlimb girdles of pterosaurs. Their voluntary activation is thought to have minimized aeroelastic flutter—those vibrations that cause uncontrolled flapping (similar to a flag caught in strong wind), which would otherwise make any flight extremely dangerous (Pittman et al. 2022). Unless we assume a geophysical and energetic state of the Earth completely different from the present one, many of nature's successful inventions made it possible for such giants to lift off the ground, conquer the ancient skies, and hunt their prey—defying the limits imposed by the *gravitational paradox* (Hurrell 2012). First and foremost, the invention of the bony *trabeculae* and the truly unique morphology of the cervical vertebrae. These giant species of pterosaurs had incredibly long necks, yet they managed to keep them straight even while soaring through the skies, supporting a head weighing over one and a half meters—not to mention the weight of their captured prey and air resistance. How they accomplished this was a mystery, finally solved by paleontologists from the University of Portsmouth and the University of Illinois at Urbana-Champaign (Irving 2021). The secret of these giant pterosaurs resided in the extraordinary internal architecture of their vertebrae—an anatomical solution never observed in other animal species. According to researchers, the neural canal was centrally placed and connected to the vertebral wall through a network of slender, rod-like trabeculae. These were distributed both radially and helically, resembling and even intersecting like the spokes of a bicycle wheel. Such a configuration reveals how evolutionary processes refined their anatomy to achieve an optimal balance of lightness and strength, enabling these prehistoric creatures to become astonishingly efficient flyers. The biomimetic researchers estimate the mechanical properties of pneumatized cervical vertebrae with a central neural canal based on X-ray computed tomography and use this to evaluate possible prey size. They show that as few as 50 trabeculae inside the vertebra increase the buckling load by up to 90%, which will promote a light and strong structure. However, there is overall little information about bone mineral content, which may impact mechanical properties alongside weight. The case of pterosaurs demonstrates how Nature's biomechanical technology, while maximizing the lightness of their anatomical structures, managed to challenge gravitational forces and achieve the best compromise between maximum strength, aerodynamics, and overall body weight. This enabled their organisms to overcome the physical limits imposed on both natural and artificial material worlds on Earth. Of course, human technology has always coexisted and confronted the consequences of gravity and its problematic static-mechanical effects. The organic innovation that evolution brings to natural organisms is, in human artifice, triggered by technology. Thus, oscillating between the frustration caused by the limitations of these constraints and the most unbridled imaginative and creative drive sparked by the constant desire to overcome any organic limits imposed on our species, we manage to evolve as a genus, conquering, for example, domains such as the sky or water which, evolutionarily speaking, were previously inaccessible. In fact, one of the most complex paradoxes of engineering problem-solving is being able, in static-mechanical design, to effectively mediate, balancing between them, the performance of lightness of a structural element (weight) with its own resistance (mass) in relation to the static-dynamic stresses caused by the application of dynamic forces (e.g., wind) and, above all, the effects of Earth's gravity. This is true whether we consider the individual physical response of basic construction materials or the far more complex and articulated structural systems, as happens in the dimensioning of beams and slabs for bridges or skyscrapers, or in dynamic mechanical components such as turbines, wings, entire bodies of flying vehicles, ships, and any other structure related to the design of functional objects subjected to static and dynamic loads, such as tools, anatomical prostheses, or simple elements of furniture and design. The quest for lightness-resistance is a technical-creative challenge that every designer must face, sometimes in a convoluted and by no means linear way, in the search for perfect harmony between form, function, and the physicality of an artifact. In the history of design, lightness and functionality are continuously balanced, as seen in the case of the 699 Superleggera chair designed in 1957 by the Italian architect Gio Ponti (1891–1979) for the Italian company Cassina. This object is recognized internationally as an icon, made famous precisely for its extreme lightness, weighing only 1.7 kg (Cassina 2025). The designer and the artisans of the company spent several years perfecting the overall design, working on the triangular section of the legs, which were only 18 millimeters thick, and inventing the notched tenon joint:

"Creating the Superleggera", said architect Gio Ponti, "I followed the eternal process of technique, which goes from the heavy to the light: removing inert material and weight, identifying the form with the structure at its limit, wisely and without virtuosity, that is, respecting at the same time both utility and exact solidity."

The chair is so lightweight that it can be lifted with two fingers. It is not just a triumph of modern design but also a tribute to the essentiality and functionality that characterized traditional Italian craftsmanship of its time. For this project, architect Ponti was inspired by the *chiavarina*, a typical chair from the Ligurian city of Chiavari. He used ash wood for the structure, a material that is both light and strong, and marsh straw for the seat, both elements commonly found in Ligurian tradition. The use of woven straw evokes the manual techniques used by basket makers and local artisans, keeping alive a craft tradition that values manual skill and knowledge passed down through generations. Despite the delicate appearance of its minimalist design, the Superleggera is robust, thanks to advanced construction techniques and the intelligent use of materials. It closely resembles the methodologies of the intricate design logic found in the lightweight structures of natural organisms, which biomimetics often draws inspiration from. Lightness, minimalism, and sobriety take on an ethical and moral dimension when a religious community like the Shakers, following their principles, must give the correct form and structure to objects that serve the functional needs of daily life. Shaker design is known for its minimalist appearance, characterized by a strict lightness both in materials and forms. This lightness reflects the spiritual and practical principles of the Shakers, who value essentiality, simplicity, and the sobriety of not wasting resources or time:

"Every force of Nature is used wisely and in moderation"- Shaker Principle (Plyroom 2025).

The famous *ladder-back chair* is made from sturdy yet lightweight woods like maple and cherry, combining durability and lightness. Its design is simple and clean, with tall backs and thin slats that give a sense of both visual and physical lightness, while the woven rush seats, like the Superleggera, help reduce the overall weight without sacrificing comfort and seat security. The introduction of advanced construction techniques, such as the use of dovetail joints and attention to assembly details, has ensured that these objects exemplify the paradox of lightness and strength. How much this design approach is still considered by design culture is demonstrated by the CAD/CAM reinterpretation of typical Shaker chairs created by designer Ankur Podder (Podder 2019). Compared to the way the theme of *lightweight-resistance* is addressed in contemporary *WunderCAD/CAMmer*, the extraordinary evolution of computer-aided design tools, combined with innovative analog biomimetic methodologies, offers designers powerful interfaces for physical simulation and structural optimization. On high-definition monitors, within the virtual space of their personal digital *Wunderkammern*, engineers, architects, scientists, and designers are now able to manipulate data and physical parameters through dedicated CAD/CAM suites, achieving a level of realistic simulation detail at a 1:1 scale. On Digital Twin models rendered through Dynamic Modeling (DM) techniques, it is possible to conduct highly accurate structural simulations. These can apply forces such as compression, shear, torsion, and bending—not only to individual materials but also to mechanical components and entire assemblies (Nezzi 2025). This allows engineers to record and verify structural behavior, even down to the molecular scale, without the need for physical prototypes. DM is a sophisticated computational method, closely integrated with current parametric and generative design suites. It is used to simulate and analyze the physical behavior of materials or systems under a variety of dynamic forces, including compression, torsion, bending, and vibrations. It is particularly valuable in the design process because it allows engineers and designers to predict how an object, material, or structure will react to different forces in real-world conditions, without the need to physically build prototypes. One of the key applications of DM is in the architecture and civil engineering industries. For example, when designing a skyscraper in an area prone to high winds or earthquakes, engineers can use DM to simulate how the building will behave under those forces. The model can replicate wind pushing against the structure or the shifting of forces during a seismic event, providing insights on how to optimize the building's design for better resistance while keeping the weight as light as possible. This not only ensures safety but also minimizes material usage, contributing to both sustainability and cost-efficiency. In the aerospace industry, DM plays a crucial role in the design of airplane wings. The dynamic forces acting on an aircraft wing during flight—such as lift, drag, and turbulence—are simulated using advanced CAD/CAM tools. These simulations help designers understand how the wing's shape and materials react under various flight conditions, allowing them to optimize the structure for both aerodynamics and strength. The result is a wing that is lightweight yet durable, enhancing fuel efficiency and overall performance. Similarly, mechanical engineering benefits greatly from digital modeling (DM) when designing complex components such as gears, turbines, and support structures. For example, engineers can simulate the stresses and vibrations a turbine will experience in service, then refine the geometry to minimize fatigue, prevent failure, and improve efficiency. Virtual testing in CAD/CAM environments lets them compare materials and shapes to strike the best balance between strength and weight—without costly physical prototypes. In biomechanics, DM is invaluable for medical implants and prosthetics. Dynamic simulations predict how a prosthetic limb behaves under different loads while walking or running, guiding designers toward devices that are lightweight yet durable. Advanced digital tools then fine-tune the prosthesis's shape and material composition, ensuring a precise fit and reliable performance for each user. Overall, DM enables precise material optimization and realistic performance prediction, delivering structures that are both robust and lightweight while greatly reducing the risk of real-world failure. Not all engineers who work with CAD are truly equipped to tackle the challenges of dynamic modeling. Even though many CAD software programs now include advanced tools like Finite Element Analysis (FEA), Computational Fluid Dynamics (CFD), and other specialized modules, that doesn't necessarily mean users can make full use of them. As noted by Australian Design & Drafting (n.d.), without a strong theoretical background in complex subjects such as solid mechanics, fluid dynamics, vibration analysis, control systems, or stochastic mechanics, a designer may struggle when facing unconventional problems.

Most importantly, Dynamic Modeling, when integrated with CAD/CAM, enables the development of more sustainable, cost-efficient, and high-performance designs across various industries. Below is a selection of powerful software suites for DM:

ANSYS® is one of the most powerful engineering simulation suites, used for a wide range of physical analyses, such as dynamics, thermodynamics, fluid dynamics (CFD), and structural analysis. Its ability to simulate physical behaviors with extreme precision at both the molecular and macroscopic levels makes it indispensable in fields such as automotive, aerospace, and energy. ANSYS® is particularly well-suited for optimizing structures and components that must withstand complex loads and dynamic forces. Another highly regarded tool is Abaqus® (Dassault Systèmes), known for its advanced nonlinear simulations. Abaqus® can model complex structures and test how they react to loads and dynamic forces. It is widely used in the automotive, construction, and industrial design sectors, where precise simulation of stress, vibrations, and material interactions is essential. COMSOL Multiphysics® is another versatile platform that allows users to integrate various physical simulations, such as mechanics, electromagnetism, and thermodynamics, into a single simulation model. It is particularly useful for multidisciplinary engineering and advanced scientific applications, where simultaneous simulation of various physical phenomena is needed. For those seeking a user-friendly solution, SolidWorks Simulation® offers powerful dynamic and structural simulation tools integrated into a popular CAD modeling platform. SolidWorks Simulation® allows for the execution of resistance, vibration, torsion, and stress analyses on components and structures. It is widely used in industrial design and mechanical engineering, particularly for its ability to integrate simulations directly into CAD design workflows. Siemens NX® is an advanced suite that combines CAD, CAM, and CAE. With specific modules for dynamic modeling, it enables testing and optimization of designs in terms of resistance, deformation, and vibration. Siemens

NX® is primarily used in sectors such as aerospace, automotive, and manufacturing, where the precision of simulations and structural optimization are crucial to project success. Finally, Autodesk Fusion 360® is a suite that, while more approachable compared to other solutions, still offers excellent dynamic simulation and structural analysis capabilities. Fusion 360® is ideal for small and medium-sized businesses or independent designers who want to combine design, engineering, and simulation in one platform. It is particularly appreciated for its ability to simulate deformations, stress, and vibrations, allowing real-time testing of CAD models before production. The transition from dynamic CAD design and simulation to 3D-CAM prototyping and digital fabrication systems completes the cycle, easily generating executable models and components. In particular, in the automotive, aerospace, robotics, and prosthetics sectors, the pursuit of lightweight materials combined with greater resistance and static-mechanical performance of components has always been of paramount importance In the field of prosthetics design, parametric applications excel and have revolutionized the orthotics industry by enabling the creation of lightweight, custom-made devices that provide improved comfort, functionality, and overall patient satisfaction. Companies like the German Ottobock® have embraced this innovation, leveraging advanced CAD/CAM systems to offer high-quality, personalized, and customized prosthetic solutions (Ottobock n.d.). Blatchford®, instead, is a UK-based company specializing in prosthetics and orthotics, known for its focus on lightweight materials and integrated systems designed to reduce metabolic effort and enhance natural movement. The Linx® system, for instance, connects knee and ankle through intelligent sensors to optimize every step (Blatchford 2025b). The process begins with a 3D scan of the patient's limb, taken by a certified prosthetist or orthotist. This scan captures the precise dimensions and shape of the limb, which are then used to create a digital model. For example, Ottobock® employs the FitSocket® platform, an advanced 3D scanning system that ensures accuracy and a tailored fit by capturing the unique characteristics of each patient's limb (MIT Media Lab n.d.). Once the scan is complete and the digital model is created, the prosthetist or orthotist can begin designing the prosthetic or orthotic. CAD/CAM technology allows for highly precise design adjustments, ensuring that the device perfectly fits the patient's individual needs. After the design is finalized, the prosthetic or orthotic is manufactured using cutting-edge 3D printing technology. Blatchford®, for instance, created the Atlas Custom TLSO®, an innovative spinal range that refines the design before using 3D printing or milling techniques to produce the final product. This technology, which in some respects involves *CAD-based X-ray CT calibration and error compensation*, enables rapid prototyping and allows for real-time adjustments to enhance the device's functionality, body's customization and safe comfort (Fragnaud et al. 2022). Compared to traditional prosthetics, these CAD/CAM models are not only lighter and more robust, enhancing the patient's comfort, but also customizable to ensure a precise fit tailored to the patient's anatomy. This is especially important in the creation of prosthetics where comfort, durability, and usability are paramount. Moreover, CAD/CAM technology is more environmentally friendly than traditional manufacturing methods like foam milling, as it generates less waste. The digital design process, paired with 3D printing, allows for the creation of prosthetics that are both sustainable and efficient. To achieve the desired performance, strength, and comfort, a range of advanced materials are used in CAD/CAM design production. These materials include:

- **Zirconium**: Known for its strength and durability, zirconium is used for components that require high mechanical properties.
- **Nanocomposites**: These materials are engineered to provide enhanced strength, stiffness, and resistance, often used for both structural components and aesthetics.
- **PMMA (Polymethyl Methacrylate)**: A lightweight material that offers good impact resistance and is often used in the design of lightweight prosthetics.
- **PEEK (Polyether Ether Ketone)**: Known for its strength, thermal stability, and biocompatibility, PEEK is commonly used in load-bearing components.
- **Lithium Disilicate**: A glass ceramic with excellent strength and aesthetic properties, used in more refined and aesthetic prosthetic applications.
- **Titanium**: With its outstanding strength-to-weight ratio, titanium is widely used for prosthetic limbs, particularly in structural elements.
- **Cobalt Chromium**: These materials provide excellent strength and corrosion resistance, making them ideal for high-stress applications.
- **Composite Ceramics**: Offering a blend of high strength and low weight, composite ceramics are used to create durable and lightweight prosthetic components.

By utilizing these advanced materials, Ottobock® and Blatchford® have developed high-performance products such as Ottobock's® C-Leg® prosthetic knee, which integrates advanced materials like PEEK and titanium, and Blatchford's® Linx® prosthetic system, which uses lightweight, durable materials to improve both the functionality and comfort of the device.[1,2]

Contemporary CAD protocols not only enable the practical and material implementation of the lightweight-resistant concept, but also support its application at the forefront of advanced, high-performance engineering. The concept of lightness goes beyond purely static and mechanical themes to take on a more nuanced meaning as an ergonomic metaphor, an optimization, and streamlining of business and production processes. Consider the strategies for functional weight reduction in handling the complexity and burdens of management, production, and marketing that companies face daily, both in the context of Industry 4.0 and its more eco-sustainable evolution, Industry 5.0. The adoption of advanced digital tools, such as CAE-CAO interfaces and AI, in coordinating and streamlining operational cycles faced daily

[1] https://www.ottobock.com/it-it/product/3C88-3~23C98-3

[2] https://www.blatchfordmobility.com/en-us/products/limb-systems/linx-microprocessor-controlled-leg/

by design and testing teams, for example, can lead to significant reductions in the *gravitational* load caused by the weight and complexity of bureaucratic-management processes that now afflict, like an energy parasite, the massive corporate structures of the largest corporations. Here is a typical example of a press release from a company offering management streamlining services:

"Industry 4.0 offers a whole range of options for optimizing processes and lightening the workload of staff – and that includes in production operations at your company. At your request, we can support you in your planning from the very outset and help you boost your production capacities. To this end, we divide your production into fractals in line with lean principles. The next step is to work with you to improve each of these fractals so that you can run at maximum efficiency and productivity across the board, whether cutting materials to size or machining them" (SEW-EURODRIVE 2025).

The concept of Industry 5.0, widely promoted for years as the next major evolution in how companies manage production, is rapidly gaining traction in the industrial landscape. Leveraging advanced technologies such as IoT (Internet of Things), artificial intelligence, additive manufacturing, and automation, this model is leading to the extinction of bureaucratic corporate dinosaurs in favor of more evolved, lighter, faster, and more flexible entities. The connection and interconnection of production resources, data analytics, and collaboration between various entities create a dynamic and responsive approach to the challenges of the modern market. In the design and construction phases of particularly complex projects, the working environment, driven by powerful algorithms, can manage large volumes of parameters and variables input into the design process. These are directed through massive flows of big data, which, through state-of-the-art software, can be easily integrated and interactively compared with each other. The integration of AI interfaces is leading to a further evolutionary leap in the entire design and industrial production system. Even shared work modes through CAD/CAM environments are generating an industrial ecosystem that is radically transforming traditional design and production methods, enabling designers and technology companies to work on cloud-based platforms and remotely handle 3D prototyping and modeling units. Industry 5.0, as an evolutionary model, promises to introduce a metaphysical form of lightweight production and management into the global economic and production system. It represents an almost ethereal industrial revolution, intangible in the very fabric of big data and algorithmic flows managing computers and CNC machines. This quantum leap seems to distance the world from the heavy mechanical traces of the early 20th-century industrial revolution. The adoption and development of ever more optimized design processes should be geared towards greater global sustainability, leading to significant improvements in the economies of scale models, which are crucial for saving fuel, energy, and CO_2 emissions during industrial cycles and material consumption. The optimization of CAE/CAO—through the application of principles like constant tension—has been learning lessons from the morphological study and the adoption of strategies from natural cycles, ecosystems, and the forms of living organisms. These protocols promise substantial improvements over previous design models, ensuring greater structural integrity for the entire artificial body. This translates into reduced wear and tear on technological and mechanical components, as well as enhanced energy and mechanical efficiency. In conclusion, Factory 5.0 is not just a technological shift; it is a holistic transformation of production processes that incorporates the best principles of nature, sustainable design, and high-performance engineering to create a future where industry becomes more agile, efficient, and responsive to both ecological and market demands.

3.2 The Sustainable Lightness of Trees

Bionics and biomimetics have introduced the concepts of technology and natural invention (Niklas 1992). Like the life sciences more broadly, when faced with the remarkable optimization strategies developed by living organisms—starting from the single cell—they almost seem to mock themselves. Their justification? Time—the countless trials of evolution across billions of years, something no laboratory or algorithm can yet replicate in full (Iouguina et al. 2014).

"The fact that biological technology has been shaped by natural selection rather than *designed* in the colloquial sense only enhances the strength of the comparison" (Labarbera 1998).

We convince ourselves that things in nature are so well made because life on Earth has had so much time to experiment, fail, and start over with clearer and better ideas to ensure evolutionary success for the most technologically capable species. Time spans of millions and millions of years, which, in the case of viruses, bacteria, eukaryotic cells, and prokaryotes, are counted in billions. But the evolutionary disputes that stir contemporary *WunderCAD/CAMmers* are far from settled, and today we are still theorizing hypotheses that fall somewhere between Neo-Darwinism and Intelligent Design (ID). And the more we think we know, the more we find ourselves deeply entangled with the enigmas of the great mystery of the birth and evolution of life (ours included) on our planet and in the Universe. One fundamental concept, however, seems to be agreed upon by all the factions in dispute:

"Nature and humans build their devices with the same earthly materials and use them in the same air and water, driven by the same gravity. So why do their designs diverge so sharply? Humans, for example, love right angles, while nature's angles are rarely right and usually rounded. Our technology runs on wheels, rotating pulleys, gears, shafts, and cams, but in nature, only the tiny flagella of bacteria rotate like true wheels. Our hinges rotate because hard parts slide over one another, while nature's hinges (for example, a rabbit's ear) more often swing by bending flexible materials" (Vogel 2000).

In consideration of the topic discussed in this chapter, within the German engineering tradition from the post-World War II era to the present day, the debate and research on the concept of *lightweight structures* is particularly vibrant and fertile. The *Institute for Lightweight Structures (IL)*, in the German language *ILEK (Institut für Leichtbau Entwerfen und Konstruieren),* was founded in 1964 and the architect Frei Otto (1925–2015) was appointed its director. With his interdisciplinary research and teaching in the field of wide-span surface structures and natural constructions, he developed the IL into a globally recognised centre of lightweight construction by 1991 (University of Stuttgart n.d.). In 1961, the German biologist Gerhard Helmcke (1908–1993) and the architect Frei Otto founded the interdisciplinary working group

Biologie und Bauen at the Technical University of Berlin (Otto 1984). The goal of their collaboration was to gain a deeper understanding of biology, technology, and architecture, and to explore the principles of *natural construction* (Staatliche Museen zu Berlin 2018).

"The transition from a purely biological and historical study of form to a hybrid one, where engineering and biological sciences worked together, culminated in the establishment of bionics and biomimetics as autonomous scientific disciplines. These were in turn considered essential components of the broader research program pursued by architects and biologists. For example, architect Frei Otto invited biologist Werner Nachtigall, the founder of the field of bionics in Germany, to formally participate in his SFBs. Additionally, biologist Helmcke, together with engineer Heinrich Hertel, professor of aeronautical engineering at the Technical University of Berlin, *created a marriage between engineering and biology* that led to the battle cry *TUB*"[3] (Hertel 1963, Tamborini 2021).

Within this German bionic tradition of structural technological lightness inspired by Nature, one of the pioneers is the German engineer Claus Mattheck—a physicist and world-renowned expert in the mechanics of trees (Mattheck and Tesari 2000). He is a true pioneer in the development of CAD/CAO (Computer-Aided Shape Optimization) and CAIO (Computer-Aided Internal Optimization) software, inspired by bionic analyses of bones and trees:

"Nature is the greatest engineer, and its design principles can be translated into tools for optimizing artificial structures. CAO design stands as a testament to the power of observing nature and applying its lessons to human technology" (Matteck 1998).

Prof. Mattheck was born in Dresden in 1947. His academic background was in theoretical physics—a discipline initially far removed from the study of the structural and mechanical behavior of trees, which he would later embrace through the development of the Visual Tree Assessment (VTA). This investigative methodology, now recognized in many countries, is used to evaluate the structural condition of trees (Mattheck and Breloer 1994).On the topic of the structural nature of plant growth, it is worth noting that some of the earliest theoretical reflections can be found in the research and drawings of Leonardo da Vinci (1452–1519). In the *Trattato della Pittura*, a 16th-century compilation based on his notes and reprinted many times over the centuries, the sixth chapter is titled *On Trees and Plants.* While Leonardo primarily addresses technical aspects of depicting vegetation in painting—such as light and perspective—he also offers insights and observations that verge on botanical analysis (Peruzzi 2019). Being deeply interested in the understanding and accurate drawing and artistic representation of trunks, branches, and crowns, Leonardo da Vinci formulates the so-called *Regola degli alberi* (*rule of trees*):

"all the branches of a tree at every stage of its height when put together are equal in thickness to the trunk."

In other words, if a tree's branches were folded upward and squeezed together, the tree would look like one big trunk with the same thickness from top to bottom

[3] Abbreviation for: *Technologie Und Biologie* (Technology and Biology).

(Jacobsmeyer 2011). Although recent studies tend to disprove the accuracy of this rule when applied to the scale of micro-lymphatic vessels, Leonardo certainly deserves credit for being the first *scientist-artist* to apply a proto-bionic method. Fig. 1 shows how, centuries later and with different intentions, Paul Klee employs the same underlying analytic concept to illustrate radial growth in his *Notebooks, Volume 2: The Nature of Nature* (Klee 1923, 1973).

Claus Mattheck specialized in biomechanics and later headed the Biomechanics Department at the Karlsruhe Research Center, where he developed the concept of *self-optimization* of biological structures, meaning adaptive growth without material waste (Matteck 2007). Before pioneering computerized CAO systems that involve the use of algorithms and 3D digital modeling, Mattheck, together with S. Heller, developed an original analog *computer-free* optimization system for topology, called the *Force Cone Method*, a new thinking tool to consider for lightweight structures design (Mattheck and Haller 2012). One of Mattheck's pioneering insights, inspired by close observation of how trees organize their growth and development, concerns the way they respond optimally to the complex array of static and mechanical stresses acting on trunks and branches. These include their own weight, wind forces,

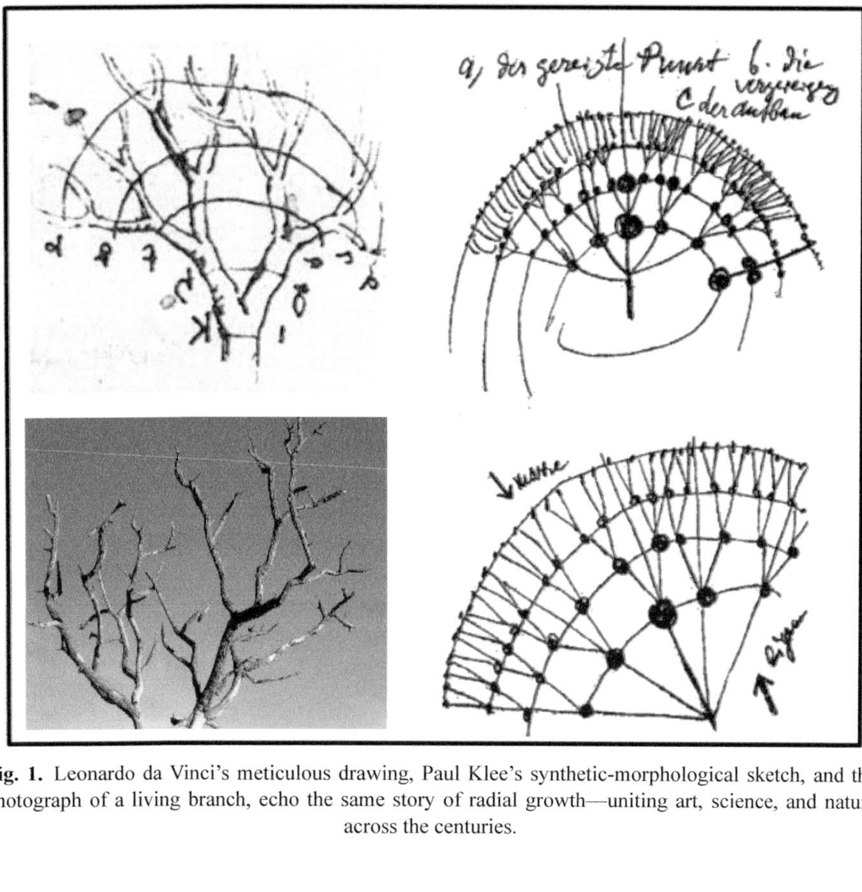

Fig. 1. Leonardo da Vinci's meticulous drawing, Paul Klee's synthetic-morphological sketch, and the photograph of a living branch, echo the same story of radial growth—uniting art, science, and nature across the centuries.

additional loads from snow or ice, and structural adaptations needed to counteract ground trauma caused by landslides or sudden soil failure. What makes Mattheck's contribution particularly significant is his ability to link these natural strategies to some of the most insidious structural problems faced by materials engineers in the design of mechanical or static components—whether in construction or in mechanical engineering. In particular, he recognized how these adaptive responses in trees directly relate to one of the most persistent challenges in structural mechanics: the notch, a critical source of weakness that often goes underestimated in both design and performance prediction. In mechanical and structural engineering, a *notch* represents a *geometric discontinuity*—such as a groove, hole, or sharp internal corner—that disrupts the uniform flow of stress within a material. It acts as a stress concentrator, a point where forces intensify and structural integrity is compromised. Even a seemingly minor notch can significantly amplify local stresses, often becoming the origin point for fatigue cracks or brittle failure. While engineers employ various methods to mitigate the effects of notches—such as fillets, reinforcements, or controlled transitions—these are corrective strategies introduced after the design has been conceived. By contrast, natural structures—and trees in particular—have evolved in such a way as to completely avoid notches. Their growth follows internal stress lines, generating organically optimized forms that distribute mechanical loads without creating sharp discontinuities. Transitions between trunk and branches, or between trunk and roots, are never abrupt: they develop as smooth, continuous curves that prevent stress concentration. The goal is not to eliminate stress, but to diffuse it across broader areas, avoiding localized overloads. This principle, derived from the observation of bones and other animal structures, suggests that nature avoids material waste by optimally adapting growth to evenly distribute static and dynamic stress. It has since been translated into algorithms that can be used in engineering design (Mattheck 2006). This approach stands apart from traditional computational engineering, as it does not use complex algorithms or 3D digital models. Instead, the method relies on an analogical and intuitive concept, focusing on understanding the flow of forces within a structure. The heart of the method lies in the graphical representation of force behavior, using force cones to visualize how stresses are distributed within a structure. This approach allows for identifying critical points where stresses concentrate and optimizing the shape and material distribution to reduce weight without compromising structural strength. In the *Force Cone Method*, the concept of the cone refers to the way forces are distributed and directed through a structure. The geometry of the force cones is such that forces are channeled naturally, minimizing the material used. This mirrors the principles of growth and optimization found in nature, where biological structures, like trees, grow in a way that maximizes efficiency with minimal resource consumption. The method is particularly useful for designing lightweight structures, such as those used in the aerospace industry, civil engineering, and innovative material design. The analysis of forces and the subsequent optimization of the structure's geometry allow for a design that adheres to principles of efficiency and sustainability, reducing material waste. In summary, the *Force Cone Method* offers a practical and visual approach to designing structures that maximize strength and lightness, reflecting the natural processes of adaptation and optimization found in living forms. This innovative design tool is a perfect example of how biological principles can be applied to technology to solve complex engineering challenges (Speck and Speck 2021). The concept of *Constant Tension* is one of the fundamental engineering axioms defined by Mattheck. This principle highlights how, in natural structures—particularly in plants—material is added only where strictly necessary to compensate for damage or to respond to environmental changes. Plants, in fact, are able to optimize the distribution of material in order to maintain structural balance, minimizing waste and maximizing efficiency. This vision has had a significant impact on Mattheck's work and has also inspired the work of German professor and engineer Lothar Harzheim, along with the Adam Opel GmbH International Development Center, in the development of *Computer-Aided Optimization (CAO)*. CAO is based on similar concepts of structural optimization, where material use is minimized through the analysis of forces and the optimal distribution of materials. Thanks to this philosophy, artificial structures can be designed more efficiently, following the principle of constant tension that guides Nature (Masselter and Speck 2011). In nature, the efficient use of body mass is a vital necessity. Biological structures—such as bones—are shaped optimally to ensure maximum strength with minimal material. Drawing inspiration from these principles, a method called the *Soft Kill Option (SKO)* has been developed. The SKO method is based on internal stress analysis, mimicking the behavior of bone cells. Using the so-called *fully stressed approach*, this method allows engineers to optimize components in order to achieve maximum stiffness with minimal material usage. In practice, the component is progressively modified by removing the least stressed areas—just as the human body eliminates excess bone tissue (Leidenfrost 2015). Building on the same principles, Mattheck developed a *tensile triangles graphical optimization theorem* that translates Nature's logic into a design tool. Starting from a 90-degree angle—the classic notch geometry—one constructs a triangle with 45-degree angles, and then, iteratively, adds progressively smaller triangles, each half the size of the previous one (Fig. 2).

The outer envelope formed by the vertices of these triangles traces a curve that replicates the way wood fibers grow in response to bending stresses. The result is a *notch-free contour*, a shape that quite literally embodies the principle that *"form follows force."* This geometry, derived from the observation of trees, can be applied directly to engineering contexts to enhance structural strength without requiring additional material. Nature does not compensate for weakness through excess, but through the *intelligence of form*—a lesson engineering would do well to embrace.

CAO (computer-aided optimization) and SKO (Soft Kill Option) software was developed by Claus Mattheck at the Karlsruhe Research Centre in Germany:

"The optimization algorithm simulates adaptive bone mineralization by varying the Young's modulus according to a calculated stress distribution. Several examples demonstrate the efficiency of this procedure, which can be done with any commercial finite element code. A combination with the shape optimization method, CAO is a complete layout procedure to design mechanical components after the rules of nature" (Baumgartner et al. 1992).

The CAO and SKO software works with FEM (Finite Element Model) used in engineering design (Li et al. 2013). FEM is a numerical tool that breaks a component of interest into finite geometrical sections, then defines the material property of each

Fig. 2. Graphical application of Mattheck's progressive triangle method to a naturally curved trunk, illustrating how stress-optimized contours emerge during tree growth.

finite element. FEM identifies areas of high stress and shows the simulated effects of adding or removing material based on CAO and SKO (Quint 2001). This imaging and simulation software mimics evolution in construction by creating a more efficient engine support structure (McKeag 2014). The software aims to secure specific elements in place while providing optimal resistance with minimal materials, using a virtual 3D model that simulates stresses. The algorithm removes all unnecessary material without weakening the structure. This process uses the same principle as evolution in living organisms.

Bones, as well as tree trunks and shrubs, are natural designs that efficiently develop their internal structural growth biologically to achieve an optimal weight-to-strength ratio, adding and removing material in response to environmental stresses, successfully addressing the paradox of lightness and strength (Fig. 3).

The CAO software by Adam Opel GmbH replicates this process: repeated simulations add material where force is needed and remove it where it is not, creating a design that maximizes strength with minimal material. The CAO program developed is able to calculate and render in a graphical model the static-dynamic stresses acting on an unoptimized structure, highlighting the greatest stresses as red spots; the structure grows until these disappear, evenly distributing the stress, analogous to the biomorphological processes in trees. According to Mattheck, the main difference between a tree and a bone lies in weight: bones are lighter to allow movement and ease of human mobility. Bone cells eliminate excess material, with

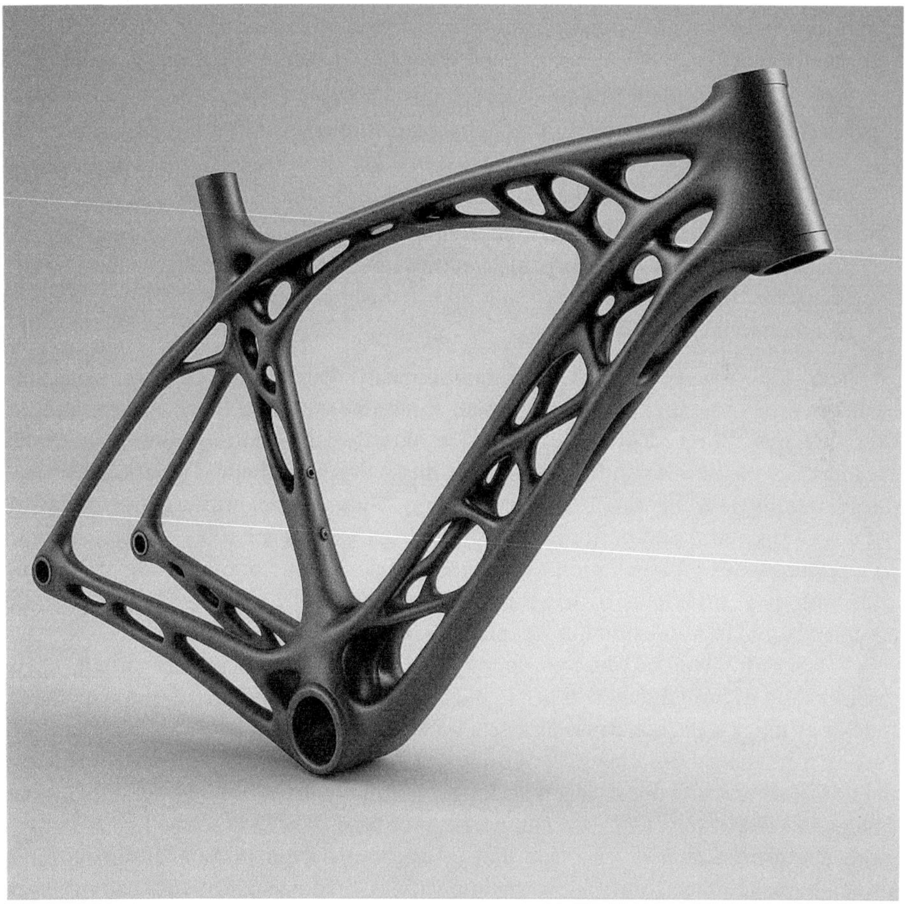

Fig. 3. A bicycle frame design inspired by bone bionics, sculpted through iterative CAD/CAE optimization under prescribed loads, refined for manufacturability, and rendered photorealistically.

internal holes determined with millimeter precision. In the software, mechanical stresses are also calculated, and surplus material is removed. This principle, applied to car and airplane bodies, has significantly reduced the material used while ensuring the strength and lightness of parts, preventing breakage with the minimum material necessary. This is the operating principle of bones. By combining CAD with Visual Tree Assessment (VTA), Mattheck has been able to develop a protocol that can enhance the accuracy and efficiency of tree inspections, especially for urban planning, landscape architecture, and construction projects. CAO is used to optimize components such as engine trees, supports, and frames, reducing the weight of components while maintaining or even improving their strength. Mattheck's CAO approach has revolutionized the way engineers tackle structural design and is now a standard adopted by popular commercial CAD routines such as Autodesk Fusion 360®, SolidWorks®, or Ansys®, to name a few (Matteck et al. 2005).

The main advantages of the CAO approach can be highlighted as follows:

- **Material efficiency**: Topological optimization allows for a significant reduction in the weight of structures, leading to a more efficient use of resources. This results in lower production costs and reduced environmental impact, in line with sustainability principles.
- **Increased durability**: Structures designed through CAO are more resilient to mechanical stresses and deformations, making them more durable and reliable over time.
- **Innovation in design**: CAO opens up new formal possibilities, enabling the creation of complex and organic geometries that are difficult to achieve with traditional methods. This approach allows designers and engineers to overcome conventional design limitations.

The applications of CAO are continuously expanding, demonstrating the transformative potential of this approach. A notable example is the *Bone Chair* by Dutch designer Joris Laarman, in which structural optimization inspired by natural evolutionary processes results in an object that merges aesthetics, functionality, and formal intelligence (Laarman n.d.). In this case, Nature is not only an aesthetic muse but also a structural and performative guide, highlighting the role of biomimicry in contemporary design innovation. By combining design and digital prototyping tools with biological principles of structural optimization, he has been able to generate complex forms in his design through evolutionary industrial processes:

"To me this was not just an animation of an optimized engine mount, but a visualization of how the industrial era in general is transforming into the digital era. Industrial times and modernist pioneers were all about assembly and standardized parts in a geometric form language dictated by the limitations of industrial machines. In our digital era, however, we are no longer bound by these limitations. With digital design and fabrication tools, we can create smarter customized forms that are much more complex. But it is also true that throughout design history, designers have striven to make objects inspired by nature: from Art Nouveau, driven by innovations in materials like cast iron, to Streamline, driven by innovations in materials like aluminum, and the organic design of the 1960s, driven by innovations in plastics and plywood. But our digital age makes it possible to not just use nature as a stylistic reference, but to actually use the underlying principles to generate shapes just like an evolutionary process."

Within the field of nature-inspired technology, the German company Synera® stands out for its innovative approach to weight reduction—both in the metaphorical, strategic sense and in its concrete engineering practice. In recent years, it has developed cutting-edge solutions that lighten industrial components by combining advanced design techniques with novel materials. This strategy not only boosts product efficiency but also promotes sustainability by lowering resource consumption and environmental impact. Applying biomimetic principles and structural optimization—and leveraging technologies such as 3D printing and computational analysis—Synera® creates components that balance strength and lightness. The company thus unites materials research with a performance-driven, engineering-excellence mindset, serving high-tech sectors such as automotive and aerospace (Synera 2025a). Synera®'s mission and vision converge in its design-optimization and Process Automation Platform, which roots the firm's philosophy firmly in biomimicry and Germany's tradition of lightweight engineering. The low-code platform automates and streamlines CAD and CAE workflows, enhancing the efficiency and precision of engineering tasks and freeing engineers to focus on creative, strategic work. By embedding automation and intuitive design tools, Synera® embodies Nature's principles of efficiency and adaptability in a holistic lightweight-design ethos. As a result, it is helping to shape the future of engineering by offering a versatile, user-friendly solution aligned with rising demands for sustainability and technological innovation in automotive, aerospace, and manufacturing:

"Bionic design is a design approach that mimics the efficient structures and functions found in nature. By combining the benefits of computer-aided engineering and bionic design, our low-code platform can help to generate lightweight structures that are not only efficient but also sustainable.

Generating efficient and sustainable bionic lightweight structures using current CAE tools and processes presents a significant challenge due to the complexity of the process, the difficulty of integrating bionic design principles, and the limitations in flexibility and ease of use provided by current CAE tools" (Synera 2023).

Synera® integrates multiple engineering tools and simulation capabilities into cohesive workflows, making complex tasks more attainable and efficient. The current platform evolved from ELISE® (Evolutionary Lightweight Structure Engineering), founded in 2010 by Dr. Moritz Maier, Sebastian Möller, and Daniel Siegel. ELISE's original mission was to transform product development with a low-code environment that automated interconnected engineering workflows (Business Wire 2022). Drawing on a strong tradition of bio-inspired research, Synera® retains ELISE's biomimetic philosophy: it uses nature-inspired algorithms and structural-optimization methods to create lighter, stronger, and more sustainable designs. By mimicking evolutionary processes, the platform helps engineers cut material usage while boosting performance. Synera® connects CAD, CAE, and optimization tools, automates repetitive steps, and offers a visual editor for design automation. This Connected Engineering approach accelerates development, fosters cross-disciplinary collaboration, and frees engineers to focus on innovation. The company aims to set the industry standard for low-code engineering by enabling rapid iteration and seamless knowledge transfer across domains. The impact is already evident. In collaborations with BMW and Airbus, Synera's bionic design workflows have delivered complex lightweight geometries unattainable with traditional methods, yielding significant weight savings and performance gains—critical advantages in automotive and aerospace. By uniting biomimetic insight with process automation, Synera® is redefining sustainable, high-efficiency engineering for the next generation of products. Its interfaces include advanced simulation tools that visualize the effects of physical forces on materials and structures. Here are some ways this is achieved:

- **Algorithm-Based Modeling**: Synera® utilizes algorithms to model and simulate physical behaviors, which helps in predicting how different forces will act on a design. This approach allows engineers to identify potential stress points and failure modes before physical prototypes are built (Synera 2025; Altair 2025; Altair Engineering Inc. 2025).
- **FEA (Finite Element Analysis) Integration**: FEA is a crucial part of Synera®'s capabilities. It allows for detailed simulations of how structures will react to various forces, such as tension, compression, and shear. By integrating FEA directly into the design process, engineers can better understand and optimize the structural integrity of their designs through Automated FE Analysis.
- **Automated Workflow and Scripting**: Synera® provides over 1,700 pre-built nodes for automating tasks and allows for custom scripting with languages like Python, Matlab, and C#. This flexibility ensures that complex simulations and data processing tasks can be handled efficiently, providing immediate feedback on engineering designs through Process Automation for Engineers.

Examples of Practical Applications:

- **Aerospace:** Synera® is used to automate repetitive design tasks, run simulations for component durability, and optimize structures for weight and strength. This capability is particularly valuable in industries where safety and performance are critical, such as aerospace. Synera®'s platform is used to design lightweight components that meet the rigorous demands of aerospace applications. This includes optimizing structures for vibration behavior, reducing mass, and supporting operational loads. Such designs help improve fuel efficiency and the overall performance of aircraft.
- **Structural Analysis**: By integrating with various CAD systems and offering powerful FEA tools, S. allows engineers to perform comprehensive structural analyses. This includes predicting how buildings, bridges, and other structures will perform under different loads and conditions.
- **Product Development**: The platform supports the entire product lifecycle from concept to manufacturing. For example, it can automate the generation of reports, handle data from multiple sources, and provide insights into material behaviors, enabling more informed decision-making throughout the development process.

- **Increased Efficiency**: By automating repetitive tasks and providing immediate feedback, Synera® reduces the time needed for design iterations and analysis, speeding up the development process.
- **Enhanced Collaboration**: Synera® democratizes gateway to advanced simulation tools, making them usable by a broader range of professionals, not just experts in simulation. This fosters a more collaborative environment where knowledge is shared and utilized more effectively.
- **Better Design Quality**: With integrated simulation and optimization tools, engineers can refine their designs to ensure they meet all necessary performance criteria before moving to physical testing, reducing the risk of costly failures.

The ELISE®–Synera® story is one of the clearest international examples of how modern technology draws strategic inspiration from the study of natural organisms to develop optimization processes for lightweight, high-efficiency design. A comparable case is GrabCAD®, an enterprise-ready software suite created to ease the workload and organizational burden in engineering—especially additive manufacturing—through connected, cloud-based tools (GrabCAD 2025). Developed by Stratasys®, a leader in collaborative design solutions, GrabCAD® addresses the needs of teams handling complex CAD projects. Its products—such as GrabCAD Print Pro® and GrabCAD Streamline Pro®—offer secure, centralized file management and real-time collaboration. Aimed at engineers, designers, and multidisciplinary teams, GrabCAD® is particularly valuable in automotive, aerospace, machinery, and other high-tech sectors where efficient sharing and version control of CAD data are critical. By integrating these capabilities, the platform accelerates additive-manufacturing workflows and reduces overhead, making it a noteworthy counterpart to Synera® biomimetic approach to optimized engineering.

Similar to the other platforms analyzed, its main features can be summarized as:

- **CAD File Management**: Allows users to upload, store, and organize CAD files in a secure cloud environment, facilitating access and sharing among team members.
- **Advanced Collaboration**: Provides tools for reviewing and annotating CAD models directly within the platform, enabling team members to give feedback and make real-time modifications.
- **Version Control**: The platform automatically manages file versions, ensuring that all team members are always working on the most up-to-date version of the data.
- **Integrations with CAD Software**: Supports integration with popular CAD software such as SolidWorks®, Autodesk Inventor®, and others, enabling users to easily synchronize their projects across different platforms.

3.3 Computational Lightness

Despite the rapid rise of platforms that simplify and optimize engineering workflows, designers still wrestle with the *gravitational paradox*: trimming mass for lightness can undermine strength, yet adding material to restore robustness re-introduces weight. At full scale, gravity dominates structural behavior; in micro- and nanoscale realms, it fades while surface tension, electrostatics, and molecular forces become the primary constraints. Natural organisms have spent billions of years resolving this light-versus-heavy dilemma. Architect Achim Menges, continuing Frei Otto's legacy at the University of Stuttgart, channels those biological strategies into contemporary design. Through the ICD (Institute for Computational Design and Construction), his team studies structures such as sea urchin shells and spider webs, distills their load-bearing logics, and re-embeds them in architecture via CAD/CAM workflows—milling, 3D printing, and experimental robotic fabrication. The result is a new generation of lightweight, resilient constructs that mirror Nature's own balance between minimal mass and maximal performance.

Some of Achim Menges' most influential technology-demonstration pavilions include:

- **Elytra Filament Pavilion**: Built in 2016 for the Serpentine Gallery in London, this pavilion is inspired by the elytra, the rigid wings of beetles like the rhinoceros beetle. The pavilion is made of fiberglass and resin elements, produced through an advanced digital fabrication process. The fibers were wound around a pre-existing wooden frame, creating a lightweight and self-supporting structure that dynamically adapts to environmental conditions (Prado et al. 2017, ICD/ITKE 2014).
- **ICD ITKE Research Pavilion 2013–14:** An innovative architectural project developed by the Institute for Computational Design (ICD) and the Institute of Building Structures and Structural Design (ITKE) at the University of Stuttgart, this research pavilion emerges from cutting-edge work that blends computational design, digital fabrication, and structural engineering while tightly integrating form, structure, and performance. It shows how advanced computational workflows can optimize architectural geometry for both aesthetic and functional goals.

The pavilion's geometry draws on the lightweight efficiency of insect exoskeletons: that biomimetic precedent informed generative algorithms that produced a shell that is simultaneously slender and robust, underscoring the dialogue between natural patterns and digital design tools. A defining feature is the project's CAD/CAM-driven coreless filament winding. Two industrial robots precisely wrap resin-impregnated fibers to build a three-dimensional composite without molds. This process achieves high accuracy, minimizes material use, and guarantees structural integrity, demonstrating how robotic fabrication can deliver complex, custom components tuned to exact performance requirements. Materially, the pavilion pioneers a novel composite strategy: local fiber orientations are varied to create a geometry that responds dynamically to external forces and loads, maximizing strength-to-weight ratio and reducing waste. The result underlines a commitment to sustainability through material efficiency. In essence, the ICD-ITKE Research Pavilion is a landmark example of how biomimicry, computational design, and robotic fabrication can converge to push architectural boundaries. By advancing parametric modeling, robotic production, and composite efficiency, it marks a significant milestone in the evolution of contemporary architecture. This project not only showcases the potential of interdisciplinary research but also highlights the increasing role of digital technology in shaping the future of architectural practice (ICD/ITKE 2018).

The computational and parametric biomimetic design approach in Menges's experimental process begins with the analysis of specific biological organisms and can be divided into several distinct phases, each well-documented in numerous videos available online, among which are recordings of the ICD-ITKE pavilions and the filament-winding processes.

- **Biomimetic Analysis & Bioinspired Principles**: In the initial phase, the design team studies the structural and functional principles found in nature, drawing inspiration from organisms such as insects, plants, and marine animals like urchins or photosynthetic unicellular organisms like diatoms. For ICD ITKE Research Pavilion 2013–14, the investigation into natural lightweight structures was carried out through an interdisciplinary collaboration between architects and engineers from the University of Stuttgart and biologists from the University of Tübingen, as part of the *Bionics of Animal Constructions* module led by Prof. Oliver Betz (Biology) and Prof. James H. Nebelsick (Geosciences). Among the various biological models examined, the elytron—the protective shell covering beetles' wings and abdomen—emerged as a particularly suitable reference for designing highly material-efficient structures. The exceptional performance of these natural lightweight systems is rooted in the geometric morphology of their double-layered architecture, as well as in the mechanical properties of their natural fiber composite. This composite, composed of chitin fibers embedded within a protein matrix, exhibits anisotropic behavior, allowing for locally differentiated mechanical characteristics. In partnership with the ANKA Synchrotron Radiation Facility and the Institute for Photon Science and Synchrotron Radiation at the Karlsruhe Institute of Technology (KIT), high-resolution 3D models of various beetle elytra were generated using micro-computed tomography. Complemented by scanning electron microscopy (SEM) from the University of Tübingen, this enabled a detailed analysis of the complex internal architecture of beetle shells. The morphology of the elytron is characterized by a double-layered structure connected through column-like, doubly curved support elements known as trabeculae. Within each trabecula, the fiber orientation merges the upper and lower shell segments via continuous fiber connections. The distribution and geometric articulation of these trabeculae vary significantly throughout the shell, contributing to the system's structural efficiency. By conducting comparative analyses across multiple flying beetle species, the underlying structural principles were identified and translated into design guidelines for architectural and engineering applications.
- **Digital Modeling**: Using advanced CAD software, the team translates biomimetic principles into three-dimensional digital models. These models allow for the exploration of different shapes and configurations for the structure. Drawing on the differentiated morphology of the trabeculae and the specific

fiber orientations observed in the beetle elytra, a modular double-layered system was developed for application in an architectural prototype. The integration of computational design tools and advanced simulation methods enabled the simultaneous incorporation of both robotic fabrication parameters and abstracted biomimetic principles into the design process. Glass and carbon fiber-reinforced polymers were selected as primary materials due to their exceptional mechanical performance—particularly their high strength-to-weight ratio—and their capacity for localized variation in material properties through controlled fiber placement.

- **Advanced Simulations**: Using simulation software, the team assesses the structural and environmental performance of the proposed design. This can include stress analysis, airflow analysis, and lighting analysis, among others. Combined with their inherent moldability, the set of materials selected offers the versatility required to replicate the complex geometries and internal organization found in natural structural systems. Traditional fabrication techniques for fiber-reinforced composites typically rely on molds to define form. However, this approach is ill-suited for the translation of natural construction logics into architectural contexts, as it limits the ability to produce non-repetitive elements and demands extensive, often unfeasible, formwork.
- **Digital Fabrication**: Once the final design is defined, the process moves to the fabrication phase. Using CNC (Computer Numerical Control) machines and advanced robotics, the components of the structure are produced with great precision. To fabricate the geometrically unique, doubly curved modules, a robotic coreless filament winding method was developed. This technique employs two synchronized six-axis industrial robots (*master & slave*), which collaboratively wind fibers between custom-designed steel frame effectors held by each robot. While the effectors define the peripheral boundaries of each module, the final form emerges through the dynamic interaction of the layered fibers themselves. Initially, the fibers are linearly tensioned between the two frames; as additional layers are wound, they begin to lie upon and tension one another, producing reciprocal deformation. This interplay generates doubly curved surfaces from originally straight fiber trajectories. The sequence in which resin-impregnated fiber bundles (rovings) are laid—referred to as the *winding syntax*— is critical to the formation process. This syntax governs the precise placement of each individual fiber, enabling a material-informed design approach where form, structure, and fabrication are intrinsically interdependent. As such, the winding syntax is embedded as a core parameter within the computational design framework. The steel frame effectors are adjustable and can accommodate a variety of component geometries, enabling the use of a single reconfigurable tool setup for all 36 unique elements. Coreless filament winding not only eliminates the need for individualized molds—thereby drastically reducing material use and tooling costs—but is also inherently efficient, producing no waste or offcuts. The robotic process consists of winding six individual layers of glass and carbon fibers. An initial layer of glass fibers defines the overall geometry of the element and functions as a temporary formwork. Subsequent layers of carbon fiber

serve as structural reinforcement and are customized through anisotropic fiber arrangements. These arrangements are determined based on the specific stress distribution of each module, as calculated through finite element (FE) analysis of the overall system. The resulting winding syntax is translated directly into robotic motion commands, enabling the automated and precise fabrication of all six fiber layers.

- **Assembly and Installation**: The fabricated components are then assembled on-site according to the project specifications. This phase requires careful attention to precision and quality of installation. A total of 36 unique components were fabricated, each derived from structural principles abstracted from the morphology of beetle elytra. Every element features an individualized fiber layout, resulting in a materially efficient, load-bearing system. The largest of these elements spans 2.6 meters in diameter while weighing only 24.1 kilograms. Collectively, the research pavilion encompasses a surface area of 50 square meters and encloses a volume of 122 cubic meters, with a total weight of just 593 kilograms. The overall geometry of the pavilion responds to the site-specific context of the public space adjacent to the university building and the nearby park. Simultaneously, it showcases the system's morphological adaptability by generating complex spatial configurations that go beyond the simplicity of a conventional shell structure. As a whole, the research pavilion exemplifies how the computational synthesis of biological structural principles—combined with a deep understanding of the reciprocal relationships between material behavior, geometric form, and robotic fabrication—can lead to the development of innovative construction methods using fiber-reinforced composites. At the same time, the multidisciplinary research approach not only enables the creation of high-performance, material-efficient lightweight structures, but also explores new spatial qualities and expands the tectonic vocabulary of architectural design.

One of the latest achievements is the ITECH Research Pavilion 2024 that:

"presents an innovative approach to bio-based hybrid architecture, featuring a first-of-its-kind system that combines timber and natural fibres. The novel process explores the architectural possibilities that arise from the complementary characteristics of both materials. In this context, timber's compressive strength and shape flexibility are strategically integrated with natural fibres' tensile abilities and versatility, enabling new directions in the design and fabrication of lightweight and performative hybrid structures. The project introduces the co-design of new computational design methods and multi-robot fabrication processes tailored to leverage the inherent characteristics of each material. The proposed timber-fibre hybrid approach not only pushes the boundaries of material innovation but also offers a regenerative pathway towards climate-positive practices in construction. The research builds upon a series of successful pavilions developed at the Institute for Computational Design and Construction (ICD) and the Institute of Building Structures and Structural Design (ITKE) at the University of Stuttgart. The project was designed and fabricated by students and researchers from the ITECH master's program at the Cluster of Excellence Integrative Computational Design and Construction for Architecture (IntCDC) at the University of Stuttgart (ICD/ITKE 2024)."

3.4 Unlocking the Aerodynamic Secrets of Lightweight Insect Wing Design

Insects' wings are an extraordinary example of Nature's high-tech, lightweight, and high-performance biodesign. These functional structures are not only incredibly light but also remarkably flexible and durable, capable of withstanding the mechanical and aerodynamic stresses of flying and hard fighting—characteristics that, when translated and contextualized into human flight technologies, seem almost unbelievable. The sophistication of insect wings is a perfect fusion of form and function, where delicate structures achieve optimal efficiency in flight. This combination of lightness and strength, along with flexibility that allows for complex maneuvers, mirrors the ideal properties that engineers seek when designing high-performance materials for aviation. The fact that such designs have evolved naturally, over millions of years, provides valuable insights for innovation in human flight, where the lessons from biology inspire the next generation of aerospace technologies. This concept also touches on the broader idea of biomimicry, where designers and engineers look to nature for solutions to complex challenges. Insects' wings, with their ability to endure extreme forces while maintaining flexibility and lightweight design, offer a perfect template for developing advanced materials and structures that could revolutionize the future of human flight, making it more energy-efficient, sustainable, and agile. By understanding and adapting these natural models, we can push the boundaries of what is possible in aerospace engineering, much like how other fields, such as medical technology and robotics, have already drawn inspiration from the animal kingdom. Just as insect wings exemplify a masterful balance between lightness, strength, and flexibility, so too does the delicate Japanese art of *origami* and *kirigami* demonstrate how the lightest material—paper—can be transformed into intricate, resilient structures. In both cases, the emphasis is on efficiency and the intelligent use of minimal material to achieve maximum performance. The thinness of the paper in origami, when carefully folded, mirrors the exceptional engineering of insect wings, where seemingly fragile elements come together to perform feats of strength and agility. The fusion of *origami and design*—now attracting growing attention in engineering fields from aerospace to soft robotics—is an area in which Biruta Kresling stands out as one of the most influential and innovative theorists. Kresling is an architect and independent German researcher exploring the intersection of biomimetics and origami design. In her projects, gathered under the title *Experimental Origami in Biology and Engineering*, she studies folded structures inspired by natural growth patterns to create stable planar elements and tensegrity models that mimic the mechanical behavior of plants and organisms (Kresling 2025). Since her pioneering study *Plant Design: Mechanical Simulations of Growth Patterns and Bionics* (1995), Kresling has demonstrated how the combination of compression and torsion in cylindrical shells—now known as the *Kresling pattern*—can generate deployable and responsive structures (Zang et al. 2024). Figure 4 illustrates how characteristic Kresling-style crease patterns arise when two thin-walled hollow cylinders are subjected to opposing axial compression combined with a slight counter-rotation. Under this coupled load, the cylinder walls buckle and snap into a regular spiraling array of mountain and valley folds.

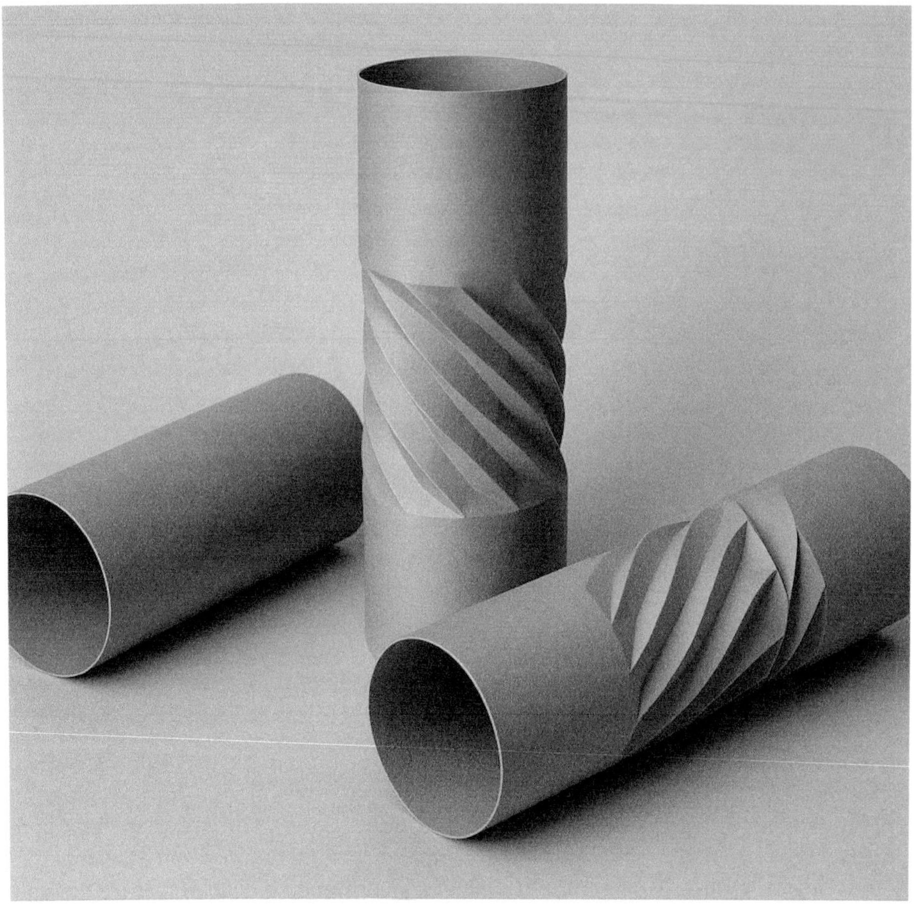

Fig. 4. Cylindrical Work Model. From a kinematic standpoint, a Kresling origami unit couples axial motion (compression or extension) with rotational twisting, resulting in a non-rigid mechanical response.

The resulting crease geometry—its pitch, amplitude and bistable response—is governed by the cylinders' diameter, wall thickness, pre-crease angle, and the magnitudes of axial displacement and rotation. This self-forming mechanism showcases a versatile strategy for creating deployable, load-bearing metamaterials directly from simple cylindrical precursors.

Her in-depth analyses of translation–rotation coupling and the mechanical properties derived from origami folds continue to inspire cutting-edge solutions in soft robotics, adaptive materials, and architectural installations, confirming origami's potential as a bridge between natural forms and engineered systems (Wang et al. 2024). Lu et al. (2022) generalize the traditional Kresling patterns—which until now could only fold into cylindrical configurations—by introducing free-form quadrilateral unit cells that enable a variety of conical folding geometries. Using a truss model, the authors analytically derive the stable states and energy landscapes of these new configurations, demonstrating that the characteristic bistability of the parallelogram patterns is preserved while expanding the design space of geometric

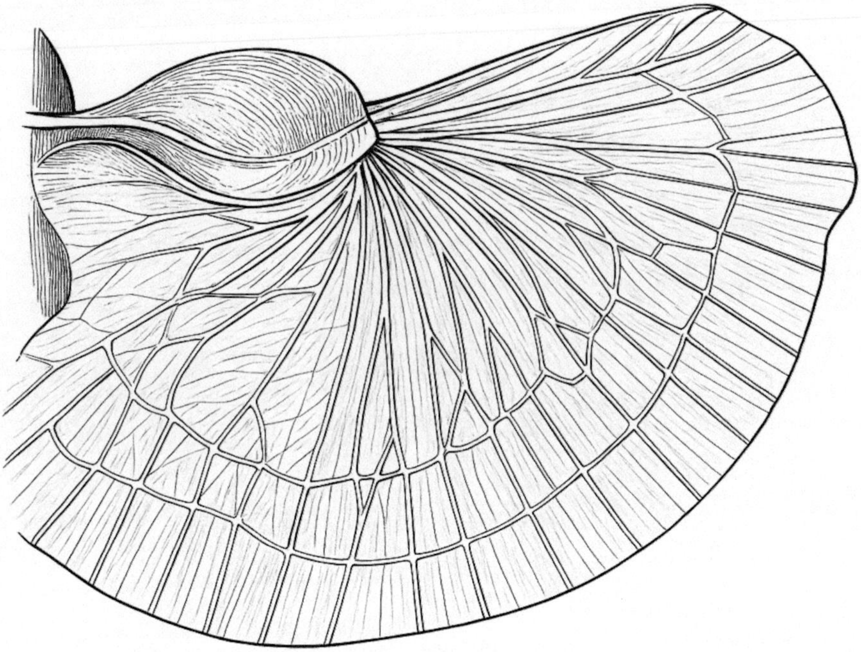

Fig. 5. Drawing of the open wing structure of a Dermapteran showing the fold architecture and radial structure.

parameters. They also propose an inverse design framework that allows conical Kresling origami to conform to arbitrary surfaces of revolution and to program predefined energy landscapes. The numerical results show excellent agreement with physical prototypes and finite-element simulations, paving the way for applications in deployable structures, shape-morphing devices, multimodal robots, and multistable metamaterials. Recent biomimetic analyses focus on studying the folding of insect wings, drawing inspiration from these extraordinary natural capabilities to develop innovative technological applications. For example, *Dermaptera*, commonly known as earwigs (Fig. 5), have wings that fold autonomously due to the presence of resilin, an elastic biopolymer that functions like a spring for rotational and linear movements (Ishiguro et al. 2023).

Researchers have analyzed these structures using computer simulations to understand the complex patterns of rotations and folds, taking inspiration from the art of origami. This understanding has led to the creation of artificial wings with remarkable folding abilities, with potential applications in foldable electronic circuits, solar sails for space missions, and geographic maps. Additionally, ladybugs (*Coccinellidae*) fold their hind wings under their elytra, utilizing a series of folds and veins that minimize the space occupied. This mechanism has been studied to develop more efficient applications such as umbrellas and solar panels, inspired by the ladybug's ability to store its wings in compact spaces through a process similar to origami. Butterflies, on the other hand, have wings with a double membrane supported by tubular veins. The surface is covered with scales arranged like shingles, creating nanometric photonic structures responsible for their iridescent colors.

This arrangement has been studied to replicate materials with similar optical properties, such as surgical tools that change color to improve visibility during operations or optical sensors sensitive to thermal variations. This analogy serves to highlight how, in both nature and human craftsmanship, the most effective designs often come from the thoughtful integration of minimal resources, where the real strength lies in their optimal configuration rather than sheer material quantity. The crucial structural challenge of lightweight-resilience design is increasingly being tackled in advanced engineering and materials science, using alternative strategies inspired by, and reinterpreted from, these refined forms of *artificialia*, enhanced by the capabilities of digital graphic computation (Massachusetts Institute of Technology 2024). Prof. Jun Mitani is a pioneer in the field of *computational origami* and *computational geometry*, known for applying the ancient art of paper folding to engineering and industrial contexts. His work has significantly influenced the use of computer-aided design for analyzing and designing complex structural folds, ensuring that stability is maintained throughout the process (Baldi et al. 2023). The algorithms developed to optimize this type of fold have been applied across various sectors, from architecture to robotics. Another important contribution by Mitani was the development of simulation software that allows users to visualize an object's behavior during the folding process in 3D (Kshirsagar et al. 2023). These tools allow users to examine how materials behave under different conditions, optimizing designs and ensuring that the folds will function as intended. The ability to test folds virtually helps avoid costly errors in the production phase and explore more innovative solutions. Mitani's work has found applications in numerous sectors, from architecture to robotics, industrial design, and even space exploration. In architecture, for example, origami has been used to design lightweight but highly resilient structures, such as foldable roofs or adaptable facades, that meet aesthetic, functional, and sustainability needs. In robotics, origami has been applied to create foldable robots that can be compact during transport and then expand to perform complex tasks, with potential applications in space missions. Mitani has explored how origami techniques can be used not only in artistic contexts but also to solve practical problems related to robotics, architecture, and industrial design. One of his main areas of research has been the automation of origami design, creating computational algorithms that generate optimized folds for applications where functionality and material efficiency are crucial. One of the most well-known software tools developed by Mitani is ORIPA® (Origami Pattern Editor), a sort of a miniCAD which allows for the automated design of origami from a basic geometry, generating crease patterns that guide the process of transforming a flat surface into a three-dimensional shape (Mitani 2012). ORIPA has revolutionized origami design, making the process more precise, fully digitalized, and widely available. This shift, according to Lumini (2022), has expanded origami's scope from pure artistry to applications in engineering and sustainable packaging (Fig. 6).

Mitani also contributed to the study of Miura-ori folds, a technique that creates foldable surfaces with surprising auxetic properties, particularly useful in contexts such as soft robotics or space design structures, where the ability of materials to expand or contract is crucial for rescue operations. In industrial design, the use of origami optimizes material usage, reducing weight and enhancing performance, while in the space sector, folding has been used to create devices like parachutes

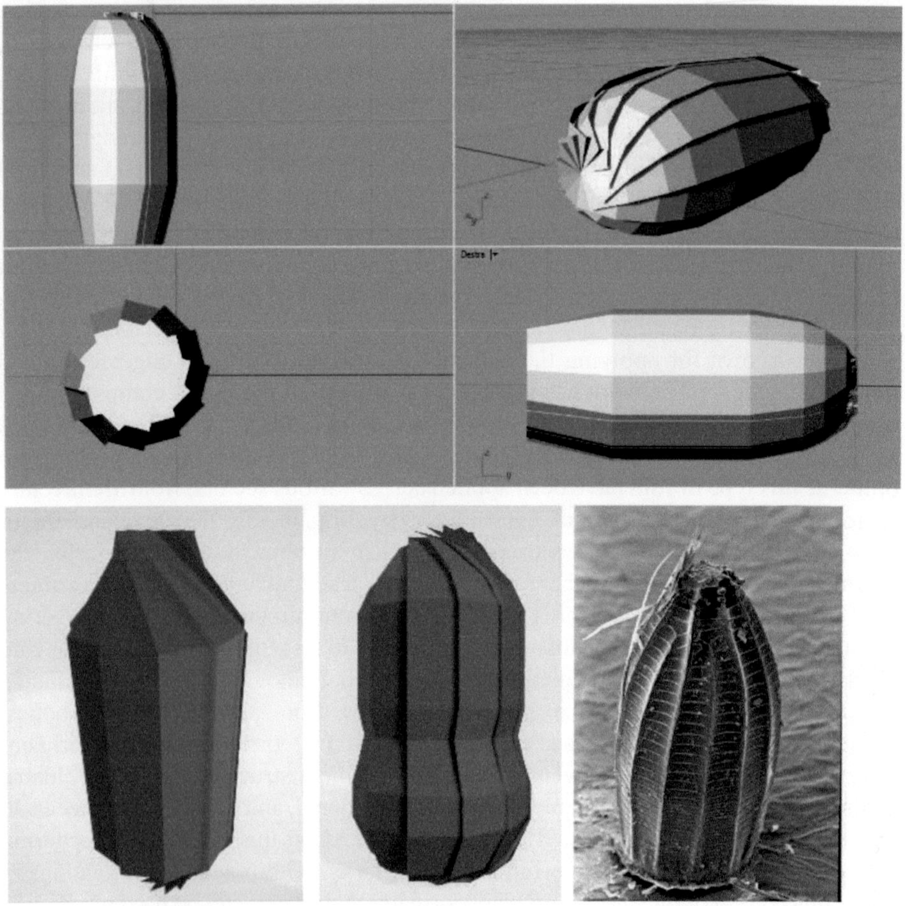

Fig. 6. Morphological analysis of the *chorion* of lepidopteran eggs and design of an origami-folding CAD-model by interpolating ORIPA® with Rhino® 3D-rendering.

or expandable structures that need to be compact for launch and then expand once at the destination (Kenney 2021). In the search for lightweight resilience tricks, here's an example of an elementary structural solution with extremely significant design consequences. Imagine holding a sheet of A4 print paper firmly between two fingers. It is observed that, under gravity's pull, a sheet of paper naturally sags; the challenge, therefore, is to improve its stiffness and capacity to support itself without increasing weight by adding material, but solely by reshaping the geometry of its surface pattern. The most strategic solution would be to fold it in half, introducing a crease along the central axis of its larger surface. This simple action, stiffening its surface with a sort of ribbing, will not only allow the sheet to resist its own weight but also enable it to support additional loads. When held at one end, it will act like a cantilever shelf. This principle of structural ribbing, widely used in the manufacturing of resistant metal sheets through surface corrugation, finds interesting inspiration in nature and its efficient design systems (Breish et al. 2024). Nature uses folded and crinkled structures to provide strength and stability to thin and lightweight structures. For example, the leaves of certain plants and the wings of insects feature folds and ribs that increase their rigidity without adding significant weight. The leaves of the *saw palmetto*, for instance, employ a system of folds and crinkles to self-support and orient themselves toward sunlight, thereby amplifying and optimizing photosynthesis. At the same time, this introduces an efficient structural solution that behaves like a refined, intricate technological gutter system, allowing the plant to collect and direct moisture and water toward the trunk and roots. It also provides extreme resistance and flexibility against dynamic forces, such as wind and rain loads during tropical storms. One of the most captivating examples of high-performance lightweight design that Nature offers is the wings of dragonflies, extraordinarily optimized structures that the animal beats at impressive rhythms and speeds during its sophisticated flight style. In fact, dragonflies are insects that exhibit some of the most advanced and sophisticated flight capabilities in the animal kingdom. Their flight system is an emblematic example of how nature has optimized biological structures to achieve efficiency, agility, and strength. Their wings are thin, transparent membranes reinforced by an intricate network of veins. These veins not only provide structural support but also play a crucial role in load distribution and wing flexibility. The forewings and hindwings can move independently, allowing for extremely precise and complex maneuvers (Di Leo and Deng 2009). Their flight ability is made possible by a combination of aerodynamic and muscular mechanisms, allowing them to perform a flapping flight, where the wings generate lift and thrust through both flapping and rotating movements. This type of flight requires fine muscular control and perfect coordination between the forewings and hindwings (Jongerius and Lentink 2010). These voracious insects, shaped by Nature into efficient and relentless aerial hunters, are capable of hovering, flying backward, and executing rapid turns to ambush and capture their prey mid-flight, thanks to these synergistic movements. Additionally, the movement of the wings creates vortices that enhance lift and flight efficiency. Researchers have observed that the forewings and hindwings can be tilted at different angles, generating vortices that further improve lift and the overall efficiency of flight (Wang 2004). The veins of a dragonfly's wings are designed to provide both strength and flexibility. Reinforced areas, such as the wing node (nodus) and the main vein (costal vein), are stiffer and more resistant, while other regions are more flexible, allowing the wings to deform and adapt during flight. This system of veins and nodes is an example of how composite structures can be optimized for specific functions (Rajabi et al. 2017). The intricate polygonal network of veins, reinforced by hemolymph, provides exceptional stability and flight control. Hemolymph, the circulatory fluid of insects, plays a crucial role in maintaining the structural integrity of the wings. One of the most distinctive features is the pterostigma, a small pigmented spot located on the leading edge of the wings. This element plays a critical role in stabilizing the flight of dragonflies, reducing vibrations, and enhancing maneuverability. The use of CAD has allowed for a deeper understanding and application of these principles in aeronautical engineering, as the pterostigma acts as a dynamic damper. Due to its additional mass and strategic position, it modifies the bending and torsional properties of the wings, reducing vibrations during flight. Experimental studies and biomechanical modeling have demonstrated that the pterostigma can lower the resonant frequency of the wings, preventing flutter phenomena that could compromise flight stability (Sun and Bhushan 2012). The effect of the pterostigma on the flight stability of dragonflies has been examined using dynamic analysis techniques and physical models. These studies have demonstrated that the pterostigma significantly reduces wing vibrations during the wingbeat. This vibration damping contributes to more stable and controlled flight, allowing dragonflies to perform complex maneuvers with precision (Kesel 2000). Through CAD simulations, engineers can create detailed three-dimensional models of the wings, including the pterostigma, and simulate their behavior under various flight conditions. These CAD models have provided valuable insights for the design of artificial wings in micro aerial vehicles (MAVs) inspired by dragonflies, showing increased stability and resistance to vibrations, enhancing their flight performance. This biomimetic approach not only boosts the efficiency of MAVs but also reduces the risk of structural damage during flight (Sun et al. 2014, Chahl et al. 2021, Zheng et al. 2023, Salami 2022, AskNature Team 2022).

References

Actlab (2018). Trabeculae Pavilion 3D printed by Delta WASP. WASP. Available at: https://www.3dwasp.com/en/trabeculae-pavilion-3d-printed-by-delta-wasp/ [Accessed: 11 May 2025].

Altair (2025). Synera. Altair. Available at: https://altair.com/synera [Accessed: 20 May 2025].

Altair Engineering Inc. (2025). *Altair® OptiStruct®: Optimization-enabled Structural Analysis*. Available at: https://altair.com/optistruct [Accessed 26 May 2025].

AskNature Team (2022). *Sensory cells give dragonflies smart wings*. AskNature.Available at: https://asknature.org/strategy/sensory-cells-give-dragonflies-smart-wings/ [Accessed: 20 May 2025].

Australian Design and Drafting (n.d.) *How CAD Technology benefits from Dynamic Modeling*. ASTCAD. Available at: https://astcad.com.au/cad-technology-benefits-from-dynamic-modeling/ [Accessed: 12 May 2025].

Baldi, A., Santucci, P. M., Carta, G., Brun, M., and Lai, D. (2023). The Miura Ori geometry: From brilliant theory to surprising auxetic properties. Available at: https://iris.unica.it/handle/11584/371803 [Accessed: 11 May 2025].

Baumgartner, A., Harzheim, L., and Mattheck, C. (1992). SKO (soft kill option): The biological way to find an optimum structure topology. *International Journal of Fatigue,* 14(6): 387–393.

Bennett, S. C. (2009). *Pterosaur* flight: The role of actinofibrils in wing function. *Journal of Vertebrate Paleontology,* 29(2): 255–284.

Bizzarri, M., Naimark, O., Nieto-Villar, J., Fedeli, V., and Giuliani, A. (2020). Complexity in biological organization: Deconstruction (and subsequent restating) of key concepts. *Entropy*, 22(8): 885.

Blatchford (2025a). Homepage. Blatchford Mobility. Available at: https://www.blatchfordmobility.com/en-gb/ [Accessed: 12 May 2025].

Blatchford (2025b). Atlas CAD-CAM. Blatchford Mobility. Available at: https://www.blatchfordmobility.com/en-gb/products/orthotics/spinal-orthotics/atlas-cad-cam/ [Accessed: 12 May 2025].

Breish, F., Hamm, C., and Andresen, S. (2024). Nature's load-bearing design principles and their application in engineering: A review. *Biomimetics,* 9(9): 545.

Business Wire (2022). ELISE raises $14.8 million Series A to scale connected engineering. Business Wire. Available at: https://www.businesswire.com/news/home/20220907005348/en/ELISE-raises-%2414.8-million-Series-A-to-scale-connected-engineering [Accessed: 20 May 2025].

Cassina (2025). Superleggera 699. Cassina. Available at: https://www.cassina.com/it/it/prodotti/superleggera.html?cas_rivestimento=X-X201#699-superleggera_26319 [Accessed: 11 May 2025].

Chahl, J., Chitsaz, N., McIvor, B., Ogunwa, T., Kok, J.-M., McIntyre, T. et al. (2021). Biomimetic drones inspired by dragonflies will require a systems-based approach and insights from biology. *Drones,* 5(2): 24.

Chatterjee, S., Lind, R., and Roberts, B. (2013). The novel characteristics of pterosaurs: Biological inspiration for robotic vehicles. *International Journal of Design & Nature and Ecodynamics,* 8(2): 113–143.

DiLeo, C., and Deng, X. (2009). Design of and experiments on a dragonfly-inspired robot. *Advanced Robotics,* 23(7–8): 1003–1021. https://doi.org/10.1163/156855309X443160.

Eiters, J., Kowalczyk, W., and Seidl, T. (2016). Simultaneous optimisation of earwig hindwings for flight and folding. *Biology Open*, 5(5): 638–644.

Fragnaud, C., Remacha, C., Betancur, J., and Roux, S. (2022). CAD-based X-ray CT calibration and error compensation. Measurement Science and Technology, 33.

Ghermandi, R., Tosini, G., Lorenzi, A., Griffoni, C., La Barbera, L., Girolami, M. et al. (2023). Carbon Fiber-Reinforced PolyEtherEtherKetone (CFR-PEEK) Instrumentation in Degenerative Disease of Lumbar Spine: A Pilot Study. *Bioengineering*. Available at: https://re.public.polimi.it/handle/11311/1257705 [Accessed: 11 May 2025].

GrabCAD (2025). GrabCAD. Available at: https://grabcad.com/ [Accessed: 20 May 2025].

Hertel, H. (1963). Struktur, Form, Bewegung (Vol. 1). Krausskopf-Verlag.

Hurrell, S. (2012). Ancient life's gravity and its implications for the expanding Earth. pp. 307–325. *In*: Scalera, G., Boschi, E., and Cwojdziński, S. (Eds.). The Earth expansion evidence: A challenge for geology, geophysics and astronomy. Aracne Editrice. ISBN 978-8854856936.

ICD/ITKE (2014). ICD/ITKE Research Pavilion 2013–14. YouTube. Available at: https://youtu.be/IMH6uGAYDps?si=DWcspnxeEjsC-AQA [Accessed: 21 May 2025].

ICD/ITKE (2018). *ICD ITKE Research Pavilion 2013–14*. YouTube. Available at: https://youtu.be/yFh8ALTjwTY [Accessed: 21 May 2025].

ICD/ITKE (2024). *ITECH Research Pavilion 2024, Stuttgart, Germany*. YouTube. Available at: https://youtu.be/LMEuf6o605E [Accessed: 21 May 2025].

Iouguina, A., Dawson, J., Hallgrimsson, B., and Smart, G. (2014). Biologically informed disciplines: A comparative analysis of bionics, biomimetics, biomimicry, and bio-inspiration among others. *International Journal of Design & Nature and Ecodynamics,* 9(3): 197–205

Irving, M. (2021). Bizarre bones helped giant *pterosaurs* straighten up and fly light. New Atlas. Available at: https://newatlas.com/science/giant-pterosaur-vertebrae-structure/ [Accessed: 11 May 2025].

Ishiguro, R., Kawasetsu, T., Motoori, Y., Paik, J., and Hosoda, K. (2023). Earwig-inspired foldable origami wing for micro air vehicle gliding. *Frontiers in Robotics and AI,* 10: 1255666.

Jacobsmeyer, B. (2011) Uncovering Da Vinci's Rule of the Trees. *Inside Science*, American Institute of Physics. Available at: https://ww2.aip.org/inside-science/uncovering-da-vincis-rule-of-the-trees [Accessed: 12 May 2025].

Jongerius, S. R., and Lentink, D. (2010). Structural analysis of a dragonfly wing. *Experimental Mechanics,* 50(9): 1323–1334.

Joris Laarman (n.d.). Bone Chair. Joris Laarman. Available at: https://www.jorislaarman.com/work/bone-chair/ [Accessed: 21 May 2025].

Kellner, A. W. A., Wang, X., Tischlinger, H., Campos, D., Hone, D. W. E., and Meng, X. (2009). The soft tissue of *Jeholopterus* (*Pterosauria, Anurognathidae, Batrachognathinae*) and the structure of the pterosaur wing membrane. Proceedings of the Royal Society B, 277(1679): 321–329.

Kenney, K. L. (2021). Folding tech: Using origami and nature to revolutionize technology. Twenty-First Century Books, an imprint of Lerner Publishing Group, Inc.

Kesel, A. (2000). Aerodynamic characteristics of dragonfly wing sections compared with technical aerofoils. The Journal of Experimental B*iology,* 203(20): 3125–3135.

Klee, P. (1923). Ways to Study Nature. Obelisk Art History. Available at: https://www.arthistoryproject.com/artists/paul-klee/ways-to-study-nature/ [Accessed: 12 May 2025].

Klee, P. (1973). Paul Klee Notebooks. Volume 2: The Nature of Nature. Lund Humphries. Available at: https://archive.org/details/PaulKleeNotebooksVol2TheNatureOfNature/page/n15/mode/2up [Accessed: 26 May 2025].

Kresling, B. (1995). Plant "Design": Mechanical simulations of growth patterns and bionics. *Biomimetics,* 3(3): 105–122.

Kresling, B. (2025). ResearchGate Profile. ResearchGate. Available at: https://www.researchgate.net/profile/Biruta-Kresling. Accessed May 27: 2025.

Kshirsagar, M., Ambike, S. D., Jaya Prakash, N., Kandasubramanian, B., and Deshpande, P. (2023). Origami engineering: Creating dynamic functional materials through folded structures. Hybrid Advances, 4: 100092.

LaBarbera, M. (1998). The Nature of Technology and the Technology of Nature. American Scientist, 86(5). Available at: https://www.americanscientist.org/article/the-nature-of-technology-and-the-technology-of-nature [Accessed: 12 May 2025].

Leidenfrost, D. (2015). Development of a Nature Inspired Hull Structure for a 46m Sailing Yacht. Master's thesis, Hochschule Bremen, Dykstra Naval Architects, Alfred-Wegener-Institut. Available at: https://epic.awi.de/40476/1/David_Leidenfrost-Master_thesis_2015.pdf [Accessed: 12 May 2025].

Li, W., Zhou, Q., Zhang, X., Xiong, X., and Zhao, J. (2013). Topology optimization design of bars structure based on SKO method. Applied Mechanics and Materials, 394: 515–520.

Lu, L., Dang, X., Feng, F., Lv, P., and Duan, H. (2022). Conical Kresling origami and its applications to curvature and energy programming. Proceedings of the Royal Society A: Mathematical, Physical and Engineering Sciences, 478(2257). Available at: https://royalsocietypublishing.org/doi/10.1098/rspa.2021.0712. Accessed May 27: 2025.

Lumini, M. (2022). The Maniola, Lycaenidae, and other Lepidoptera eggs as an inspiration source for food storage and packaging design solutions. *In*: Palombini, F. L., and Muthu, S. S. (Eds.). Bionics and sustainable design (Environmental Footprints and Eco-design of Products and Processes). Springer.

Manzini, E. (1989). The Material of Invention: Materials and Design. MIT Press.

Massachusetts Institute of Technology (2024). Using Kirigami to Make Ultrastrong, Lightweight Structures. Tech Briefs. Available at: https://www.techbriefs.com/component/content/article/49997-using-kirigami-to-make-ultrastrong-lightweight-structures [Accessed: 21 May 2025].

Masselter, T., and Speck, T. (2011). Materials. In Biomimetics: Learning from Nature (pp. 1–25). IntechOpen.

Mattheck, C., and Breloer, H. (1994). The body language of trees: Encyclopedia of visual tree assessment. Kessel Publishing House.

Mattheck, C. (1998). Design in nature: Learning from trees. Springer-Verlag.

Mattheck, C., and Tesari, I. (2000). Design in nature. Development & Application of Computer Techniques to Environmental Studies VII, 218–226.

Mattheck, C., Scherrer, M., Bethge, K., and Tesari, I. (2005). Shape optimization: An analytical approach. WIT Transactions on The Built Environment, 80, 10.

Mattheck, C. (2006). Teacher tree: The evolution of notch shape optimization from complex to simple. Engineering Fracture Mechanics, 73: 1732–1742.

Mattheck, C. (2007). Design and growth rules for biological structures and their application to engineering. Fatigue & Fracture of Engineering Materials & Structures, 13(4): 535–550.

Mattheck, C., and Haller, S. (2012). The Force Cone Method: A new thinking tool for lightweight structures. WIT Transactions on Ecology and the Environment, 160: 1–11.

McKeag, T. (2014). Shape means strength, from a Boeing Dreamliner to a Bone Chair. Trellis. Available at: https://trellis.net/article/shape-means-strength-boeing-dreamliner-bone-chair/ [Accessed: 12 May 2025].

MIT Media Lab (n.d.). Tools for Measuring Limb Tissue Biomechanics. Biomechatronics Group. Available at: https://biomech.media.mit.edu/portfolio_page/socket-fit/ [Accessed: 12 May 2025.]

Mitani, J. (2012). ORIPA: Origami Pattern Editor. Available at: https://mitani.cs.tsukuba.ac.jp/oripa/ (Accessed: 21 May 2025).

Naboni, R., Breseghello, L., and Kunic, A. (2019). Multi-scale design and fabrication of the Trabeculae Pavilion. Additive Manufacturing, 27: 305–317.

Nezzi, C. (2025). Kinematic and dynamic modeling of mechanical systems towards digital twins. Procedia CIRP, 128: 123–128.

Niklas, K. J. (1992). Plant biomechanics: An engineering approach to plant form and function. University of Chicago Press.

Otto, F. (1984). Die Forschungsgruppe Biologie und Bauen. pp. 225–238. *In:* Burkhardt, B., and Klotz, H. (Eds.). Schriften und Reden. Schriften des Deutschen Architekturmuseums zur Architekturgeschichte und Architekturtheorie. Vieweg+Teubner Verlag.

Ottobock (n.d.). Homepage. Ottobock. Available at: https://www.ottobock.com/it-it/Homepage [Accessed: 12 May 2025].

Peruzzi, L. (2019). Degli alberi e verdure: Leonardo da Vinci e la botanica. Scienza in rete. Available at: https://www.scienzainrete.it/articolo/degli-alberi-e-verdure-leonardo-da-vinci-e-botanica/lorenzo-peruzzi/2019-02-24 [Accessed: 12 May 2025].

Pittman, M., Kaye, T. G., Campos, H. B., and Habib, M. B. (2022). Quadrupedal water launch capability demonstrated in small Late Jurassic pterosaurs. Scientific Reports, 12: 6540.

Plyroom (2025). Finding Inspiration in the Shaker Design Principles. Plyroom. Available at: https://www.plyroom.com.au/blogs/journal/finding-inspiration-in-the-shaker-design-principles [Accessed: 11 May 2025].

Podder, A. (2019). User-Centered Design-Production for Assembly & Disassembly: A Shaker Chair. Medium. Available at: https://medium.com/@ar.ankurpodder/user-centered-design-production-for-assembly-disassembly-a-shaker-chair-b5358b6585e3 [Accessed: 11 May 2025].

Prado, M., Dörstelmann, M., Menges, A., Solly, J., and Knippers, J. (2017). Elytra filament pavilion: Robotic filament winding for structural composite building systems. In Proceedings of the 2017 Design and Technology of Structures (pp. 224–231).

Quint, M. (2001). Adaptive Biological Growth: Introduction to Shape (CAO) & Topology (SKO) Optimization Derived from the Growing of Trees. Structural Design, Reutlingen. Available at: https://www.xperteez.de/publikationen/docs/NAFEMS/NAFEMS_2001.pdf [Accessed: 12 May 2025].

Rajabi, H., Schroeter, V., Eshghi, S., and Gorb, S. N. (2017). The probability of wing damage in the dragonfly Sympetrum vulgatum (Anisoptera: Libellulidae): A field study. Royal Society Open Science, 4(7): 170467.

Salami, E., Montazer, E., Ward, T. A., Nik Ghazali, N. N., and Badruddin, I. A. (2022). Aerodynamic performance of a dragonfly-inspired tandem wing system for a biomimetic micro air vehicle. Frontiers in Bioengineering and Biotechnology, 10: 787220.

SEW-EURODRIVE (2025). Smart Factory Assembly and Production. SEW-EURODRIVE. Available at: https://www.sew-eurodrive.it/automation/factory-automation/industry-40/smart-factory-assembly-production/smart-factory-assembly-production.html [Accessed: 12 May 2025].

Singh, B., Kumar, R., and Chohan, J. S. (2020). Polymer matrix composites in 3D printing: A state of art review. Materials Today: Proceedings, 33: 4258–4263.

Speck, O., and Speck, T. (2021). Functional morphology of plants – a key to biomimetic applications. New Phytologist.

Staatliche Museen zu Berlin (2018). Bildräume: Biologie und Bauen. Available at: https://www.smb.museum/ausstellungen/detail/bildraeume-biologie-und-bauen/ [Accessed: 12 May 2025].

Sun, J., and Bhushan, B. (2012). The structure and mechanical properties of dragonfly wings and their role on flyability. Comptes Rendus Mécanique, 340(1–2): 3–17.

Sun, J., Ling, M., Pan, C., Chen, D., Tong, J., and Li, X. (2014). Biomimetic structure design of dragonfly wing venation using topology optimization method. Journal of Mechanics in Medicine and Biology, 14(4): 1450078.

Synera (2023). Elise is now Synera. Synera. Available at: https://www.synera.io/news/elise-is-now-synera (Accessed: 20 May 2025).

Synera (2025a). Bionic lightweight design. Synera. Available at: https://www.synera.io/workflows/bionic-lightweight-design [Accessed: 21 May 2025].

Synera (2025b). Automated FE analysis. Synera. Available at: https://www.synera.io/automated-fe-analysis [Accessed: 21 May 2025].

Tamborini, M. (2021). Morphogenesis – The Riddles of Form, in Twenty-First Century Science. Perspectives on Science, 29(5): 559–567.

University of Stuttgart. (n.d.). History of the Institute. Institute for Lightweight Structures and Conceptual Design (ILEK). Available at: https://www.ilek.uni-stuttgart.de/en/institute/history/ [Accessed: 12 May 2025].

Vogel, S. (2000). Cats' paws and catapults: Mechanical worlds of nature and people. (K. K. Davis, Illus.). W.W. Norton & Company.

Wang, X., Qu, H., Zhao, K., Yang, X., and Guo, S. (2024). Kresling origami derived structures and inspired mechanical metamaterial. Smart Materials and Structures, 33.

Wang, Z. J. (2004). The role of drag in insect hovering. Journal of Experimental Biology, 207(23): 4147–4155.

Williams, C. J., Pani, M., Bucchi, A., Smith, R. E., Kao, A., Keeble, W. et al. (2021). Helically arranged cross struts in azhdarchid pterosaur cervical vertebrae and their biomechanical implications. Scientific Reports, 11(1): 11374.

Zang, S., Misseroni, D., Zhao, T., and Paulino, G. (2024). Kresling origami mechanics explained: Experiments and theory. Journal of the Mechanics and Physics of Solids, 188: 105630.

Zheng, H., Mofatteh, H., Hablicsek, M., Akbarzadeh, A., and Akbarzadeh, M. (2023). Dragonfly-inspired wing design enabled by machine learning and Maxwell's reciprocal diagrams. Advanced Science, 10(13): 2207635.

4

Make it Cleaner

"The only useful part of medicine is hygiene;
and even hygiene, after all,
is more a virtue than a science."

Jean Jacques Rousseau

4.1 Pure as a Lily

In the heart of the murkiest ponds of the Asian marshes, since time immemorial, a small miracle of Nature repeats itself every day: from the dark mud emerges the miracle of lotus flower (*Nelumbo nucifera*), whose fleshy petals unfold toward the sky, while its green leaves remain perfectly dry, glossy, and spotless. Dust, impurities, organic waste, microbes, and bacteria are captured and swept away by microscopic spheres of water created by raindrop splashes and by the humidity that collects on the leaf surface. This passage highlights the lotus's exceptional self-cleaning ability—so striking to ancient observers that it became a powerful symbol of purity and rebirth in Indian and Eastern spiritual traditions (MeMeraki 2024). Although other plants such as *Colocasia esculenta* share similar hygienic traits, the lotus represents their maximum efficiency, embodying an ideal of untarnished beauty emerging immaculate from the silt (Srisuwan 2020, Koch et al. 2009). Only recently has materials science focused its research, experimental applications, and the creation of functional patents on natural surfaces exhibiting hydrophobic properties—studying not only plants but also bird feathers, lepidopteran wings, and other specific morphological structures of living organisms (Barthwal et al. 2024). Since H. Ollivier first introduced the phenomenon of superhydrophobicity in 1907, over 27,000 documents—articles, patents, and books—have been published in the Web of Science, with a rapidly growing trend (Ollivier 1907, Huang et al. 2025). Thanks to botanical studies integrated with biomimetic research, more than 200 plant species with characteristics similar to those of the lotus have been identified, including cauliflower (*Brassica oleracea*), nasturtium (*Tropaeolum majus*), prickly pear (*Opuntia ficus-indica*), and lady's mantle (*Alchemilla vulgaris*). Researchers have discovered that many of the textures present on biological surfaces interact with the main terrestrial physical

DOI: 10.1201/9781003438946-4

forces, such as surface tension, enabling the formation of micro-air cushions that act as barriers against water (Nosonovsky and Bhushan 2008, Tang et al. 2024). It is important to emphasize that natural superhydrophobicity results from different technological strategies developed by evolution. Both bird feathers and lotus leaves exhibit high water-repellency properties, but for different reasons. Lotus leaves possess a complex hierarchical structure with interacting roughness levels, while feathers adopt a pseudo-hierarchical structure in which the roughness levels are independent. This configuration allows them to remain dry both in flight and when submerged, requiring high pressures (around 100 kPa) to become completely wetted. Furthermore, an energy barrier prevents the transition from the superhydrophobic state to the wet state, thus ensuring stability. In general, when a water droplet comes into contact with a solid surface, its behavior depends on surface tension and energy. If the surface has a high contact angle and low *hysteresis*,[1] water does not adhere and slides off easily (Johnson et al. 1977). The contact angle—formed between the surface and the tangent to the droplet at the point of contact—depends on the material's physicochemical properties and affects water adhesion. In the *Cassie regime*,[2] the water droplet rests on a structured surface, trapping air bubbles, assuming an almost perfectly spherical shape that promoting droplet sliding, can move freely in all directions, simultaneously removing dust from the surface and thus producing an effective self-cleaning effect. In contrast, in the *Wenzel regime*[3]—as observed on mango leaves—the liquid penetrates the microstructures, increasing adhesion. Measurements have shown that the water contact angle on a lotus leaf is approximately 160°, while the roll-off angle is around 2°. For a visual demonstration of superhydrophobic droplet behavior, see the educational videos produced by Durham University (Durham University n.d.). Returning to the lotus plant, one of the most successful cases in superhydrophobicity—both scientifically and commercially—was the discovery and subsequent industrial application of the so-called Lotus Effect®. First described and analyzed by Barthlott and Neinhuis (1977) at the University of Bonn in Germany, this phenomenon refers to the extremely efficient control of surface wettability exhibited by lotus leaves. Later, Koch et al. (2009) deepened understanding of the effect by examining in greater detail the ultrastructure and multifunctionality of such surfaces (Vonna 2023). From that same period, the first studies emerged hypothesizing the applicability of these observations to the materials technology sector (Balani et al. 2009). For the first time, materials science

---

1 Contact angle hysteresis is the difference between the advancing contact angle (θ_a), measured as the liquid front advances, and the receding contact angle (θ_r), measured as the liquid front retreats. A small hysteresis ($\theta_a - \theta_r$) means drops roll off easily, promoting self-cleaning, whereas large hysteresis causes pinning, trapping droplets and reducing mobility.

2 In the Cassie regime, a liquid droplet rests atop the peaks of a textured surface, with air pockets trapped beneath it. This composite solid–air interface results in very high apparent contact angles and minimal adhesion, so droplets remain nearly spherical and can roll off easily, carrying away dirt and contaminants.

3 In the Wenzel regime, a liquid droplet fully wets the textured surface, filling its grooves and asperities. The liquid–solid interfacial area increases, so the apparent contact angle is modified by the surface roughness factor (r), amplifying the material's inherent wettability. As a result, droplets adhere more strongly and resist rolling off, limiting self-cleaning performance.

researchers proposed that the lotus leaf's hierarchical structure could be replicated to design materials with advanced water-repellent and dirt-resistant properties, including resistance to bacterial contamination. The European patent resulted from extensive research led by Professor Wilhelm Barthlott, director of the Botanical Institute at the University of Bonn. STO SE & Co. KGaA (2025), headquartered in Stühlingen, Baden-Württemberg, Germany, launched Lotusan® façade paint in 1999 as the first biomimetic market product. Since then, it has been regarded worldwide as a prime example of how a natural principle (Lotus Effect®) can be translated into a technical product. Although lotus leaves achieve extreme superhydrophobicity (contact angles > 150°) through a complex, hierarchical natural structure—yielding near-perfect self-cleaning—Lotusan® mimics this principle with a simpler, controlled micro-roughness (contact angles ~100–110°). This affords excellent water repellency and rain-driven cleaning on building façades. The same patent was later used for Aeroxide self-cleaning surfaces, developed by an interdisciplinary team from Creavis Technology & Innovation alongside Degussa's Aerosil & Silanes Business Unit (now Evonik). Biomimetic research combined with materials science enabled superhydrophobic coatings that produce self-cleaning surfaces, prevent corrosion, and improve the durability of metal materials exposed to atmospheric agents (Lafuma and Quéré 2003). New applications exploit lotus-leaf–inspired textures—using advanced deposition techniques or nanostructured films—to achieve durable, self-cleaning coatings for sectors such as aerospace, construction, automotive, glass, ceramics, and textiles (Lotus Leaf Coatings n.d.). Li et al. (2023) analyzed lotus leaf surfaces via scanning electron microscopy (SEM), identifying micro- and nanostructures as key to their hydrophobic and self-cleaning properties. Their review explains why superhydrophobic (SH) coatings remain largely confined to the lab—due chiefly to poor durability, water impalement, environmental degradation, and scale-up challenges—and proposes solutions such as co-optimized material–texture design, standardized testing protocols, and academia–industry partnerships. With these strategies and shared performance metrics, they project that commercially viable SH surfaces could emerge by 2035. Their studies revealed the lotus surface to be covered in countless waxy, tubiform micro-emulsion structures of micrometric dimensions. This finding led to the hypothesis that the lotus effect—characterized by superhydrophobicity—arises from combining a low-surface-energy waxy material with a complex network of branched nanostructures across each leaf lobe. This micro–nano composite configuration, akin to the *fakir effect*,[4] drastically reduces droplet–surface contact area, endowing the leaf with exceptional hydrophobicity. Because superhydrophobicity requires micro-/nanoscale surface roughness (contact angle >150°), experimental modification techniques are employed. Among these, 3D printing (additive manufacturing) has emerged as a cost-effective, versatile method: a CAD model is digitally sliced and reproduced layer by layer, embedding specific

[4] The *fakir effect* in superhydrophobicity arises from the interplay of surface tension and capillary forces acting over a hierarchical micro–nano roughness, where low surface energy and entrapped air pockets minimize adhesive forces and contact angle hysteresis—yielding extreme water repellency and facile droplet roll-off.

functionalities at predetermined locations within the geometry. Numerous studies have demonstrated this approach's effectiveness in generating highly rough, superhydrophobic structures (Asmatulu et al. 2022). Below is Table 1 with a concise comparison of key physicochemical and performance properties for the natural lotus leaf versus the biomimetic Lotusan® paint coating.

Drawing on the work of Domen et al. (2014) on fog water as an alternative and sustainable resource, a Biomimetics workshop (2017)—led by personnel from *BIONIKONLab* (Lumini 2023) in Iglesias—adopted a bottom-up strategy centered on ultrastructural analyses of the leaves and stem of *Cotula fallax*. This small perennial herb of the *Asteraceae* family bears finely divided pinnate leaves covered in dense glandular hairs. Its surface microtopography excels at harvesting moisture

Table 1. Comparison of Lotus Leaf vs. Lotusan® Paint.

Property	Lotus Leaf (Natural)	Lotusan® Paint (Technology)
Application Process	Biological growth—no human intervention	Industrial dip–pad–dry finishing followed by packaging and QC
Contact Angle	> 150° (extreme superhydrophobicity)	≈ 100–110° (high hydrophobicity, but lower than lotus)
Contact Angle Hysteresis	< 5°	10–15°
Durability Over Time	Limited by the leaf's seasonal lifecycle	High—resistant to UV, freeze/thaw cycles, washing, and weathering
Entrapped Air Layer	Continuous air cushion beneath the droplet	Discontinuous air pockets
Fields of Use	Specific to the plant's leaf function	Protective and decorative façades in construction
Manufacturing Complexity	Natural growth process	Simple spray application with nanoparticle additives
Mechanical Durability	Low (delicate biological tissue)	High (mineral-based, UV-stable)
Mechanism	Micro- and nanostructures trap air, minimizing liquid–leaf contact	Microparticles form a rough layer that promotes droplet bouncing
Regeneration	New leaves emerge with intact surfaces each growth cycle	Microstructure persists as long as the coating remains undamaged
Self-Cleaning	Exceptional—each drop efficiently removes dirt and dust	Good—rain washes away most deposits (≈ > 80 %)
Solid–Liquid Contact Fraction (f_s)	~ 0.02–0.05	~ 0.2–0.3
Sustainability	Renewable and biodegradable	Biocide-free and low-VOC, though requires energy and materials to produce
Surface Structure	Irregular, hierarchical roughness (nano-on micro-papillae)	Controlled silica microparticles creating a uniform rough texture
Typical Applications	Plant leaves	Exterior coatings for buildings

Figs. 1-2. Detail of the leaves and stem of *Cotula fallax* and schematic diagram of the fog water collection (from Domen et al. 2014).

from saturated air, driving local condensation to the dew point and nucleating microscopic droplets that merge into ever-larger drops (Figs. 1-2). When these drops reach a critical size, gravity carries them down to the plant's base, securing hydration even in extreme environments.

The proposal of designer teams is to emulate this natural moisture-collection mechanism in engineered panels designed to pull water from humid air (Fig. 3).

Research on superhydrophobic surfaces is advancing rapidly, driven by their potential in critical sectors—from civil and military aviation (where wing icing disrupts lift via Bernoulli's principle) to aerospace, satellites, and beyond. Preventing ice formation and adhesion not only enhances safety but also reduces weight and drag (Ma et al. 2020). Achieving these goals requires precise control of surface structure, contact angle, and droplet behavior—tasks now explored via CAD/CAM workflows and multiphase CFD (Computational Fluid Dynamics) simulations with OpenFOAM®, which models contact angles, adhesion forces, and Cassie–Baxter stability (OpenFOAM Foundation Ltd. n.d.). Beyond computational design, 3D fabrication methods such as laser-assisted microfabrication and additive manufacturing enable creation of bio-inspired micro-/nano-textures that trap air, minimize water–surface contact, and promote self-cleaning and anti-fouling performance (Afshar-Mohajer et al. 2023). These ideas are highly patentable for applications like superhydrophobic fabrics. As suggested by Rousseau, "hygiene is a virtue." Despite these innovations, human cleanliness has historically relied on cultural, social, and religious practices that have undergone significant historical transformations. Unlike many animals that secrete protective oils, humanity's cleaning capabilities evolved more slowly—marked by alternating eras of sanitation and neglect, and by devastating pandemics such as the Black Death and Spanish Flu. This contrast underscores the promise of engineered superhydrophobic materials to augment our innate defenses with nature-inspired self-cleaning surfaces. The natural oils found on the skin and fur of many animal species play a crucial role in protection and physiological functionality. Animal species have evolved specific adaptations to

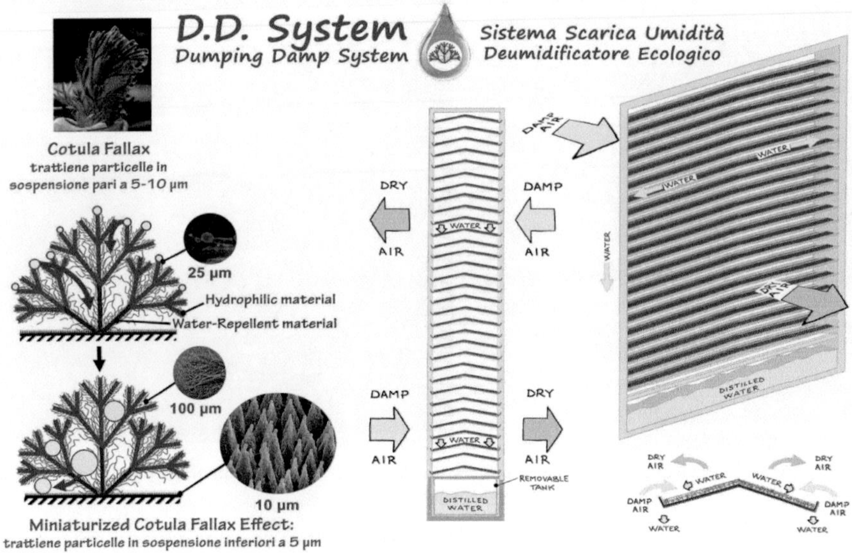

Fig. 3. D.D. System (Dumping Damp System) (designer team: Castangia, C., Mura, A., Mura, L., Porru, S., Ruggeri, T., Stivaletta, M., Uda, F., & Zaru, F. Sci. Coord. Lumini, M. 2017).

produce natural oils that respond to environmental and behavioral needs. These adaptations have been the subject of numerous studies, linking the chemical composition of the oils to protection against environmental factors and predators. The use of natural oils for skin protection also has biomedical implications, such as in the development of new therapies for human skin disorders, based on understanding the protective properties of these oils in animals. Hygiene is a crucial function for evolutionary success, as neither humans nor natural organisms are immune to the risks associated with poor cleanliness of bodies and habitats. Pathogenic viruses and bacteria have always threatened life, and even today, billions of people live in precarious sanitary conditions, highlighting dramatic socio-anthropological disparities. Even in advanced societies, hygiene and health management pose technological, economic, and environmental challenges, making innovative solutions indispensable. In this context, studying nature-based technologies may offer valuable inspiration for improving human-made artifacts and technological systems. However, many current biomimetic self-cleaning surfaces still rely on chemical impregnation rather than replicating the sophisticated micro- and nano-textures of leaves that naturally modulate water surface tension. Astonishingly, biomimetics has uncovered examples of self-cleaning technologies in the dark depths of the oceans—where time seems suspended, legends blur with reality, and sharks stand as living relics. These creatures bear witness to distant geological eras and guard evolutionary secrets rooted in a mysterious and primordial past. This family of fish showcases a remarkable series of successful evolutionary designs. The remains of the oldest known articulated shark specimen, *Doliodus problematicus*, date back to the Early Devonian period—approximately 410 million years ago—and were discovered in what is now New Brunswick, Canada (Davis n.d.). Although sharks are often regarded as primitive creatures, they include some of the largest and most highly adapted marine predators alive today. Unlike many other fish, their skin does not possess mucus or typical scales to reduce friction. Instead, it is covered by densely packed microscopic structures known as dermal *denticles*. These are small, tooth-like scales with a chemical and mineral composition remarkably similar to that of teeth (*denticle*), giving the shark's skin[5] a rough and highly functional surface (Domel et al. 2018, Lloyd et al. 2023). The three-dimensional *microgeometry* of natural surfaces not only optimizes the hydrodynamic efficiency of aquatic animals—reducing drag and enhancing swimming speed—but also plays a crucial role in protection against microorganisms, including algae and crustacean fouling. *Microtextured* surfaces, such as those found on shark skin, hinder biofilm formation and make it difficult for bacteria and other pathogens to adhere and colonize, thereby preventing infections and tissue damage (Baeten et al. 2024). In human contexts, the growing threat of ultra-resistant bacteria—capable of evading both antibiotics and conventional sanitizers—has become a serious healthcare emergency. These microorganisms are responsible for hospital-acquired infections, post-operative complications, and, in some cases, fatalities. Pathogens such as *Escherichia coli*, *Klebsiella* spp., Pseudomonas *aeruginosa*, and *Clostridioides difficile* can survive for months on inanimate surfaces, complicating treatment and sanitation efforts. Of particular concern is the spread of *Clostridioides difficile*, whose highly resistant spores persist in healthcare environments despite rigorous cleaning protocols. While disinfectants such as chlorine, hydrogen peroxide, and peracetic acid have proven effective, their efficacy depends on proper application, with adequate contact times and correct usage to avoid unwanted side effects, such as material degradation or skin irritation (Kramer et al. 2006). For these reasons, the shark skin effect generated by dermal denticles—whose microtextures vary across species—has attracted the attention of biomimetic research for years (Gabler-Smith et al. 2021). In this context, Sharklet Technologies™ is recognized for its innovation inspired by the dermal denticle structure of sharks (Sharklet Technologies Inc. 2023). It replicates the microscopic pattern to create microstructured surfaces through 3D printing, consisting of millions of micro-elements arranged in a precise diamond pattern. The aim is to prevent bacterial and germ adhesion, as well as the formation of biofilms for colonization and growth, all without the use of chemical additives. This innovative approach is used in a wide range of applications, from healthcare to boat surface protection. To design and create these types of microstructured surfaces, CAD/CAM protocols are employed, allowing for precise modeling of geometries at the micro and nano levels. Thanks to advanced software, it is possible to create detailed and customized models of Sharklet™ surfaces, adapting the size and shape of the pattern to specific needs.

---

[5] Although often considered primitive, sharks comprise some of the largest and most highly adapted marine predators alive today. Unlike many other fish, their skin lacks mucus and typical overlapping scales used to reduce friction. Instead, it is covered with densely packed microscopic structures known as dermal denticles—tiny, tooth-like scales with a chemical and mineral composition strikingly similar to that of human teeth. This unique adaptation gives shark skin a rough texture and exceptional functional properties.

In the healthcare sector, for example, CAD/CAM enables the design of medical devices, such as catheters and surgical instruments, with Sharklet™ geometry to maximize protection against bacterial growth (Damodaran and Murthy 2016). Similarly, in boat surface protection, CAD/CAM protocols allow the design of coatings that prevent the formation of fouling and algae, improving durability and efficiency without the use of toxic chemicals. The integration of Sharklet™ technology systems offers customizable, effective, and sustainable solutions, addressing the challenges of microbiological protection across various sectors. However, despite decades of research and numerous studies on the properties of these bio-structures, practical applications remain quite limited, with only a few exceptions. Wong and Mak (2021) employed Computational Fluid Dynamics (CFD) and 3D printing to compare the fouling resistance and self-cleaning performance of two patterned surfaces against an unpatterned control, while assessing the feasibility of biomimetic shark-skin topographies for real-world use. Using SolidWorks 2020® and ANSYS Fluent® under laminar flow conditions, they modeled both shark-skin and smooth surfaces, analyzing fluid behavior through contour maps and vector fields of hydrodynamic variables such as wall shear stress. Their results indicate that the shark-inspired microtopography increases shear stress at the wall, suggesting an antifouling effect. In the project's second phase, they explored fabrication methods to reduce costs: by leveraging Fused Deposition Modeling (FDM) 3D printing instead of traditional photolithography, they demonstrated a more user-friendly approach to producing antifouling microstructures.

Despite the efforts of the scientific and technological community, the gap between the knowledge produced by basic biomimetic research and its potential technological transfers into functional artifacts is still apparent in many other cases (Bell et al. 2013, Ciulla et al. 2020).

This gap can be synthetically attributed to several factors:

1. **Complexity of Translation from Nature to Technology**
    - Nature develops extremely complex and multifunctional solutions that are difficult to replicate with current technologies.
    - The dynamic and adaptive properties of many biological structures do not easily correspond to conventional synthetic materials.
2. **Technological Limitations**
    - Despite the adoption of techniques such as additive manufacturing and programmable materials, many biomimetic innovations remain prototypes due to production complexity and high costs.
    - The ability to design and simulate biomimetic materials accurately requires advanced computational models that are not yet fully developed or integrated into current industrial sectors.
3. **Long Patent and Technology Transfer Times**
    - The patenting process for new biomimetic solutions can be lengthy and complex, often not synchronized with productive development cycles.

- Companies may be reluctant to invest in technologies that are not yet established or have uncertain industrial applications.

4. Collaboration Between Research and Industry

- The transfer of academic discoveries to the industrial sector requires close collaboration between universities, research centers, and companies, which is not always effective or well-structured. For example, the territorial BIOKON project is a national network, launched in 2001 and expanded in 2004, that connects the main stakeholders—companies, research institutes, and universities—in the field of bionics and biomimetics in Germany. Funded by the Federal Ministry of Education and Research, BIOKON aims to translate nature-inspired solutions into industrial applications by facilitating scientific exchange, training, and technology transfer to both the production sector and the general public (BIOKON 2025). Also, CEEBIOS (Centre d'Études et d'Expertises en Biomimétisme), founded in 2015 as a cooperative of collective interest, is a French network that unites universities, research institutes, SMEs, large corporations, and public bodies to promote the adoption of biomimicry in industrial and territorial projects through training, workshops, consulting, and the establishment of thematic working groups (CEEBIOS 2025).
- There is often a lack of intermediary figures who can translate research findings into practical solutions for industry.

An example of bridging the technological gap is represented by projects which are helping to accelerate technology transfer through industrial partnerships. These collaborations allow scientific discoveries to be quickly translated into practical solutions, fostering innovation and the application of developed technologies (Wyss Institute for Biologically Inspired Engineering at Harvard University 2025).

Here is how their eight major Focus Areas are defined:

- **Bioinspired Therapeutics & Diagnostics**

Therapeutic discovery and diagnostics development enabled by microsystems engineering, molecular engineering, computational design, and organ-on-a-chip *in vitro* human experimentation technology.

- **Computational Design & Discovery**

Combining predictive bioanalytics and machine learning with physical and mathematical modeling and simulation.

- **Diagnostics Accelerator**

Developing new diagnostic technologies that solve important healthcare challenges through collaboration at the Wyss Institute with clinicians and industry partners.

- **Immuno-Material**

Material-based systems capable of modulating immune cells *ex vivo* and in the human body to treat or diagnose disease.

- **Living Cellular Devices**

Re-engineered living cells and biological circuits as programmable devices for medicine, manufacturing and sustainability.

- **Molecular Robotics**

Self-assembling molecules that can be programmed like robots to carry out specific tasks without requiring power.

- **3D Organ Engineering**

Highly functional, multiscale, vascularized organ replacements that can be seamlessly integrated into the body.

- **Synthetic Biology**

Breakthrough approaches to reading, writing, and editing nucleic acids and proteins for multiple applications, varying from healthcare to data storage.

The Wyss Institute for Biologically Inspired Engineering has had a significant impact on research and production worldwide, transforming the way we approach design and technological innovation. Its discoveries and technologies, which combine biology, engineering, and advanced design, are profoundly influencing sectors such as biomedicine, sustainable materials production, and complex systems engineering. Through its ability to translate biological concepts into engineering solutions, the institute has created a bridge between academic research and industrial applications, accelerating the realization of technologies that not only improve quality of life but also promote a more sustainable and innovative future for the entire world (Humphries 2011, Tolikas et al. 2018).

In the case of Hydrophobic Surfaces (HS), when commercializing materials inspired by these surfaces, it is essential to consider the application contexts during their preparation and performance optimization (Khajuria et al. 2023).

The main issues regarding the applications of SH surfaces, along with their solutions, are:

- **Evaluation with Different Liquids**:

Most studies use pure water to test SH surfaces, but it is clear that in reality, SH surfaces come into contact with a variety of liquids (solutions, suspensions, emulsions, oils, etc.) that differ in composition, viscosity, and other physicochemical parameters. The evolution of CAD/CAM in this field allows for the design of customized SH surface models to accurately simulate interaction with different types of liquids. For example, using computational fluid dynamics (CFD) simulation software integrated with CAD, it is possible to model and simulate the interaction of SH surfaces with viscous liquids or suspended particles. Additionally, advanced CAD systems allow for the design of variable microstructures, optimizing the geometry based on different viscosities and surface tension of liquids. A concrete example could be the design of SH coatings for pharmaceutical containers or industrial cooling systems that come into contact with very different fluids.

- **Real Testing Conditions:**

Laboratory tests often do not reflect real environmental conditions, creating a gap between what is studied in the lab and the practical use of SH surfaces. CAD/CAM technology can be revolutionary in this regard, as it enables the design of models that replicate real environmental conditions, such as extreme temperatures, variable humidity, or surfaces subjected to dynamic stress. With the integration of thermal and mechanical simulations into parametric software, SH surfaces can be tested not only in controlled environments but also in realistic conditions, such as protection against ice at very low temperatures with high humidity, or resistance to abrasion in industrial environments. For example, an SH surface designed for an airport or automotive components can be tested to resist real atmospheric conditions, such as heavy rain or contact with oil and other contaminants.

- **Objective Performance Evaluation**:

To obtain a comprehensive evaluation of SH surface performance, it is essential to use more complex testing methods that go beyond simple static simulation. Digital models can be used to design SH surfaces that can be tested in various scenarios, such as corrosion resistance, and to integrate simulations of salt spray or tests for resistance to chemical agents. The use of CAD/CAM for prototyping and subsequent production through 3D printing technologies can enhance objective evaluation, allowing physical testing of various samples before large-scale production. For example, SH surfaces designed for the naval industry could be optimized to resist saltwater corrosion in marine environments. Huang et al. (2024) developed an extrusion-based 3D printing method to fabricate a soft composite material—Ecoflex 00-30 (a silicone elastomer) loaded with SiO_2 particles—with high superhydrophobicity and programmable anisotropic wettability. The degree of hydrophobicity and wettability anisotropy is achieved by controlling the geometry and orientation of the deposited filaments and can be further tuned by applying mechanical deformations (directional stretching and bending). Thanks to these features, the printed surfaces enable efficient liquid transport, microdroplet manipulation, and the operation of soft grippers in wet environments, demonstrating promising applications in wearable devices, healthcare, and human-interactive robotics (Fig. 4).

- **Exaggeration and Misinformation**:

Many studies on SH materials exaggerate the potential of the technology, creating unrealistic expectations. The integration of CAD/CAM allows for precise design and simulation, reducing the possibility of misinformation. Advanced software enables testing and validating every modification to the SH surface in a detailed and scientifically rigorous manner, improving transparency in performance statements. For example, through parametric modeling and real-time behavior simulation, researchers can avoid misleading promises and focus on real solutions for specific application areas.

- **Future Perspectives**:

SH surfaces have made significant progress in recent decades, but to fully exploit their commercial potential, strong commitment to research and development

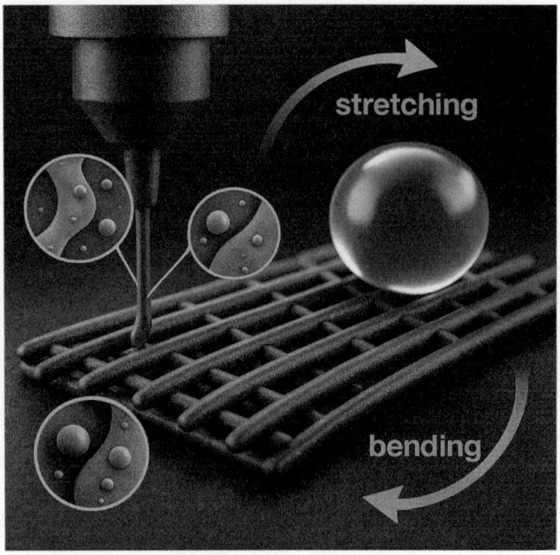

Fig. 4. 3D-Ecoflex Printing.

is required. The use of CAD/CAM techniques is crucial for creating customized solutions, analyzing performance in complex scenarios, and optimizing production. In particular, the evolution of CAD software, which now supports advanced simulation of fluid-structure interactions and integration with artificial intelligence (AI) and machine learning technologies, promises to accelerate progress in SH surfaces. A practical example could be the use of CAD/CAM to design SH surfaces at the molecular level, improving effectiveness against biofouling in marine environments or creating nanostructured coatings for high-precision medical devices. Significant developments in this direction of controlling micro-physical parameters are enabled by research applying specific design and digital manufacturing protocols, such as stereolithographic (SLA) 3D printing or Digital Light Processing (DLP) (Kaur et al. 2020).

DLP is a 3D printing technology through which it is possible to create surfaces with complex micro and nanostructures, used to quickly produce photopolymer parts. It is very similar to SLA with one significant difference: while SLA machines use a laser to trace one layer at a time, DLP machines use a projected light source to polymerize the entire layer at once. DLP printing can be used to print highly intricate design objects in resin, such as toys, jewelry molds, dental molds, figurines, and other finely detailed items. Since it polymerizes the entire layer at once, it is much faster than SLA. The direct light processing 3D printing technology was originally developed in 1987 (Markforged, Inc. 2025). Table 2 below provides a comparative chart between SLA (Stereolithography) and DLP (Digital Light Processing), highlighting the main technological parameters, material characteristics, and materials science.

Table 3 offers an overview of companies producing SLA 3D printers, while Table 4 presents an overview of companies producing DLP 3D printers.

In more recent times, Computer Aided Biomimetics (CAB) design technologies, along with digital manufacturing integrated with AI-Aided Design (AIAD), are driving a further evolution in technological processes for the production of materials featuring superhydrophobic and self-cleaning properties (Kruiper et al. 2016, 2017, Diaz-Lantada et al. 2020). Researchers and industries worldwide are mining nature's *design tricks*—from carnivorous plants like *Dionaea muscipula* and Nepenthes—to inspire advanced materials (Kruiper et al. 2020, Zhang et al. 2023). Nepenthes' peristome, with its hierarchical nano/microstructure that traps a water film and creates extreme slipperiness, has guided the creation of biomimetic slippery surfaces for applications such as self-cleaning robotic skins (Bohn and Federle 2004, Han et al. 2023). FsLDW-fabricated LRGO/GO surfaces combine tunable nano-/micro-topography and chemistry to deliver SLIPS-like controllable wettability, friction, and lubricant infusion, alongside inherent moisture-responsive deformations for dynamic actuation. This rapid, CAD/CAM-compatible approach enables bespoke, stimuli-responsive materials ideal for soft robotics, biomedical devices, microfluidics, and humidity-driven actuators. In practice, Nepenthes-inspired, self-cleaning, water- and contaminant-repellent coatings can be integrated into robots, drones, and automated cleaning systems to enhance performance and durability in humid or dirty environments, while cutting maintenance costs and improving hygiene in industrial, domestic, and medical applications. Building on this, Cai et al. (2022) developed a humidity-responsive, oil-infused surface with hybrid bionic capabilities, while Femtosecond Laser Direct Writing (FsLDW)—a CAD/CAM-compatible technique using ultrashort pulses—enables precise micro-/nano-fabrication of materials to mimic nature's wetting and adhesion control (KLOÉ France n.d.). When the laser strikes the material with ultrafast pulses of high intensity, multiphotonic ionization and two-photon absorption processes occur, enabling localized modifications without significant thermal damage to the surrounding areas. Due to its unique characteristics, FsLDW can be considered an advanced form of CAD/CAM applied to micromachining and nanofabrication, as the production process involves (Bioengineer 2025). FsLDW is revolutionizing the creation of bioinspired surfaces, offering unprecedented precision and remarkable versatility. Below is a summary of the main characteristics and applications of this technology:

1. **High Precision and Resolution**
   - Thanks to femtosecond laser pulses, FsLDW allows the creation of micro and nanometric structures with extreme detail, down to the molecular level. This enables the replication of complex characteristics of natural surfaces, such as those of carnivorous plants.

2. **Assisted Digital Design**
   - Design is done through CAD software, which allows the creation of customized geometries. The information is then translated into instructions for the laser through CAM systems, ensuring precise and repeatable production.

3. **Direct Material Processing**
   - The laser acts directly on materials without damaging the surrounding areas. This minimizes defects and material loss, ensuring efficiency in production.

Table 2. A Comparative Chart Between SLA (Stereolithography) and DLP (Digital Light Processing.

Parameter	SLA (Stereolithography)	DLP (Digital Light Processing)
Accuracy	Very high precision, especially in detailed models	High precision, but dependent on projector and material
Applications	High-detail prototyping (medical, jewelry, etc.)	Rapid prototyping, general industrial applications, dental, jewelry
Build Size	Generally smaller build volumes	Can offer larger build volumes depending on the projector used
Complexity of Prints	Excellent for fine details and high complexity	Excellent for rapid prints but may struggle with highly intricate features
Environmental Considerations	Some resins may be toxic; requires proper ventilation	Similar concerns with resins but more energy-efficient
Layer Thickness	Typically between 25–100 microns	Typically between 25–100 microns
Light Source	UV Laser (focused)	Digital Light Projector (DLP chip, LED-based)
Material Innovation	Strong in bioresins and specialized formulations (e.g., dental, medical)	Rapid advancements in resin types, including flexible, transparent, multi-material
Material Strength	Suitable for precise, strong models	Suitable for a range of materials, often less strong than SLA prints
Material Types	Photopolymer resins, bioresins, castable resins	Photopolymer resins, bioresins, castable resins, flexible materials
Material Variety	A wide range of resins available, including castable, tough, flexible, etc.	Broad selection of resins, including rigid, flexible, tough, transparent, etc.
Operational Cost	Higher due to laser maintenance and resin cost	Generally lower due to less complex hardware and lower material cost
Post-Processing Requirements	High, requires cleaning, curing, and finishing	Moderate to high, requires cleaning and curing
Print Speed	Slower due to laser point-by-point exposure	Faster (whole layer exposure in one go)
Printer Cost	Generally more expensive due to laser technology	Generally more affordable, especially with LED-based DLP printers
Resolution	High resolution (down to microns)	High resolution (down to microns)
Resolution of Print Layers	Can achieve higher resolution and finer detail, particularly at micro levels	Similar resolution to SLA but may vary depending on DLP projector resolution
Speed of Setup	Slower due to precision of laser and layer-by-layer focus	Faster, as the projector exposes an entire layer at once
Surface Finish	Smooth finish, especially at higher resolutions	Smooth finish, but may have a slight pixelated appearance
Technology Type	Photopolymerization, Layer-by-Layer 3D Printing	Photopolymerization, Layer-by-Layer 3D Printing

Table 3. Companies Producing SLA 3D Printers.

Company	Models	Features	Materials	Website
3D Systems®	ProJet MJP 2500, SLA 7000	Pioneer in 3D printing, high-precision for automotive, medicine, and jewelry	Resins for prototyping and production	www.3dsystems.com
Anycubic®	Photon Mono X, Photon Ultra	Good balance between performance and cost, popular for rapid prototyping	Resins for high-precision printing	www.anycubic.com
Asiga®	Asiga MAX X, Asiga PRO 4K	Specializes in high-precision printing for dental and industrial applications	Dental resins, biocompatible, prototype resins	www.asiga.com
Carbon®	M1, M2	Advanced SLA solutions for industrial applications	Advanced resins for production and prototypes	www.carbon3d.com
DWS Systems®	XPRO, XFAB 3500	Italian company offering SLA solutions for jewelry, automotive, and medical	Resins for rapid prototyping, jewelry, and medical	www.dwssystems.com
Elegoo®	Mars 2 Pro, Saturn	Leader in DLP printing for hobbyists and small businesses	DLP resins, high-detail, elastic, and transparent resins	www.elegoo.com
EnvisionTEC®	Perfactory, Micro Plus	High-precision SLA printers for medicine, jewelry, and automotive	Resins for prototypes, medical, dental, and jewelry	www.envisiontec.com
Formlabs®	Form 3, Form 3B	High-quality SLA printers, popular for professional and prototyping applications	Rigid, elastic, castable, and bioresins	www.formlabs.com
Peopoly®	Moai, Moai 130	Affordable, high-quality SLA printers for small businesses and hobbyists	Standard resins and castable resins	www.peopoly.net
Prusa Research®	SL1S	Known for FDM printers, also offering high-quality SLA printing	Standard and special resins	www.prusa3d.com
XYZprinting®	Nobel 1.0A, Nobel 1.0	Entry-level SLA printers for small businesses and hobbyists	Standard resins and castable resins	www.xyzprinting.com

Table 4. Companies Producing DLP 3D printers.

Company	Models	Features	Materials	Website
Anycubic®	Photon Ultra, Photon DLP	DLP printers combining speed and high resolution	DLP resins, including rigid and flexible resins	www.anycubic.com
Asiga®	Asiga MAX X, Asiga PRO 4K	High-precision DLP printers used in dental and industrial applications	Dental resins, biocompatible, and prototype resins	www.asiga.com
Elegoo®	Mars 2 Pro, Saturn	Leader in DLP printing for hobbyists and small businesses	DLP resins, including high-detail, elastic, and transparent resins	www.elegoo.com
FlashForge®	Hunter DLP	Reliable DLP printers with superior print quality	DLP resins for high-resolution and prototyping	www.flashforge.com
MoonRay®	MoonRay S	Professional DLP 3D printing, high resolution and excellent surface quality	Resins for prototypes, jewelry, and medical devices	www.moonray3d.com
Photocentric®	LC Pro, LC Magna	DLP printers with high-speed and precision, ideal for rapid prototyping	Resins for industrial applications and prototypes	www.photocentricgroup.com
PicoFab®	Pico 3D, PicoPro	Affordable DLP printers for small businesses and rapid prototyping	Resins for prototypes and high-resolution models	www.picofab.com

4. High Versatility in Materials

- FsLDW allows for the processing of a wide range of materials, including metals, polymers, ceramics, and even biological materials. This makes it ideal for applications in fields like microelectronics, biomedicine, and photonics.

5. Low Thermal Deformation

- Since the laser pulses are extremely brief, the amount of heat generated is limited, minimizing thermal deformation effects. This is crucial for preserving the properties of the materials.

6. Customization and Rapid Prototyping

- FsLDW is ideal for rapid prototyping, enabling the quick creation and iteration of highly specific micro- and nano-scale models in short timeframes. When paired with advanced materials like GO and LRGO, FsLDW's ultrashort pulses produce bioinspired, highly functional surfaces—such as those mimicking the adaptive slipperiness of Nepenthes—that combine precision, versatility, and nature-inspired performance.[6]

A highly effective fabrication method for producing these biomimetic micro- and nano-textured surfaces is Laser Interference Ablation (LIA), which uses the overlap of multiple coherent laser beams to generate precise interference patterns and achieve micrometer- and nanometer-scale material removal (Lasagni et al. 2018). By integrating CAD-guided design with CAM-controlled ablation, LIA ensures the constructive interference zones ablate material to form uniform micro-reliefs or nanostructures, while destructive zones remain intact, allowing scalable production of consistent, slippery patterns (Böhlen et al. 2003, Mutapcic et al. 2004, Ackerle et al. 2019). These patterned surfaces find applications across photonics, sensing, nanotechnology, and electronics, enabling advanced solutions in a wide range of industrial and technological sectors.

Table 5 provides a comparative overview of various technologies, including DLIP, LIA, FsLDW/FEM laser, and LBIM.

Biomimetic research continues to discover how Nature has designed ingenious, zero-impact strategies to maintain high levels of cleanliness and hygiene in living organisms. This stands in contrast to human solutions based on industrial chemistry for manipulating hydrocarbons over the past hundred years, which continue to introduce highly polluting substances harmful to both ecosystems and humans. The ideal Nature's Patent Office gathers numerous examples:

Table 6 presents a comprehensive list of these bio-inspired surface design solutions, highlighting the diversity and ingenuity found in nature.

4.2 The Zero Impact Cleaning Syndrome

CAD/CAM and related 3D printing processes are continuously evolving, and the field of superhydrophobic materials is experiencing significant scientific and industrial advances (Murariu et al. 2022, Liu et al. 2023). The design optimization of self-cleaning surfaces will reduce the environmental impact of cleaning processes, minimize the use of detergents harmful to human health and the environment, and cut water consumption, energy usage, and cleaning costs. They can also be applied across a wide range of industrial sectors—such as construction, electronics, and transportation—where surface contamination undermines material performance and reliability, thereby improving overall process efficiency. The ability to engineer functional surfaces at micro- and nanoscales is becoming increasingly feasible, further advancing the concept of Zero Impact Cleaning. Traditionally, self-cleaning

[6] GO stands for Graphene Oxide, which is a form of graphene that contains hydroxyl (-OH), carboxyl (-COOH), and epoxy (-O-) groups on its surface, making it hydrophilic and more reactive compared to pure graphene. LRGO stands for Low-Reduced Graphene Oxide, which is a form of graphene oxide that has undergone controlled reduction, removing some of the oxidized groups present in GO, but without returning it completely to pure graphene. Together, these two materials are combined in a bilayer structure to exploit the specific properties of each, such as the ability to respond to humidity or enhance conductivity.

Table 5. Comparative Table of Laser Technologies.

	DLIP	**LIA**	**FsLDW/ FEMlaser**	**LBIM**
Main Purpose	Creating micro- and nanostructured periodic surfaces	Selective ablation for periodic structures	Ultra-precise micromachining with ultrashort pulses	Measurement and characterization using laser interference
Operating Principle	Direct laser interference modifies the surface without ablation	Laser interference removes material selectively	Ultrashort laser pulses for high-precision processing with minimal thermal effects	Laser interference detects dimensional variations or optical properties
Pulse Duration	Nanoseconds (ns) or picoseconds (ps)	Nanoseconds (ns) or picoseconds (ps)	Femtoseconds (fs, 10^{-15} s)	Continuous or pulsed, depending on the technique
Thermal Damage	Low, but present	Moderate to high (thermal ablation)	Almost none (cold ablation)	None (optical measurement technique)
Resulting Structures	Regular patterns (lines, grids, dots)	Complex scalable structures	Ultra-precise engravings, 3D micromachining	No physical modification of the material
Resolution	Sub-micrometer (100 nm - μm)	Micrometer-scale (μm)	Nanometer-scale (10–100 nm)	Sub-nanometer for measurements
Main Applications	Biomimetics, optics, tribology, electronics	Structured surfaces, microfluidics, photonics	Biomedical, optoelectronics, nanotechnology	Metrology, quality testing, advanced materials

of surfaces was achieved through chemical coatings that promoted low surface energy and created high roughness on the material's surface. This allowed contaminating particles or droplets to float rather than adhere. While these chemical methods worked well for textiles produced with traditional weaving technologies, the advent of 3D printing in textile production has introduced new challenges. In fact, chemical coatings are not as effective on 3D-printed fabrics, which have a different morphology and structure, thus requiring more innovative solutions to ensure self-cleaning functionality (Chan et al. 2023). In relation to the application of advanced CAD/CAM protocols to the research of active materials, a 2022 study published in *Polymers* explores the control of microscopic characteristics in self-cleaning fabrics using typical 3D additive printing technologies (Atwah et al. 2022). The research investigates how these technologies can be leveraged to create fabrics with optimized microstructures that enable self-cleaning properties, providing a more efficient and sustainable solution for textile materials. By manipulating parameters during the 3D printing process, the study aims to enhance the functional properties of fabrics, such as their ability to repel dirt and water, mimicking natural self-cleaning phenomena. Self-cleaning fabrics inspired by Nature offer significant advantages, including aesthetic appeal, energy savings, and a reduction in washing needs. The evolution of CAD/CAM 3D printing technology now offers a new

Table 6. Animal's Surface Characteristics.

Species	**Surface Characteristics**
Androctonus australis (Scorpion)	Superhydrophobic surface due to specific microstructures
Anopheles gambiae (Mosquito)	Superhydrophobic legs with setae and microstructures
Aptenodytes forsteri (Emperor Penguin)	Microstructures on feathers
Blaberus giganteus (Cockroach)	Hydrophobic exoskeleton with microscopic ridges
Cataglyphis (Desert Ant)	Hydrophobic microstructures on exoskeleton
Collembola (Springtail)	Nanostructured cuticle
Culex pipiens (Mosquito Larva)	Hydrophobic cuticular surfaces
Cyclocephala (Neotropical Beetle)	Microstructures on exoskeleton
Cylindera germanica (Golden Beetle)	Micro- and nanostructures on exoskeleton
Drosophila (Fruit Fly)	Setae on legs creating superhydrophobic surface
Dolomedes (Water Spider)	Hydrophobic hairs on legs
Gekko gecko (Tokay Gecko)	Microscopic setae on feet
Litoria caerulea (White's Tree Frog)	Superhydrophobic microstructures on skin
Lycosidae (Wolf Spider)	Hydrophobic hairs on legs
Morpho menelaus (Morpho Butterfly)	Nanostructures on wings
Notonectidae (Backswimmer)	Superhydrophobic surfaces on body
Polistes fuscatus (Paper Wasp)	Superhydrophobic properties of wings
Reticulitermes flavipes (Eastern Subterranean Termite)	Nanostructures on wings
Salmo salar (Atlantic Salmon)	Scales with hydrophobic structures
Salticus scenicus (Jumping Spider)	Microscopic setae on legs
Scalopus aquaticus (Eastern Mole)	Hydrophobic fur

opportunity to regulate these parameters during fabric production, creating structural self-cleaning characteristics. Specifically, the fabrication of 3D-printed materials has been achieved using the low-cost Fused Filament Fabrication (FFF) technique (Galvagno et al. 2024). Various printing parameters, such as orientation angle, layer height, and extruder width, have been studied to control the microscopic self-cleaning characteristics of the fabrics. Among these, aspects like surface roughness, wettability angle, and porosity were analyzed by varying the printing parameters. From the research conducted, the optimal self-cleaning surface was obtained by combining three parameters: layer height (LH) (0.15, 0.13, 0.10 mm), extruder width (EW) (0.5, 0.4, 0.3 mm), and two different printing orientation angles (O) (45° and 90°). The materials used for the flexible thermoplastic filaments were three: thermoplastic polyurethane (TPU 98A), thermoplastic elastomers (TPE felaflex), and thermoplastic copolyester (TPC flex45). The self-cleaning properties were quantified using a predefined criterion. Finally, the printing parameters were modeled to optimize the self-cleaning characteristics of the obtained samples, demonstrating the importance of precise control over printing parameters to achieve the best results. The advanced textile materials industry is rapidly growing, with innovations in fabric characteristics such as waterproofing, antimicrobial, antistatic, and self-cleaning properties. The search *for Zero Impact for Cleaning systems* proposes solutions applicable to various materials and artifacts, such as small household appliances, a category of objects that have become an integral part of our daily lives and have radically changed how we live and work, thanks to unprecedented innovations and high performance. However, a negative aspect of this technological revolution is the bacterial colonization of the surfaces of such devices, a phenomenon that can have serious consequences for our health and well-being. To address this issue, a recent study explored the use of composite resins to enhance the antifouling and antimicrobial properties of household appliances, utilizing photopolymer 3D printing technology (Xu 2024). Composite resins were developed by adding various proportions of sulfobetaine methacrylate (SBMA) and 12-methacryloyloxydodecylpyridinium bromide (MDPB). The innovation of this approach lies in the ability of these resins to combine hydrophilic and antimicrobial properties, resulting in surfaces that reduce bacterial adhesion and other contaminations. Specifically, thanks to the natural hydration of the zwitterionic SBMA, these resins effectively suppressed protein adsorption and bacterial adhesion, creating surfaces resistant to fouling. The results obtained are remarkable: the presence of MDPB significantly enhanced the antimicrobial properties of the resins, demonstrating notable effectiveness against bacteria such as *S. aureus* (Gram-positive) and *E. coli* (Gram-negative), with reductions of 98.3% and 97.1%, respectively. These resins not only proved effective in fighting bacteria but also maintained low cytotoxicity, demonstrating good biocompatibility, which is essential for use in devices that come into contact with skin and domestic environments. An additional step forward was made with the development of baby nebulizers, which, by using these composite resins, achieved surfaces with greater resistance to protein adsorption and reduced bacterial accumulation. This demonstrates how photopolymerizing 3D printing technology can not only improve device performance but also meet growing health and hygiene needs, significantly contributing to the creation of healthier and cleaner domestic environments. This innovation fits perfectly within the concept of zero polluted syndrome, a goal that is increasingly central to modern technologies, where improving surfaces and materials tends to minimize the accumulation of harmful substances, promoting a healthier and safer environment for humans. Through the use of advanced materials and technologies like 3D printing, it is possible not only to address cleaning and hygiene needs but also to prevent the proliferation of contaminants that could compromise our health and that of others. An interesting study mixed three natural sources of inspiration: the rice leaf, the well-known lotus leaf, and the already widely described shark skin. By combining their different surface properties, the material developed on a three-level hierarchical structure exhibits exceptional superhydrophobic characteristics that motivate its biomimicry. For the first time, 3D-printed channels via stereolithography (SLA) with a width of 100 μm were used, characterized by intrinsic roughness due to the printing filaments (10 μm), and then coated with TiO_2 nanoparticles (22 and 100 nm). This structure shows a maximum advancing contact angle of 165°, with lower anisotropy and hysteresis compared to other 3D-printed surfaces, thanks to the presence of air pockets at the surface/water interface (Cassie-

Baxter state). Dynamic tests with water droplets demonstrate that the biomimetic surface has self-cleaning properties, although these properties decrease under UV-A irradiation. Furthermore, the biomimetic surface gives increased buoyancy to 3D-printed objects, reducing friction thanks to the smaller contact area between solid and liquid. Numerical simulations of a channel with biomimetic walls confirm that the presence of air is crucial to explain the results: it increases the average velocity and reduces the friction coefficient due to a slip effect on the wall. Our results show that SLA 3D printing is an effective approach for developing superhydrophobic biomimetic surfaces, with potential applications in antifouling devices and reducing fluid dynamic resistance (Barraza et al. 2022). At the heart of the most advanced technology, in the clean rooms of microelectronics and bioengineering laboratories, an invisible battle is fought against an ancient enemy: dust. It is an almost metaphysical obsession, a desire for absolute control over an elusive substance that slips away from us, much like atoms eluded the eyes of early philosophers. Clean room design aims to eliminate every particle, every speck suspended in the air, as if technological perfection must emerge from an absolute void—a contemporary horror vacui that echoes the ancient struggle between Eleatic thought and Greek atomism. Yet, if we return to the origins of scientific thought, we discover that dust itself was one of the first tools of philosophical intuition. In *De Rerum Natura*, Lucretius invites his readers to observe the dust particles suspended in a sunbeam: those seemingly chaotic trajectories, which today we can associate with Brownian motion, offer an intuitive image of the ceaseless movement of the indivisible in the void (Kyprianou 2012).

The incessant motion of dust in suspension, once symbolizing chaos, has now revealed the invisible presence of energy and vitality within the particles of atoms and molecules through our nanometric vision. Dust thus became a cosmological revelation, a visible trace of the invisible. In a curious paradox, the very dust that is obsessively expelled from laboratories today is what, in ancient times, allowed for the imagining of the fundamental structure of matter. In the attempt to eliminate dust, Zero-CleaningRoom seems to chase an ideal of absolute order, which, however, clashes with the irreducible tendency of matter to disperse, fragment, and move relentlessly. Perhaps, upon closer inspection, these sterile rooms are not just technological spaces, but modern manifestations of an ancient philosophical tension: the dream of separating form from chaos, order from disorder, structure from the primordial dust from which everything originates. Thus, in recent decades, scientific research and the high-tech industry have witnessed the spread of a sort of total cleanliness syndrome, a phenomenon that reflects the ever-increasing demand for contamination-free operational environments. From microelectronics to biotechnology, including pharmaceuticals and aerospace, multiple sectors require conditions of absolute purity, where even the tiniest particle of dust or impurity can compromise the functionality of devices or alter sensitive processes. To meet these needs, cleanrooms have been developed—controlled environments where temperature, humidity, pressure, and especially the concentration of suspended particles in the air are maintained within rigid parameters. These sterile spaces are indispensable, for example, in the production of microchips, where even a particle just a few nanometers in size can compromise the efficiency of a circuit. Similarly, in pharmaceutical research and bioengineering, cleanrooms are crucial to ensure the safety of drugs, cell cultures, and advanced medical devices (Allied Cleanrooms 2025). The design and maintenance of a cleanroom require sophisticated tools for simulating and optimizing spaces and airflows, avoiding contaminant buildup and ensuring optimal environmental conditions. In this context, CAE (Computer-Aided Engineering) and CAO (Computer-Aided Optimization) technologies play a crucial role.

- **CAE** allows for simulating aerodynamic flows within the cleanroom, assessing the distribution of suspended particles, and optimizing the use of HEPA and ULPA filters to maintain the environment's sterility. Computational Fluid Dynamics (CFD) software, such as OpenFOAM® or ANSYS Fluent®, is used to model and predict particle behavior in the air, minimizing turbulence zones that could promote contaminant deposition on critical surfaces.
- **CAO** is used to design optimal solutions for the cleanroom layout, reducing energy consumption and improving operational efficiency. With advanced optimization algorithms, it's possible to define the ideal arrangement of laminar airflows, filtration points, and access zones, minimizing the risk of cross-contamination.

An economical way to ensure contamination control comes through the CFD analysis solution of SimScale® is a cloud-based CAE software for cleanroom design simulation, allowing testing, validation, and optimization of environmental and functional layouts, including computational fluid dynamics (SimScale 2025). Well-designed cleanroom structures are essential for activities in controlled environments with low pollutant levels, a critical requirement for many manufacturing, pharmaceutical, and scientific research applications (14644.dk 2025). One way to ensure proper cleanroom design is through engineering simulation, particularly computational fluid dynamics. Depending on the application, contaminating particles are often present in cleanrooms, and if not designed correctly, they could negatively impact the product and the people within the laboratory. Cleanrooms can consume up to 50 times more energy than non-classified areas, which can be costly. With high regulatory standards, optimizing energy efficiency is not only necessary but can also have additional financial benefits. In aerospace manufacturing, by using simulation software as part of the design process, engineers can assess the performance of their design, optimize airflow and air change effectiveness, reduce contamination risk, and improve energy efficiency (Lund 2025). Guided simulation is the future of product development. Engineering companies are increasingly using simulation at every stage of the design cycle to improve outcomes. They aim to make simulation open to all users, not just specialists, and cloud-based and SaaS solutions offer this flexibility at a lower cost than traditional software. This helps overcome obstacles such as limited computational resources and high on-premise infrastructure costs, allowing engineers to be more efficient and creative in their design. By using simulation software as part of the design process, engineers can assess the performance of their design, optimize airflow and air change effectiveness, reduce contamination risk, and increase energy efficiency (SimScale 2017).

The design of *cleanrooms* also places primary importance on research into advanced materials related to the development of self-cleaning and antibacterial surfaces, inspired by natural mechanisms of impurity repulsion. These surfaces can significantly reduce the need for frequent chemical cleaning, thereby lowering the risk of releasing potentially harmful residues that could affect sensitive processes within the *cleanroom*. It was noted that this pursuit of absolute cleanliness also prompts wider concerns about whether, in an ever more sterilized world, we might be disturbing the natural balance of biological systems (Moskvitch 2015). While the absence of contaminants is essential in technological and healthcare settings, it may also affect our relationship with the environment and the evolution of bio-integrated materials. The future of cleanrooms will depend not only on continuous technological innovation but also on the ability to balance operational efficiency with sustainable solutions that are compatible with biological systems. While the Zero-*CleaningRoom* represents the pinnacle of technological purification, perhaps the contemporary digital *WunderCAD/CAMmer* constitutes its mirror, its conceptual alter-ego: not a place of subtraction, but a virtual room of accumulation; not a sterile empty room, but a digital cabinet full of wonders, where disorder is not an enemy, but a creative resource. Yet, both respond to the same tension: the human desire to dominate matter, to predict its evolution, to design the invisible. In Zero-*CleaningRoom*, every particle of dust is an intruder, an anomaly that threatens the extreme precision required for the fabrication of microchips and biotechnological devices. The goal of CAE programs is to create a perfectly clean environment, free of impurities. It aims to create an unrealistic technological *vacuum*, far from the swirling atomic chaos that permeates all space, in order to guarantee absolute control over the production processes of the most sophisticated and high-tech artificial materials. The *WunderCAD/CAMmer*, on the other hand, is a virtual space, a representation, a technological stage that embraces and showcases chaos and complexity. Just as the ancient *Wunderkammer* gathered natural and artificial mirabilia to create a layered visual encyclopedia of the world, the *WunderCAD/CAMmer* serves as a design laboratory where apparent disorder is tamed and transformed into a wellspring of inspiration. Here, dust is not eliminated but observed; error is not a malfunction but an opportunity; design is not only prediction, but also discovery. The analogy between these two seemingly opposite environments reveals itself in their common aspiration: to create a space where matter is transformed into information, where the visible gives glimpses of the invisible. But while the Zero-*CleaningRoom* pursues an ideal of unnatural purity, the *WunderCAD/CAMmer* embraces unpredictability as part of the dirty organic work of creative life process; not imposing order a priori, but discovering it in the emerging dynamics drifting of matter and information. Perhaps the future of design lies precisely in the meeting point between these two models, in the delicate boundary between subtraction and accumulation, order and disorder, human and natural. A clean room that is not only a place of subtraction, meaning fewer particles, fewer surprises, fewer variables, but also a place of wonder, and a *WunderCAD/CAMmer* where the idea of neo-accumulation turns into an intelligent selection process. It does not fear dust, but accepts it as a design principle. If the first seeks emptiness, the second welcomes accumulation. In a Zero-*CleaningRoom*, dust is an enemy to be eradicated, but in the *WunderCAD/CAMmer*, it is the very metaphor of creativity: a chaotic aggregate that, when observed from the right perspective, reveals unexpected structures. It is not a random sum of chaotic events, but a collection of meaningful elements capable of generating unprecedented connections, a place that can recognize, in the extreme precision of its digital analyses, a form of poetic matter. If ancient *Kunstkammer* gathered natural and artificial wonders to fuel new visions of the world, the contemporary digital wonder rooms operates in the same way, but within a digital and design context: data becomes the new dust, a formless material that must be ordered to generate knowledge. However, just as dust settles on an object revealing its texture, it is in the unpredictable detail that the deepest intuition hides. Perhaps the future of design lies neither in the Zero-*CleaningRoom* nor in the *WunderCAD/CAMmer*, but in their imaginative fusion: a room where dust is not only a problem to eliminate but a trace to interpret. A sterile room that is, at the same time, a laboratory of wonder, capable of recognizing that precision is not just rigor, but also the ability to observe the unexpected and transform it into knowledge. In this vision, the designer is not just a purifier of impurities, but an explorer of the visible and invisible, capable of reading in the microscopic chaos the hidden geometries of Nature and human innovation (Nunes 2023).

References

14644.dk (2025). Navigating the complexities of cleanroom design and construction. Available at: https://www.14644.dk/navigating-the-complexities-of-cleanroom-design-and-construction [Accessed: 20 May 2025].

Ackerle, N., Gysel, J., Warhanek, M., and Wegener, K. (2019). CAM solution for quasi-tangential laser ablation of complex 3D workpieces. Lasers in Manufacturing Conference 2019.

Afshar-Mohajer, M., Yang, X., Long, R., and Zou, M. (2023). 3D printing of micro/nano-hierarchical structures with various structural stiffness for controlling friction and deformation. Additive Manufacturing, 62: 103368.

Allied Cleanrooms (2025). Cleanrooms vs. Laboratories. Available at: https://alliedcleanrooms.com/cleanrooms-vs-laboratories/ [Accessed: 20 May 2025].

Asmatulu, E., Rajakaruna, R. A. D. N. V., Subeshan, B., and Uddin, M. (2022). 3D printed superhydrophobic structures for sustainable manufacturing benefits: An overview. Journal of Management and Engineering Integration, 15(1): 45–56.

Atwah, A. A., Almutairi, M. D., He, F., and Khan, M. A. (2022). Influence of printing parameters on self-cleaning properties of 3D printed polymeric fabrics. Polymers, 14(15): 3128.

Baeten, S. R. R., Kochovski, A., Jovanova, J., and Sakes, A. (2024). Characterization of shark skin properties and biomimetic replication. Bioinspiration and Biomimetics, 19(5): Article 051002.

Balani, K., Batista, R. G., Lahiri, D., and Agarwal, A. (2009). The hydrophobicity of a lotus leaf: A nanomechanical and computational approach. Nanotechnology, 20(30): 305707.

Barraza, B., Olate-Moya, F., Montecinos, G., Ortega, J. H., Rosenkranz, A., Tamburrino, A. et al. (2022). Superhydrophobic SLA 3D printed materials modified with nanoparticles biomimicking the hierarchical structure of a rice leaf. Science and Technology of Advanced Materials, 23(1): 300–321.

Barthlott, W., and Neinhuis, C. (1997). Purity of the sacred lotus, or escape from contamination in biological surfaces. Planta, 202(1): 1–8.

Barthwal, S., Uniyal, S., and Barthwal, S. (2024). Nature-inspired superhydrophobic coating materials: Drawing inspiration from nature for enhanced functionality. Micromachines, 15(3): 391.

Bello, O. S., Adegoke, K. A., and Oyewole, R. O. (2013). Biomimetic materials in our world: A review. IOSR Journal of Applied Chemistry (IOSR-JAC), 5(3): 22–35.

Bioengineer.org (2025). Femtosecond laser pulses: a new frontier in cross-scale micro and nanofabrication. Available at: https://bioengineer.org/femtosecond-laser-pulses-a-new-frontier-in-cross-scale-micro-and-nanofabrication/ [Accessed: 20 May 2025].

BIOKON (2025). Network of BIOKON. Available at: https://www.biokon.de/en/network/network-of-biokon/ (Accessed: 20 May 2025).

Böhlen, I., Fieret, J., Holmes, A., and Lee, W. (2003). CAD/CAM software for an industrial laser manufacturing tool. Proceedings of SPIE - The International Society for Optical Engineering.

Bohn, H. F., and Federle, W. (2004). Insect aquaplaning: Nepenthes pitcher plants capture prey with the peristome, a fully wettable water-lubricated anisotropic surface. Proceedings of the National Academy of Sciences, 101(39): 14138–14143.

Cai, Z., Chen, F., Tian, Y., Zhang, D., Lian, Z., and Cao, M. (2022). Programmable droplet transport on multi-bioinspired slippery surface with tridirectionally anisotropic wettability. Chemical Engineering Journal, 449: 137831.

CEEBIOS (2025). Centre d'Études et d'Expertises en Biomimétisme. Available at: https://www.ceebios.fr [Accessed: 20 May 2025].

Chan, K. P., He, F., Atwah, A. A., and Khan, M. (2023). Experimental investigation of self-cleaning behaviour of 3D-printed textile fabrics with various printing parameters. Polymer Testing, 119: 107941.

Ciulla, M. G., Massironi, A., Sugni, M., Ensign, M. A., Marzorati, S., and Forouharshad, M. (2020). Recent advances in the development of biomimetic materials. Materials Science and Engineering: R: Reports, 140: 1–24.

Damodaran, V. B., and Murthy, N. S. (2016). Sharklet technologies mimicking the micropatterns on Shark skin. ResearchGate. Available at: https://www.researchgate.net/figure/Sharklet-technologies-mimicking-the-micropatterns-on-Shark-skin-a-Skin-of-a-Bull-Shark-C_fig6_303904711 [Accessed: 20 May 2025].

Davis, J. (n.d.) Shark evolution: a 450 million year timeline. Natural History Museum. Available at: https://www.nhm.ac.uk/discover/shark-evolution-a-450-million-year-timeline.html [Accessed: 20 May 2025].

Desisa, S. T., Dalbaşi, E. S., and Çay, A. (2022). Production of superhydrophobic cotton fabric by layer-by-layer deposition of SiO2/TiO2-polydimethylsiloxane. Journal of Natural Fibers, 20(1).

Diaz Lantada, A., Franco Martinez, F., Hengsbach, S., Rupp, F., Thelen, R., and Bade, K. (2020). Artificial intelligence aided design of microtextured surfaces: Application to controlling wettability. Nanomaterials, 10(11): 2287.

Domel, A. G., Domel, G., Weaver, J. C., Saadat, M., Bertoldi, K., and Lauder, G. V. (2018). Hydrodynamic properties of biomimetic shark skin: Effect of denticle size and swimming speed. Bioinspiration & Biomimetics, 13(5): 056014.

Domen, J. K., Stringfellow, W. T., Camarillo, M. K., and Gulati, S. (2014). Fog water as an alternative and sustainable water resource. Clean Technologies and Environmental Policy, 16(2): 235–249

Durham University. (n.d.). Functional Surfaces: Superhydrophobic Droplet Behavior [Video]. YouTube. Available at: https://youtu.be/jiEKUi42r1I?si=iPmaROYSnEUBMuS0 [Accessed 19 May 2025].

Huang, F., Zhang, Y., Gong, M., Lin, X., Wang, D., Zhang, L. et al. (2024). 3D printing of soft materials with superhydrophobicity and programmable anisotropic wettability. 3D printing of soft materials with superhydrophobicity and programmable anisotropic wettability, Surfaces and Interfaces, Volume 46: 103953.

Galvagno, S., Tammaro, L., Portofino, S., Loffredo, F., De Girolamo Del Mauro, A., Villani, F. et al. (2024). Fabrication of 3D micropatterned hydrophobic surfaces by fused filament fabrication printing technology. Applied Research. Available at: https://onlinelibrary.wiley.com/doi/10.1002/appl.202400003 [Accessed: 20 May 2025].

Han, D.-D., Zhang, Y.-L., Chen, Z.-D., Li, J.-C., Ma, J.-N., Mao, J.-W. et al. (2023). Carnivorous plants inspired shape-morphing slippery surfaces. Opto-Electronic Advances, 6: 210163.

Huang, G., Guo, Y., Lee, B., Chen, H., and Mao, A. (2025). Research Advances and Future Perspectives of Superhydrophobic Coatings in Sports Equipment Applications. Molecules, 30(3): 644.

Humphries C. (2011). Designing from Life. Harvard Magazine, May–June 2011. Available at: https://www.harvardmagazine.com/2011/04/designing-from-life [Accessed: 20 May 2025].

Johnson, R. E., Dettre, R. H., and Brandreth, D. A. (1977). Dynamic contact angles and contact angle hysteresis. Journal of Colloid and Interface Science, 62(2): 205–212.

Kaur, G., Marmur, A., and Magdassi, S. (2020). Fabrication of superhydrophobic 3D objects by Digital Light Processing. Additive Manufacturing, 36: 101669.

Khajuria, D., Kaur, K., and Singh, V. (2023). Biochar based self-cleaning superhydrophobic surface with aqueous DESphobic properties. Journal of Molecular Liquids, 380: 121736.

Koch, K., Bhushan, B., and Barthlott, W. (2009). Multifunctional surface structures of plants: An inspiration for biomimetics. Progress in Materials Science, 54(2): 137–178.

KLOE France (2025). Photolithography Equipment – Direct Laser Writing. Available at: https://www.kloe-france.com/en/photolithography-equipment/direct-laser-writing?gad_source=1&gclid=CjwKCAiAneK8BhAVEiwAoy 2HYQITJAsKYxt86AVjTeErqsy17Xk9ddzEmz GC7qLks1avLCR5OHNj0xoCs9MQAvD_BwE [Accessed: 20 May 2025].

Kramer, A., Schwebke, I., Kampf, G. (2006). How long do nosocomial pathogens persist on inanimate surfaces? A systematic review. BMC Infectious Diseases, 6(130).

Kruiper, R., Chen-Burger, J., and Desmulliez, M. P. Y. (2016a). Computer-Aided Biomimetics. In Lecture Notes in Computer Science (Vol. 9793: pp. 131–143). Springer.

Kruiper, R., Vincent, J. F., Chen-Burger, J., and Desmulliez, M. P. Y. (2017b). Towards identifying biological research articles in computer-aided biomimetics. In Lecture Notes in Computer Science (Vol. 10384). Springer.

Kruiper, R., Vincent, J. F., Chen-Burger, J., Konstas, I., and Desmulliez, M. P. Y. (2020c). A scientific information extraction dataset for nature inspired engineering. arXiv.

Kyprianou (2012). A. E. Kyprianou: Brownian motion or Lucretian motion? Available at: https://people.bath.ac.uk/ak257/LvB.html?ref=josephnoelwalker.com [Accessed: 20 May 2025].

Lafuma, A., and Quéré, D. (2003). Superhydrophobic states. Nature Materials, 2(7): 457–460.

Lasagni, A. F., Alamri, S., Aguilar-Morales, A. I., Rößler, F., Voisiat, B., and Kunze, T. (2018). Biomimetic surface structuring using laser-based interferometric methods. Applied Sciences, 8(8): 1260.

Li, L., Wei, J., Zhang, J., Li, B., Yang, Y., and Zhang, J. (2023). Challenges and strategies for commercialization and widespread practical applications of superhydrophobic surfaces. Science Advances, 9(42): eadj1554.

Liu, H., Zhang, Z., Wu, C., Su, K., and Kan, X. (2023). Biomimetic superhydrophobic materials through 3D printing: Progress and challenges. Micromachines, 14(6): 1216.

Lloyd, C. J., Mittal, K., Dutta, S., Dorrell, R. M., Peakall, J., Keevil, G. M. et al. (2023). Multi-fidelity modelling of shark skin denticle flows: Insights into drag generation mechanisms. Royal Society Open Science, 10: Article 220684.

Lotus Leaf Coatings (n.d.). Lotus Leaf Coatings. Available at: https://lotusleafcoatings.com [Accessed 19 May 2025].

Lumini, M. (2023). BIONIKONLab FABNAT 14: Una Bottega tra Natura, Arte e Tecnologia. Chap. on Map Research Bionics: Transdisciplinary Ecosystems in Bionics, Biodesign and Biomimicry – Scientific and Technological Aspects for a Design Culture. São Paulo: Blucher. pp. 243–266.

Lund (2025). The Role of Cleanrooms in Aerospace Manufacturing. Available at: https://www.14644.dk/the-role-of-cleanrooms-in-aerospace-manufacturing [Accessed: 20 May 2025].

Ma, L., Zhang, Z., Gao, L., Liu, Y., and Hu, H. (2020). Bio-inspired icephobic coatings for aircraft icing mitigation: A critical review. Reviews of Adhesion and Adhesives, 8: 168–199.

Markforged, Inc. (2025). What is Digital Light Processing (DLP)? 3D Printing Processes. Available at: https://markforged.com/resources/learn/3d-printing-basics/3d-printing-processes/what-is-digital-light-processing-dlp [Accessed: 20 May 2025].

MeMeraki (2024). The Role of Lotus in Indian Mythology and Arts: Symbolism and Iconography. MeMeraki, 18 December 2024. Available at: https://www.memeraki.com/blogs/posts/the-role-of-lotus-in-indian-mythology-and-arts-symbolism-and-iconography?srsltid=AfmBOop-h1fugy9K1S8_482eA0ttsIiqWRl1NFmGxGMRPDKj_phZGK-V [Accessed: 18 May 2025].

Moskvitch (2015). Can you be too clean? Available at: https://www.bbc.com/future/article/20151118-can-you-be-too-clean [Accessed: 20 May 2025].

Murariu, A. C., Sirbu, A., Cocard, M., and Duma, I. (2022). Influence of 3D printing parameters on mechanical properties of the PLA parts made by FDM additive manufacturing process. Engineering Innovations, 2: 7–20.

Mutapcic, E., Iovenitti, P., and Hayes, J. (2004). A novel CAD/CAM tool for excimer laser ablation-machining process. Pacific International Conference on Applications of Lasers and Optics, M304.

Nosonovsky, M., and Bhushan, B. (2008). Multiscale Dissipative Mechanisms and Hierarchical Surfaces: Friction, Superhydrophobicity, and Biomimetics. pp. 1809–1819. In: Encyclopedia of Nanotechnology. Springer, Dordrecht. Available at: https://link.springer.com/doi/10.1007/978-94-017-9780-1_158 [Accessed 19 May 2025].

Nunes (2023). Self-cleaning Surfaces: Harnessing the Power of Nature and Technology. LinkedIn. Available at: https://www.linkedin.com/pulse/self-cleaning-surfaces-harnessing-power-nature-technology-tiago-nunes [Accessed: 20 May 2025].

Ollivier, H. (1907). Recherches sur la capillarité. J. Phys. Theor. Appl., 6(1):757-782. Available at: https://doi.org/10.1051/jphystap:019070060075700 [Accessed: 18 May 2025].

OpenFOAM Foundation Ltd. (n.d.) OpenFOAM® – Official home of The Open Source Computational Fluid Dynamics (CFD) Toolbox. Available at: https://www.openfoam.com/ [Accessed: 20 May 2025].

Sharklet Technologies Inc. (2023). Home – Sharklet Technologies, Inc. Retrieved May 20, 2025, from https://www.sharklet.com/ [Accessed: 20 May 2025].

SimScale (2017). How to improve clean room design using simulation [Video]. YouTube. https://youtu.be/pRMeSfvtPTY?si=aVEkaI1EwOT641uZ

SimScale (2025). SimScale. Available at: https://www.simscale.com [Accessed: 20 May 2025].

Srisuwan, T. (2020). Phyto-Inspired Design: Innovative Solutions for Architecture. International Journal of Building, Urban, Interior and Landscape Technology, 16: 7–22.

STO SE and Co. KGaA (2025). Lotusan® – Bionic Products. Available at: https://www.sto.de/s/inspiration-information/bionische-produkte/lotusan [Accessed 19 May 2025].

Tang, Z. Q., Tian, T., Molino, P. J., Skvortsov, A., Ruan, D., Ding, J. et al. (2024). Recent advances in superhydrophobic materials development for maritime applications. Advanced Science, 11(16): 2308152.

Tolikas, M., Antoniou, A., and Ingber, D. E. (2018). The Wyss institute: A new model for medical technology innovation and translation across the academic-industrial interface. Nature Biomedical Engineering, 2(5): 276–280.

Vonna, L. (2023). The Barthlott effect. Quantitative Plant Biology, 4, e16.

Wong, F. Y. M., and Mak, M. S. (2021, June 30). A study to compare the fouling resistance and self-cleaning properties of two-patterned surfaces with an un-patterned control surface. In 15th International Engineering and Computing Research Conference (EURECA 2021), Journal of Physics: Conference Series, 2120(1): 012015.

Wyss Institute for Biologically Inspired Engineering at Harvard University (2025). Wyss Institute for Biologically Inspired Engineering. Available at: https://wyss.harvard.edu/ [Accessed: 20 May 2025]

Xu, H. (2024). Antifouling and antimicrobial 3D printed small household appliances with incorporation of zwitterions and quaternary ammonium compounds. Applied Materials Today, 38: 102201.

Zhang, Y., Wu, D., Zhang, Y., Bian, Y., Wang, C., Li, J. et al. (2023). Femtosecond laser direct writing of functional stimulus-responsive structures and applications. International Journal of Extreme Manufacturing, 5.

Zheng, Y., Gao, X., and Jiang, L. (2007). Directional adhesion of superhydrophobic butterfly wings. Soft Matter, 3(2): 178–182.

5

Make it Smarter

"It takes something more than intelligence to act intelligently."

Fëdor Dostoevskij

"It has yet to be proven that intelligence has any survival value."

Arthur C. Clarke

5.1 Artificial Smartness vs. Natural Wit

Nature's ingenuity never ceases to amaze: only recently have we discovered that slime molds, such as *Physarum polycephalum*, are incredibly intelligent organisms, capable of solving problems, learning, and retaining information (Awad et al. 2023). These unicellular organisms (*plasmodia*), visible to the naked eye, can tackle complex challenges—such as solving mazes or traveling long distances in search of food—always choosing the most efficient path, despite having neither a brain nor a central nervous system. This demonstrates that *intelligence* can emerge even in very simple life forms, capable of learning from past experiences and recalling previously successful strategies. Their behavioral responses to environmental changes are such that *plasmodia* can be used as *biomachine*s, programmable via gradients of attractants and repellents to perform tasks in computational geometry, logic, and even arithmetic. Recently, Ehrlich, Hakimshafaei, and Elek (2022) leveraged a 3D simulation of *Physarum polycephalum* to design topologically optimal scaffolding structures for CAM 3D printing. Their approach is based on a *Monte Carlo algorithm* inspired by the foraging behavior of *Physarum polycephalum*.[1] As a case study, the

[1] The *Monte Carlo* method is a computational technique based on repeated random sampling to solve mathematical problems or simulate complex systems. It is named after the famous Monte Carlo casino, due to its association with the element of chance, much like gambling. Widely used across various fields—from statistical physics to economics, computational biology, and engineering design—this method is particularly effective when analytical solutions are too complex or impossible to obtain. In the case of *Physarum polycephalum*, the Monte Carlo algorithm simulates the organism's exploratory and adaptive behavior, modeling its optimized foraging strategy and translating it into a computational system capable of generating functional and high-performance structures.

DOI: 10.1201/9781003438946-5

researchers designed a biologically inspired bicycle helmet, using this computational simulation to generate a structure capable of effectively absorbing impact. A physical prototype of the helmet was also produced, demonstrating the potential practical applications of this methodology in the design of everyday objects. The project was presented during the demo session at the ACM Symposium on Computational Fabrication 2022, held in Seattle on October 27, 2022. The authors also propose further studies to evaluate and validate the effectiveness of the designs generated through this biomimetic method. Another original example of computational use in the CAD/CAM field—likewise inspired by slime mold *plasmodia*—is the Dreamcatcher® system by Autodesk®, a *generative design solutions in CAD system,* that uses algorithms to create optimized solutions based on user-defined parameters (AskNature 2016). The Dreamcatcher® lets users set goals and constraints, then uses optimization algorithms to explore countless design options and select the most efficient solution—much like slime-mold *plasmodia* navigate complex environments to find optimal paths to food (Nagy et al. 2017). By dynamically adapting to specifications (performance, strength, sustainability), Dreamcatcher® mirrors the emergent, *distributed intelligence* of slime molds, solving complex problems through simple rules rather than explicit programming (Wei et al. 2024). Slime molds and the Dreamcatcher® system both employ efficient resource-based optimization to tackle complex problems—one in a biological context, the other digitally. Recent advances in biology, ecology, life sciences, and biomimetics are fundamentally reshaping how we view Nature's creative force (Bensaude-Vincent 2019). Walking through a forest, we recognize each tree, plant, and blade of grass as the product of ingenious evolutionary problem-solving—a form of ancestral *biotechnological* wisdom. The growing field of plant neurobiology is uncovering this hidden natural intelligence, challenging our conventional understanding of cognition and adaptation (Mancuso 2023). Trees—long viewed as silent, motionless giants—actually participate in a vibrant underground *Wood Wide Web,* a silent and mysterious intelligence via fine *mycorrhizal fungal* networks, that sinks its roots deep into the soil and across geological eras. First characterized by Suzanne Simard and colleagues (1997), this subterranean system lets trees exchange nutrients and information, revealing a hidden web of communication and collective intelligence beneath the forest floor. *Mycorrhizal fungi* form vast underground networks—often spanning entire forests—that link plant roots and enable trees to share nutrients, warn one another of threats, and even bolster weaker individuals. Beyond this *bionetwork*, trees themselves sense and respond to their surroundings—detecting gravity to direct root growth, orienting leaves toward light for optimal photosynthesis, and remembering repeated physical stimuli to adapt their behavior (Gagliano 2017). The *Wood Wide Web* is a fascinating term and represents one of the most revolutionary discoveries in ecology and plant biology, demonstrating how plants are not isolated entities but part of a complex system of relationships and exchanges. From a biological perspective, *mycorrhizae* are classified as mutualistic symbioses between fungi and plant roots:

- ***Ectomycorrhizae***: Where the fungus surrounds the plant roots without penetrating the root cells.
- ***Endomycorrhizae*** (or arbuscular): Where the fungus penetrates the plant root cells, forming structures called arbuscules.

Mycorrhizal fungi colonize plant roots and extend fine hyphae through the soil, accessing nutrients (e.g., phosphorus, nitrogen) beyond the roots' reach. These nutrients travel back along the *hyphae* to the plant, which in return supplies the fungi with photosynthetic sugars (Simard et al. 2012). Beyond this mutual nutrient trade, these fungal networks also enable plants to *communicate*, sharing signals and resources—a phenomenon documented in detail by Suzanne Simard (2018) and other forest ecologists Plants under attack send chemical alerts through the Wood Wide Web, triggering neighbors to bolster their defenses and even share nutrients. Mycorrhizal fungi also boost seedling survival by supplying key nutrients and protecting against soil pathogens. The discovery of the *Wood Wide Web* not only reveals previously unknown aspects of plant *intelligence* but also has profound implications for our understanding of forest community dynamics. Recognizing that plants are part of an interconnected system radically changes the way we perceive and manage forests. This integrated strategy—combining sustainable forest management, biodiversity conservation, and ecosystem resilience—demonstrates how human and natural intelligences operate through distinct yet interconnected modes, as evidenced by *mycelial* networks that first evolved in the Neoproterozoic Era (Anderson 2020, Nace 2019, Heckman et al. 2001, Smith and Read 2008). Originating perhaps a billion years ago, these primordial *mycelia* drove nutrient cycling, soil formation, and early *mycorrhizal* symbioses, and today are harnessed in biodegradable biomaterials that promise eco-friendly replacements for plastics and conventional construction materials (Stamets 2005, Alima et al. 2021). Furthermore, this material is also being explored for innovative textiles and packaging, which could contribute to reducing polymers waste in our environment (HBBE, Hub for Biotechnology in the Built Environment).[2] One of the most interesting aspects is that mycelium composites can be cultivated through a circular-economy framework using agricultural waste, thereby reducing overall environmental impact. As seen, they also find interesting applications in the innovative CAD/CAM technologies of 3D printing biomaterials (3Dnatives 2024). Severi (2023) introduces MyCera®, a project by 3D WASP that explores the use of *mycelium* to reinforce 3D-printed structures. *Mycelium*-reinforced composites enhance the mechanical performance of 3D-printed materials while providing a sustainable, biodegradable alternative to conventional construction methods, aligning with the shift toward eco-friendly building practices. Mitchell Jones et al. (2020) review engineered *mycelium* composites from fungal biorefineries, underscoring their biodegradability, low energy demands, valorization of agricultural by-products, and the advances and challenges

[2] The *Hub for Biotechnology in the Built Environment (HBBE)* is a unique research centre held jointly between Northumbria University and Newcastle University. The HBBE intends to establish the world's first research Hub for Biotechnology in the Built Environment, aiming to create a new generation of *Living Buildings* which are responsive to their natural environment, grown using living engineered materials to reduce inefficient industrial construction processes, metabolise their own waste, reducing pollution, generate energy and high-value products and modulate their microbiome to benefit human health and wellbeing. The HBBE is funded through Research England's Expanding Excellence in England (E3) Scheme. The HBBE is organised around four core research themes: - Building Metabolism; Living Construction; Microbial Environments; Responsible Interactions.

in their application for construction. *Mycelium* acts as a negative-carbon biomaterial by sequestering more CO_2 than it emits. ARUP® and MOGI® leverage this property in the FORESTA system—acoustic panels made from *fungal mycelium* grown on textile-industry residues and framed with local beech wood. These panels are biodegradable, require low-energy to produce, sound-absorbing, and customizable in color and texture, embodying circular-economy principles by reducing waste and promoting material reuse (Adlington 2021, Mogu 2024). Architects, designers, and engineers are embracing carbon-sequestering materials to replace petroleum-based products. Leaders in construction, composites, and biomaterials partnered with Autodesk® researchers to create a carbon-negative façade for a California housing project. Swapping petroleum-based foam in fiberglass panels for *mycelium* composites reduced embodied carbon, while using Forma® and Fusion 360® accelerated design and installation. The study highlights the potential of novel biomaterials, methods for assessing embodied-carbon, and strategies to streamline façade assembly (Autodesk 2023).

Design Key Learnings:

- Discover how novel sustainable materials are being researched and developed to meet building code requirements.
- Learn about the leading building products for reducing a building envelope's total embodied carbon.
- Learn about analyzing embodied carbon data to make trade-off decisions with net zero building materials.
- Learn how you can improve sustainability, habitability, and construction performance by using novel building facades.

Plants, fungi, and roots not only fascinate ecologists, botanists, and life-science researchers but also inspire architects and designers using digital platforms and biomimicry—as demonstrated by the University of Westminster's Diploma Studio 10 *Fungal Incubator* project (WeWantToLearn.net 2023). Since 2011, University of Westminster professors Toby Burgess and Arthur Mamou-Mani have run WeWantToLearn.net and co-managed Diploma Studio 10, a biomimetic design platform that leverages parametric CAD/CAM tools—Grasshopper®, Kangaroo®, RhinoScript®, Ecotect®, Processing®—alongside robotics and 3D printing to generate complex, nature-inspired geometries. Every project begins by mapping a precise biomimetic concept (e.g., fungal and mycelial systems) into a generative CAD/CAM/CAE process, uniting experimental craftsmanship with cutting-edge computational design. Their workflow blends hand sketching, rapid prototyping, rendering, and DIY experiments with sustainable materials—organic plastics and plant-based fabrics—to produce modular structures like shelters and the *Living Fabric Tower* that embody natural forms and functions (Figs. 1-3).

It is highly recommended to consult Studio 10's work—particularly their aforementioned detailed design documents and project boards—which provide a fascinating and comprehensive illustration of how parametric CAD/CAM

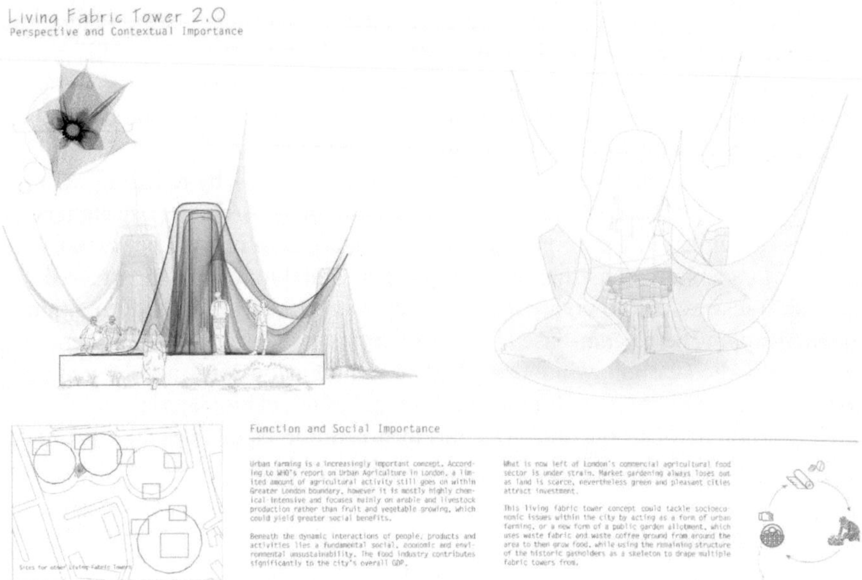

Fig. 1. Living Fabric Tower 2.0. Perspective and Contextual Importance.

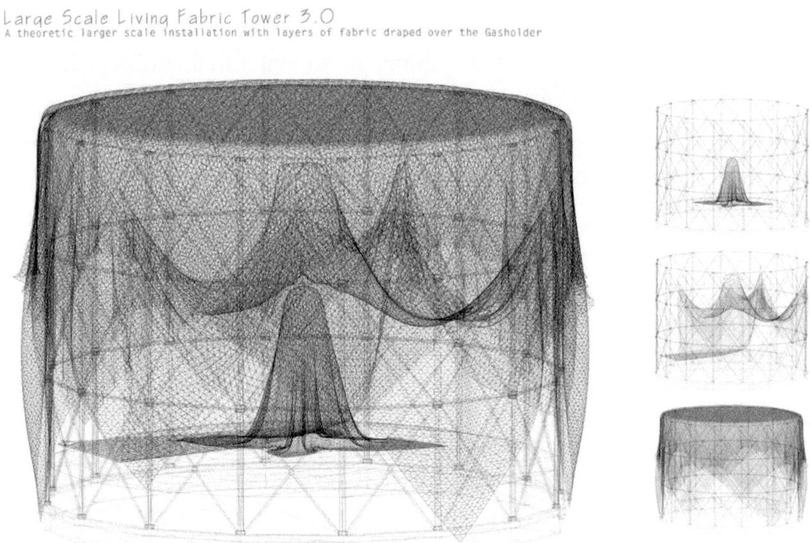

Fig. 2. Living Fabric Tower 3.0. Theoretical Large-scale Installation

Fig. 3. Living Fabric Tower 2.0 Digital Rendering of Installation.

computational tools are used to develop the supply chain for this complex experimental eco-biodesign process:[3]

"How do natural structures and organisms interact with their environment? We are seeking an architectural language that relates to the natural world, rather than remaining separate from it, by designing a modular, performative urban artifact that brings living nature into the city. The artifact will be highly site-specific, half human-made and half grown by nature. Fungi absorb nutrients through vast underground systems of white filaments called *mycelium*. Although hidden in the soil and sometimes confused with roots, *mycelium* is actually the true body of a fungus. Fungi fruit only when conditions are favorable for the spread of their spores. *Mycelium* plays a key role in decomposing plant matter, but it can also form a symbiotic relationship with the roots of some plants, called *mycorrhiza*. Most plants rely on *mycorrhiza* to absorb phosphorus and other nutrients. In return, fungi have continuous access to the carbohydrates of plants. Often, neither the fungus nor the plant will grow without a mycorrhizal partner" (WeWantToLearn.net 2021).

Bio-inspired experimental designs of this caliber demonstrate that although humans have developed conceptual and technological tools to understand and transform the world, nature has—over billions of years of evolution—refined adaptation and optimization strategies that often surpass human ingenuity in efficiency, sustainability,

[3] For an in-depth analysis of the *WeWantToLearn* project — including objectives, methodology, and preliminary results — see paragraph 2.03, titled *A Virtuous Ring*, which outlines the development phases, key collaborations, and future prospects.

and self-organization, and that attention to biomimetic processes in design can make a real difference.

In summary, this leads us to compare two distinct paradigms:

- **Artificial Smartness**, or computational and algorithmic intelligence, which relies on processing large amounts of data, machine learning, and predictive optimization typical of human-made systems.
- **Natural Wit**, or adaptive and biological intelligence, which is not programmed but emerges from the very properties of living matter and its ability to evolve through self-organization mechanisms and responses to stimuli.

In this context, human intelligence stands out for its capacity for abstraction, prediction, and modeling, expressed through:

- **Language and symbolism**, which enable the transmission of knowledge and the creation of complex systems (mathematics, art, science).
- **Design and technology**, through which humans extend their capabilities and modify the environment to meet their needs.
- **Creativity and problem-solving**, which allow the generation of innovative solutions even in unexpected situations.

Meanwhile, natural design is characterized by:

- **Self-repair and regeneration**, as many living systems can restore functionality and structural integrity after damage or stress.
- **Hierarchical and modular organization**, with structures composed of repeated modules (from cells to tissues to organs), providing high flexibility and adaptability without compromising the entire system.
- **Resource-use efficiency and waste minimization**, since biological processes recycle materials and make the most of available energy, minimizing losses and by-products.

Human intelligence often hits cognitive, cultural, and resource-use limits, struggling to embed truly long-term adaptive solutions. In contrast, Nature's *intelligence* unfolds as recurring geometric and spatial patterns—spirals, meanders, fractals, the golden ratio—across scales from microscopic to macroscopic. Far from mere aesthetics, these forms are the product of evolutionary optimization, embedding deep order and efficiency into every aspect of living systems. Nature's geometrical patterns transcends human conceptual and rational logic, embodying one of the deep archetypal forms of the universe's intelligence—and it's no accident that, in vernacular traditions, geometry has always been regarded as *Sacred Geometry* (Schneider 2025). The *Fibonacci spiral*—derived from the *Fibonacci sequence*—recurs in plant leaf arrangements, shell shapes, and even galactic structures, offering an optimal solution for packing and light capture. Likewise, the *Golden Ratio* (or Divine Proportion) governs perfect natural forms like the *Golden Spiral*, which guides both botanical growth and celestial formations (Pietroluongo 2022). Many fruits and flowers, such as sunflowers, arrange their seeds following this geometry, maximizing the efficiency of their development. Fractals, with their self-similar nature at any scale, are another form of natural geometric intelligence, and Soo and Yu (2003) apply rapid prototyping to designs characterized by self-similarity, using the CAD process to optimize the creation of complex, self-similar models in advanced materials contexts, representing one of the earliest experiments in generative and parametric bio-computation. Fractal patterns—from tree branching to river networks—demonstrate Nature's use of simple, self-similar rules to model complexity and enable efficient growth and resource distribution. River meanders likewise emerge from an optimization between erosion, sedimentation, and flow resistance, minimizing erosion and regulating water speed. These recurring geometries (spirals, fractals, meanders) reveal that what appears random in nature actually follows precise mathematical laws that optimize resource use, reduce *errors*, and maximize efficiency—all the while producing forms of striking beauty. Viewed this way, natural systems perform a kind of innate computation, embedding problem-solving directly into their structure rather than relying on abstract symbols or conscious calculation. Bio-inspired engineering aims to harness these geometric and mathematical principles—emulating Nature's optimization strategies to create technologies that are efficient and in harmony with the environments in which they operate. In addition to those already highlighted, here are other significant aspects that deeply distinguish natural technological intelligence from human intelligence. They are the intrinsic capabilities of its systems—both organic and inorganic—to:

- **Self-organize** without an apparent control center, as seen in protein folding at the nano and micro scales, but also in insect swarms, bacterial colonies, or mycelial networks at the ecosystem scale.
- **Create emerging biochemical hierarchies** that organize materials and organs with exceptional technical and functional performances, such as fibers, filaments, calcium shells, tendons, or bones, elaborating basic elemental materials available in their vital environment (for example, oceanic waters).
- **Adapt dynamically to environmental stimuli**, developing survival strategies that minimize waste of matter and energy, and maximize functional efficiency, building vast self-regulating ecosystem networks.
- **Evolve through natural selection**, a process that finds functional solutions without an evident manifestation of intentional and centralized design.

Biological organisms and ecosystems are examples of what we could define as **distributed smartness**, where complexity emerges from the interaction between simple components, without a central command unit. *Intelligence*, in the context of design, is often associated with computational processing capacity, the precision of algorithms, and optimization based on data and simulations (Yang et al. 2023). This perspective broadens intelligence beyond computation to include matter's innate ability to adapt, transform, and respond to its surroundings. Traditional design focuses on foreseeing and solving specific problems, but by embracing adaptability and real-time interaction with context, the line between computational and bio-inspired design fades. The pivotal question ahead is whether we'll advance smart solutions primarily through ever-more powerful algorithms or by drawing deeper lessons from nature's own material-based technologies.

Today, the most advanced engineering design is moving by seeking intersections and connections between these two complementary directions:

Computational Intelligence:

- **CAD, CAE, CAM, and AI** are transforming design, enabling predictive simulations that optimize structures and functions.
- **Deep learning algorithms** analyze data and generate new solutions, improving design in real time.
- **Generative design** allows exploring thousands of variants, finding optimal configurations based on defined constraints.

***Intelligence* of Matter:**

- **Bio-inspired materials**, such as self-healing ones, are changing the concept of maintenance and longevity over time.
- **Reactive surfaces and biomaterials** can adapt to environmental conditions without the need for computational inputs.
- The use of **molecularly programmable materials** paves the way for designs incorporating emergent and self-organizing capabilities.

Most likely, the evolution of design will not proceed exclusively towards greater dependence on *Evolutionary Algorithms (EA)*, nor will it simply imitate biological processes (Evolv AI 2019). The real challenge lies in integrating these approaches: a design that not only calculates but also understands, and a material that not only reacts but also predicts and optimizes—concepts anticipated by Renner and Ekárt (2003) and Case et al. (2003). In this vision of an unprecedented *olistic technology*, the designer is no longer just an engineer of algorithms or a manipulator of matter, but becomes a holistic mediator between *computational intelligence* and *adaptive intelligence*, between the digital and the biological (Wright and McLaughlin 2023). Thus, the design of the future will not only be a matter of computing power and speed but of an intimate and fruitful dialogue with the matrices of the structural complexity of the material and natural world.

"Smart is one of the buzz words of the digital age, and in this context, refers to connected, data-driven, tech that is sensor-based, and increasingly autonomous, making greater use of artificial intelligence. Everyone seems to understand and agree that the new world that we are creating will be *smarter* in this sense. And yet, many of us aren't really aware of the character and consequences (both good and bad) of this new and smarter world" (Vermeulen 2018).

The term is also associated with predictive and analytical capabilities, typical of technologies that leverage advanced artificial intelligence algorithms and machine learning to optimize processes or predict behaviors, as seen in smart analytics or smart grids. In the context of contemporary urban design, the integration of Computer-Aided Design (CAD), Building Information Modelling (BIM), and Geographic Information Systems (GIS) is leading to an epistemological and operational evolution of the city itself, now viewed not merely as a social and spatial structure, but as a *high-tech synthetic organism,* governable through predictive virtual models. These tools no longer just represent the city: they model it as a dynamic and simulable system, where every element—from material composition to energy flows, individual behaviors to environmental policies—can be encoded, visualized, and predicted. BIM, in particular, functions as a sort of *digital twin, a doppelgänger* built environment, enabling continuous management of buildings and infrastructure life cycles. In an advanced smart city, this model extends to include urban-scale data, integrated with real-time information from IoT sensors and dynamic GIS flows. In the history of urban design, there has always been a need to build scaled models of cities to understand their spaces and functions. This practice dates back to Renaissance humanist modeling and even to the analog-CAD-like precision of Leonardo da Vinci's drawings of ideal cities. In philosophical and theoretical culture, the notion of the "ideal city" already anticipates today's Building Information Modeling (BIM): both seek a holistic representation that unites form, function, and context. Utopian thought manifests itself in the geometric perfection of forms and proportions designed to regulate the spontaneous irrationality and ungovernability of human behavior—as seen in the 18th- and 19th-century ideal cities:

- **Saline de Chaux** (Claude-Nicolas Ledoux, 1775): A rigorously symmetrical saltworks complex whose architecture reflects an almost mathematical order.
- **Phalanstères** (Charles Fourier, 1820s): Modular, self-contained communal buildings conceived to harmonize work, leisure, and social relations.
- **New Harmony** (Robert Owen, 1825): A grid-based town in Indiana, laid out to foster cooperation, equitable land use, and moral improvement.

While Renaissance maquettes and Leonardo's meticulous plans provided a visual and spatial framework for urban ideals, modern BIM goes further by embedding geometric, material, functional, and temporal data into a single digital model. BIM enables architects, engineers, and planners to:

1. **Visualize and simulate** urban dynamics (traffic flow, daylighting, energy use) before construction
2. **Coordinate** multidisciplinary inputs in real time, reducing clashes and errors
3. **Manage** a building's entire lifecycle—from conception through operation—within the same dataset
4. **Engage** stakeholders through interactive 3D environments and data dashboards.

In this way, BIM extends the Renaissance and utopian ambitions of comprehensively understanding and perfecting the city, transforming analog representations into a *living digital twin* that informs more resilient, efficient, and human-centered urban environments. From here emerges a theoretical hypothesis: the city is treated as an ultimate, an *evolving synthetic material*, a complex *in vitro* manmade artifact that, through its virtual representation, can be pre-understood, controlled, and manipulated, much like what happens in the field of advanced materials science. In the design of synthetic materials, CAD/CAM simulations and machine learning are used to predict behaviors, deformations, and structural responses, aiming to produce materials that can adapt to environmental conditions and withstand external stresses. Similarly, in urban design, generative algorithms, parametric models, and artificial intelligence seek to anticipate the evolution of the city and make controllable and manipulable complex phenomena such as traffic, energy consumption, and resource management. This ambition raises critical questions:

- How far is it possible to model the unpredictability of social phenomena?
- Is it ethically and culturally legitimate to treat an urban environment as a programmable organism?
- What are the epistemological limits of the predictive approach when applied to emerging complex systems like cities?

Indeed, the risk of a technocentric approach might lead to a reductive view of urbanity, where the complexity of the city is reduced to a series of predictive models and algorithms that exclude the richness of nonlinear interactions and social dynamics. As scholars like Antoine Picon (2015) and Bernadette Bensaude-Vincent (2019) argue, hyper-rationality in the use of technologies risks overshadowing the more human dimension of the city, which is made of social diversity, unpredictable changes, and ecological resilience. In this framework, the *smart city* becomes a computational *WunderCAD/CAMmer*, a digitalized *chamber of wonders*, where the illusion of total control and predictive accuracy may obscure what makes the urban ecosystem vital: its capacity for nonlinear adaptation, creative divergence, and unpredictable resilience. Some of the most advanced smart cities, such as Singapore and Amsterdam, already use CAD, CAE, CAO, BIM and IoT systems for the dynamic management of urban flows, integrating real-time data to optimize mobility, water resource management, and security (TAAL Tech 2025). These models are used to predict the future behavior of the city and to plan urban expansion sustainably. However, as critics point out, these digital cities may overlook the social dynamics that escape modeling, such as local conflicts of interest or social inequalities. Another example is the *London Digital Twin project*, a virtual replica of the city that collects data from IoT sensors and uses it to simulate various scenarios, such as climate change or traffic evolution (Middlesex University 2025). This methodology enables detailed forecasting of urban development outcomes but also challenges the credibility of models that aim to govern every facet of city life. To address these concerns, smart-city strategies must blend technological precision with the adaptability and resilience intrinsic to living systems—embracing uncertainty, fostering creativity, and supporting civic engagement. The pervasive deployment of wireless networks, IoT devices, and powerful algorithms promises enhanced efficiency in private, public, and commercial services (Chipilska 2024), yet raises legitimate privacy and data-security issues as cities increasingly rely on continuous data streams. Ultimately, 3D digital twins, BIM workflows, and smart-city visions are tightly coupled and represent an urgent pathway for urban innovation (GimmiGIS 2014, Philipsen 2013).

- **Simulation and planning:** They allow decision-makers to preview traffic flows, energy consumption, and infrastructure layouts, optimizing efficiency, sustainability, and livability through CAD and BIM, which integrate building lifecycle data.

- **Real-time monitoring:** Leveraging sensors and IoT, virtual cities ingest data on air quality, water networks, and power grids, enabling dynamic adjustments and data-driven decisions powered by AI and machine learning.
- **Resource optimization:** By simulating energy distribution, water systems, and waste management, authorities can test low-impact strategies before real-world deployment.
- **Testing innovative technologies:** In a controlled environment, autonomous vehicles, smart grids, and renewable energy systems can be tested risk-free using detailed prototypes created with CAD and BIM.
- **Community engagement:** Interactive 3D models let citizens explore proposed urban changes, give feedback, and participate actively in decision-making, strengthening transparency and trust between governments and residents.

Virtual cities use IoT, AI, and related technologies to create digital twins that optimize energy, traffic, and infrastructure by leveraging BIM's data-rich models (Sharma 2025). These platforms offer crucial insights for planning, testing, and refining urban systems—helping cities address population growth, climate change, and rapid technological shifts (Abugu 2025). Success depends on: (1) accurately mirroring real-world dynamics, (2) integrating emerging technologies, and (3) fostering collaboration among governments, businesses, and residents. Beyond emissions reduction, smart cities embrace advanced transport networks, smarter water and waste systems, intelligent building controls, interactive governance, and services for aging populations. The global smart-city market is projected to reach $300 billion by 2032, driven by a 27.1% annual growth in smart energy (from $122.2 billion in 2020 to $652 billion by 2026) (Pvcase Team 2024). Common initiatives include adaptive street lighting, energy-storage solutions for peak loads, and smart grids that integrate renewables. Yet, the exaltation of hyperconnectivity—a "connectivity mystique"—can override other social and ethical values, raising privacy and well-being concerns (Fernández 2014, Settembre 2012).

In the contemporary debate on the future of cities, two visions stand in contrast: on one hand, the Smart City, focused on connectivity, data, and automation; on the other, the notion of a more organic city inspired by the mechanisms of nature. In this context, the concept of the Bionic City shifts attention from predictive and control technologies to processes of self-regeneration and adaptation intrinsic to living organisms. Bridging these two worlds is the work of Melissa Sterry, who offers a model of a city capable of integrating biological resilience with advanced urban design. Melissa Sterry, designer and researcher, in her article *The Bionic City: A natural blueprint for future cities* (2010), explores how the principles of self-repair, modularity, and optimization typical of living systems can inspire urban design. The Bionic City relies on biomimetic structures able to regenerate and adapt, emphasizing resilience and active interaction with the natural environment. In contrast to the Smart City, which prioritizes connectivity, data, and automation to maximize efficiency and control, the bionic approach centers on flexibility, organic sustainability, and the capacity for self-repair. While the Smart City seeks to manage every aspect of urban life through sensors and algorithms, the Bionic City aims to integrate the intelligence inherent in natural systems, fostering a more organic urban evolution less bound by technology:

"The bionic city provides a natural blueprint for future urban environments, capable of self-repair and dynamic adaptation through biomimetic design" (Sterry 2010).

Table 1 presents a comparative overview of Smart City vs. Bionic City approaches.

5.2 Toward a Conscious Materiality (I)

Material smartness is linked to the optimization of the dialogue between CAD/CAM and CAE/CAO operating systems, and it is a process considered an essential standard for improving the design and production workflows in contemporary industries (Kreis et al. 2024). The fusion of computer-aided systems with advanced AI—such as Generative Design in Autodesk Fusion 360®—is transforming engineering, architecture, and industrial design by enabling real-time simulation, parametric optimization, and intelligent data management; for instance, Fusion 360®'s AI algorithms explore millions of material- and process-aware design permutations to auto-generate structurally and manufacturably optimized shapes (Autodesk 2025). AI-driven CAD/CAE workflows—such as Autodesk Fusion 360®'s Generative Design—automatically propose and refine optimized geometries (e.g., lightweight yet robust automotive components) via real-time simulation (Autodesk 2025). Meanwhile, Rhino® + Grasshopper®—enhanced with Ladybug Tools® and Honeybee®—supports environmentally conscious design by integrating real weather data into a parametric process to optimize building orientation, façade configuration, and overall energy performance in high-efficiency architecture (Sadeghipour Roudsari and Pak 2013). Parametric CAD/CAM tools are revolutionizing multiple industries: in STEM educational robotics, the VEX Robotics® protocol (e.g., VEXcode VR®) enables fully parametric robot design with real-time CAM simulation of physical behavior (VEX Robotics 2025); in soft industrial robotics, Autodesk Inventor® models elastic and piezoelectric material responses under stress to refine component geometries; and in automotive engineering, Siemens NX®'s unified CAD/CAE/CAM workflow—including CFD (Computational Fluid Dynamics), structural simulation, and CAM optimization—produces engine parts capable of withstanding extreme thermal and mechanical loads (Siemens Software 2025). Finally, Materialise Magics® supports advanced 3D-print data preparation paired with parametric design to minimize defects and boost efficiency, a capability critical for creating micrometric aerospace parts and functional prototypes (Materialise 2025). COMSOL Multiphysics® enables advanced bionic-design simulations that integrate dynamic material behavior—particularly in soft robotics for piezoelectric devices and biocompatible hydrogels—by analyzing responses to stimuli such as pressure and humidity, thus allowing designers to predict real-world performance and optimize functionality and longevity (Bionic Surface Technologies GmbH 2025, Banerjee et al. 2018). By fusing CAD/CAM/CAE/CAO systems with these smart-material protocols, industries from manufacturing and architecture to robotics and medicine can create responsive, optimized designs that leverage materials capable of self-organization and adaptive responses—suggesting

Table 1. Comparative Overview of Smart City vs. Bionic City Approaches .

Aspect	Smart City	Bionic City
Primary Focus	Connectivity, data collection, automation	Biomimetic processes, self-repair, organic adaptability
Core Technology	IoT sensors, big data analytics, AI/ML algorithms	Shape-memory materials, neuromorphic systems, living materials
Adaptability	Predictive optimization based on real-time data	Innate self-organization and regeneration in response to environmental stimuli
Resilience	Redundancy through digital backups and fail-safe protocols	Structural and functional self-repair mechanisms derived from living systems
Design Principle	Modular digital twins (BIM), rule-based control systems	Hierarchical, modular structures mirroring cellular to organism scales
Governance Model	Centralized monitoring platforms, data-driven decision-making	Distributed, participatory processes integrating community and ecological feedback
Resource Use	Dynamic resource allocation via smart grids and adaptive lighting	Closed-loop material cycles, waste-to-resource strategies
Privacy & Ethics	High data collection raises privacy and surveillance concerns	Emphasizes natural cycles and minimal intrusive monitoring
Performance Metrics	KPIs like energy saved, traffic flow improvements, reduced emissions	Indicators of ecosystem health, self-repair rate, biodiversity integration
Illustrative Example	Barcelona's smart traffic management using sensor networks	Living façade buildings that heal minor damage via microbial or plant-based coatings

a nascent form of awareness (Kantorovitch 2019). This technological convergence, including additive manufacturing and transient electronics, supports a transition from linear consumption to a circular economy, illustrating how sustainability and advanced smart functionality can reinforce one another. With the continuous development of AI, bioengineering, and material science, we find ourselves in an era where the boundaries between intelligence and matter are becoming increasingly blurred. While AI will likely never achieve true human consciousness, our ability to design and simulate adaptive, learning materials brings us closer to understanding the deep connections between mind, matter, and nature. Perhaps the consciousness of the material world does not reside in silicon circuits, but in the principles of self-organization that govern both living and inanimate systems. An awareness that, if fully understood, could radically change our perception of the universe and our role within it. Biology provides us with extraordinary examples, such as the molecular mechanisms within cells, from self-folding proteins to motor proteins like kinesin, which operate with precision and efficiency that suggests some form of intelligence and intrinsic awareness capable of creating order and functionality from a primordial soup of chaotic and random phenomena (Fan and Lai 2022). Biomimetic bioengineering takes inspiration from these phenomena to develop materials and systems capable of learning and adapting over time. However, despite the progress, artificial intelligence remains fundamentally different from biological intelligence (Shevlin et al. 2019). While AI can predict and simulate the behavior of materials with remarkable accuracy, its decisions do not derive from subjective experience. AI can enhance the functional adaptability of materials, but it will struggle to give them true self-awareness. The natural world operates with a level of organization and response that often appears indistinguishable from conscious intelligence. It has been repeatedly emphasized that plants communicate through chemical signals, fungi create vast mycelial networks to regulate ecosystems, and even inanimate systems like fluid dynamics and crystal growth exhibit principles of self-organization. From this perspective, consciousness may not be exclusive to the human brain, but a phenomenon distributed in different ways throughout Nature. This idea challenges our anthropocentric view of intelligence and urges us to consider a broader definition of awareness, one that extends far beyond our species to encompass the entire material world. With the continued development of AI, bioengineering, and material science, we are entering an era in which the boundaries between intelligence and matter are becoming increasingly fluid. An exploration of embodied intelligence and its implications points toward a theory of intelligence in general; with case studies of intelligent systems in ubiquitous computing, business and management, human memory, and robotics. Science asks how the body could influence our thinking when it seems obvious that the brain controls the body. In *How the Body Shapes the Way We Think*, Rolf Pfeifer and Josh Bongard demonstrate that thought is not independent of the body but is tightly constrained, and at the same time enabled, by it (Pfeifer and Bongard 2006). They argue that the kinds of thoughts we are capable of have their foundation in our embodiment—in our morphology and the material properties of our bodies. This crucial notion of embodiment underlies fundamental changes in the field of artificial intelligence over the past two decades, and Pfeifer and Bongard use the basic methodology of artificial intelligence—*understanding by building*—to describe their insights. If we understand how to design and build intelligent systems, they reason, we will better understand intelligence in general. The industrial sector of so-called smart materials has seen rapid development in recent decades, with applications ranging from robotics to medicine, from construction to biosensing. These materials respond dynamically and adaptively to a variety of stimuli, such as light, temperature, pressure, and electrical, chemical, or acoustic forces. Thanks to advances in digital design and manufacturing technologies—particularly through CAD/CAM—it is now possible to model and customize materials in a precise and versatile way, leading to innovations that often aim to mimic natural phenomena. Here is a compendium of the main artificial smart materials and their applications, integrated with references to the use of CAD/CAM technologies to optimize their design and production, along with functional parallels to the performance offered by natural smart materials:

- **Response to Pressure and Vibrations**

Piezoelectric polymers and elastomeric compounds—used in sensors, actuators, and monitoring devices—change their properties under pressure or vibration (Tazwar et al. 2025). CAD/CAM techniques enable designing complex geometries and optimizing force distribution to calibrate sensors for precise pressure thresholds. In nature, the *thigmonastic* folding of *Mimosa pudica's* leaves illustrates pressure-driven movement (Félix and Costa 2021), while the Venus flytrap (*Dionaea muscipula*) snaps shut via action-potential–triggered lobes. Inspired by these mechanisms, materials science aims to replicate such rapid, predictable mechanical responses in synthetic surfaces for enhanced environmental sensitivity.

- **Response to Electric Impulses**

Materials such as EAPs (Electroactive Polymers), conductive polymers, and shape memory materials respond to electrical impulses by changing their structure (ScienceDirect 2018).

EAPs are generally divided into two main categories:

- **Electromechanical (dielectric) EAPs**, which operate through electrostatic forces and require high voltages to activate. *Examples:* Dielectric Elastomer Actuators (DEA).
- **Electrochemical EAPs**, which rely on electrically induced chemical reactions and operate at lower voltages. *Examples:* Conductive polymers such as polyaniline and PEDOT.

Application Examples:

- **Micromanipulation and Robotics:** Worm-inspired soft robots enable adaptable pipeline and tunnel inspection (Liu et al. 2021).
- **Prosthetics and artificial muscles:** DEA (Dielectric Elastomer Actuators) are used to build bioinspired actuators *in artificial neuromuscular prostheses* (Bruschi et al. 2021).
- **Soft robotics:** Harvard's *Octobot* uses EAPs to move without rigid components (Wyss Institute for Biologically Inspired Engineering 2016).
- **Tactile interfaces:** EAPs are integrated into haptic gloves for virtual reality.
- **Medical devices:** EAP-based sensors can monitor physiological parameters with high sensitivity.

CAD/CAM enables the creation of devices with complex geometries and ultra-precise motion control, boosting their performance and lifespan, mirroring how electrical impulses drive neurons and muscle fibers—and how electric fish like *Electrophorus electricus* and *Torpedo torpedo* generate fields for communication and hunting (Neveln et al. 2013).

- **Electroactive Polymers (EAPs)**

Active electropolymers deform under an electric field—ranging from slight bending to complex movements—and are employed in actuators, sensors, soft robotics, haptic devices, and smart prosthetics, with CAD/CAM processing, enabling the design of intricately shaped, electricity-activated structures (Bar-Cohen and Anderson 2019). Inspired by cephalopod *muscles* that morph for hunting and defense, these biomimetic materials promise breakthroughs across industrial design, medicine, and renewable energy. Integrating metamaterials into CAD/CAM, supported by advanced simulation and parametric design, heralds an era of intelligent, adaptable manufacturing driven by the convergence of biology, engineering, and AI (Engel et al. 2021).

- **Response to Light**

Photoresponsive or phototropic materials change their structure or color in response to light and are used in displays and solar shading to optimize energy use. CAD/CAM tools allow the design of photonic materials with tailored geometries that selectively respond to specific wavelengths, boosting the efficiency of optical and photovoltaic systems. In nature, phototropism is seen in sunflowers and other plants, orienting toward the sun—an effect now digitally replicated in renewable energy applications.

- **Photonic crystal (PC) materials**

This class of materials—inspired by structural colors found in natural organisms, such as those on butterfly wings—is gaining increasing attention due to advanced applications in biomedicine. When integrated with smart materials or developed through advanced fabrication techniques, these materials offer significant advantages in fields such as biosensing, bio-probing, biological screening, and tissue engineering. Their ability to manipulate light at the nanoscale makes them extremely promising tools for developing new diagnostic and therapeutic technologies (Xu et al. 2021).

- **Photochromic materials**

Photochromic materials change color in response to UV light. This characteristic is utilized in applications such as self-adjusting sunglasses, adaptive screen materials, and adaptive optical devices. CAD/CAM and 3D printing technologies allow the design of materials that dynamically respond to lighting conditions, enhancing the interaction between the user and the device. A 2020 study used tungsten oxide (WO_3) nanoparticles as reversible photothermal sensitizers (Powell et al. 2020). These nanoparticles, transparent under normal conditions, become highly absorbing in the infrared when exposed to UV light. By incorporating them into polyamide powders, it was possible to produce 3D-printed objects with high color purity and controllable color changes, demonstrating the potential of photochromic materials in advanced manufacturing. Photonic components that respond to light are generating great interest because they allow for rapid, precise, and remote control of a material's properties. New 3D printing technologies offer advanced possibilities, but printing photoactive materials remains a challenge: it is difficult to preserve their optical properties and minimize light losses. Szukalski et al. (2019) present advanced 3D printing techniques capable of embedding photo-reactive compounds into optical devices, which can be activated by UV or visible light, with response times between 1 and 10 seconds. A natural parallel can be found in marine isopods, which are able to change color in response to ambient light variations, thereby adapting to their surroundings (Hultgren and Mittelstaedt 2015).

- **Shape Memory**

Shape Memory Alloys (SMA) and Shape Memory Polymers (SMP) are smart materials capable of recovering their original shape after being deformed, thanks to a crystalline phase transition triggered by temperature or mechanical stress. These materials are used in medical applications such as self-healing sutures and soft robotics. With the integration of CAD/CAM, it is possible to design objects and devices that can deform in response to environmental changes and then restore their original shape (Khalid et al. 2024). A parallel can be drawn in the natural world with jellyfish, which contract or expand in response to environmental stimuli, similar to the behavior of shape-memory materials designed to adapt to external conditions (Machado et al. 2023).

- **Self-Healing**
Self-healing materials can autonomously repair damage. Applications in self-healing coatings, protective layers for electronic devices, and construction materials are becoming increasingly important. CAD/CAM enables the creation of precise models that allow materials to heal specific areas without compromising overall functionality. A natural example of this phenomenon can be seen in organisms like cacti and certain plants that regenerate damaged surfaces to survive harsh conditions (Mylo et al. 2020). A natural parallel can also be found in the skin repair process of some amphibians, such as frogs, which are able to rapidly heal damaged skin through cellular regeneration (Liu et al. 2023).

- **Response to Temperature**
Thermosensitive materials, which expand or contract in response to temperature changes, are used in thermal control devices, smart textiles, and construction materials. The use of CAD/CAM technologies enables the design of devices capable of managing and optimizing their response to thermal variations. A natural example is the duckweed (*Lemna minor*), which changes its structure to adapt to temperature fluctuations—a process replicated in various artificial materials through computational design (Sender and Grzywna 2025).

- **Thermochromic Materials**
Thermochromic Materials (TCMs) change color depending on temperature. They are used in temperature labels, environmental condition monitoring devices, and materials for interactive design. TCMs are of significant interest due to their potential in energy-efficient buildings, textiles, thermal storage, sensor technologies, and antique maintenance (Nawade et al. 2020). These materials are categorized into *inorganic, organic, polymeric*, and *hybrid systems*, based on their properties. TCMs are synthesized using various techniques, often combined to improve yield, stability, and efficiency. However, exposure to external factors like UV radiation, temperature, pressure, and humidity can degrade their performance. Microencapsulation and coating methods are used to protect TCMs and enhance their thermochromic properties. The development of microencapsulated TCMs with nanostructural features shows promise for future applications in inks, paints, and coatings for sustainable energy efficiency and beyond. Hakami et al. (2022) provide a thorough review of thermochromic materials, detailing their development processes, characterization methods, and diverse applications across fields such as smart coatings, temperature sensing, and adaptive building technologies. Thermochromic membranes fabricated via extrusion-based 3D printing—combining a biocompatible polymer with thermochromic pigments—demonstrate rapid, reversible color change at fever-relevant temperatures, showcasing how CAD/CAM-enabled additive

manufacturing can produce stimuli-responsive devices for medical diagnostics (Jeon et al. 2021). An interesting example of natural thermochromism is provided by the chameleon, which can change the color of its skin not only for defensive purposes but also for thermoregulation. From the study of the chameleon's skin, Wu et al. (2021) drew inspiration to create a photonic structure. Using ZnS nanospheres and polymers, they designed non-close-packed photonic crystals through a two-step technique. The resulting structure is highly sensitive to external stimuli, similar to the chameleon's skin, due to the refractive contrast and its structure. This structure showed a color change (blue-shift) when exposed to temperature changes or compression, demonstrating ultrasensitivity useful for many applications.

- **Response to Humidity**

Some polymers and composite materials are sensitive to humidity, changing their mechanical and thermal properties accordingly. These materials are used in environmental monitoring devices and adaptive textiles. With the aid of parametric and generative computational processes, it is possible to design humidity control systems that respond to varying levels of exposure, as in certain specialized devices that are 3D-printed and capable of accumulating different levels of airborne humidity while maintaining proper environmental comfort standards for both humans and plants (Daneluzzo et al. 2023). Some carnivorous plants and mosses exhibit a similar dynamic response to humidity, changing their morphology according to water availability. In *Aldrovanda vesiculosa* (an aquatic carnivorous plant), an ultra-fast snap mechanism functions similarly to that of *Dionaea muscipula* (Venus Flytrap), but is based on a rapid variation in cell turgor pressure. The trap movement is influenced by humidity and the surface tension of the surrounding water (Biomimetics in Botanical Gardens—Educational Trails and Guided Tours 2023). In *Drosera capensis* (round-leaved sundew), its mucus-secreting glands move slowly in response to humidity and the presence of prey, gradually closing around the insect (Qiu et al. 2021). *Funaria hygrometrica* (hygroscopic moss) has capsules (*sporophytes*) that open and close in response to humidity, optimally regulating spore dispersal (Reyssat and Mahadevan 2009). The biomimetic inspiration drawn from the study of plant species that exhibit humidity-sensitive structures in terms of bio-evolutionary functionality (such as the scales of pine or larch cones that open in dry air to release seeds and close with humidity to protect them) was the foundation that inspired German architect Achim Menges, in collaboration with Oliver David Krieg and Steffen Reichert from the Institute for Computational Design—University of Stuttgart—to design and construct, through an advanced CAD/CAM system and robotic digital fabrication, the experimental meteorosensitive pavilion named *HygroSkin (ArchDaily 2013, Krieg et al. 2014). This* design "explores a novel approach to climate-responsive architecture," particularly "as the computational process integrates the material's capacity to physically compute form in the elastic bending process, the cumulative structure of the resulting building components, the computational detailing of all joints, and the generation of the required machine code for fabrication with a 7-axis industrial robot" *(Menges 2012). The plant Tortula ruralis* (Resurrection moss) is capable of completely drying out and then resuming photosynthetic activity when moisture levels increase, just like *Selaginella lepidophylla* (resurrection plant), which curls into a compact ball under arid conditions and unfurls to reactivate photosynthesis when humidity returns (Rafsanjani et al. 2015). Although better known for its response to mechanical touch, the leaves of *Mimosa pudica* (sensitive plant) can also react to changes in humidity by folding up to prevent excessive water loss. Finally, in the case of *Avena sterilis* (rain grass), the seeds of these grasses possess hygroscopic appendages that twist and move in response to humidity changes, helping the plant to self-bury into the soil *(Fiorello et al. 2024)*.

- **Self-Assembly**
Self-assembly is an inherently smart mechanism by which materials autonomously organize into complex, emergent structures, as demonstrated in a building-block 3D printing process based on molecular self-assembly monolayers with self-healing properties (Hamoudi et al. 2022). Technologies such as nanotechnology and polymer self-assembly are used in advanced electronic devices and biosensors. The CAD/CAM approach, integrated with a range of computational biology tools, enables the molecular-level design of materials—optimizing their performance and bridging the gap between nanoscale organization and macroscale fabrication (Requicha and Arbuckle 2006, Sola et al. 2023). Organic–inorganic self-assembly is a process in which an ordered structure spontaneously emerges from the local interaction between individual components. When these elements are molecules, the phenomenon is called molecular self-assembly (Mendes et al. 2013). Even on a more macroscopic level—such as the formation of beehives by bees and coral reefs by corals—these are examples, for structural engineers, of how Nature can organize complex structures without external intervention, a process that nanotechnology is trying to replicate in artificial materials (Chen et al. 2015, Yahia et al. 2021).

- **Acoustic Properties**
Acoustic metamaterials—engineered structures whose unique internal geometries, often at subwavelength scales, manipulate sound waves in ways unattainable by conventional materials—are used in applications ranging from soundproofing and noise cancellation to advanced sensors and sonar (Choi et al. 2020, Li et al. 2023). CAD/CAM tools enable precise modeling, simulation, and fabrication of these complex architectures, optimizing properties like sound redirection, absorption, cloaking, and beam forming before production (Zhang et al. 2024). Inspired by biological echolocation—seen in bats (*Chiroptera*; Tech 2021) and cetaceans (Vishnu et al. 2022)—researchers have developed underwater acoustic metamaterials capable of negative refraction, cloaking, and super-resolution imaging, despite challenges posed by impedance boundaries in marine environments (Dong et al. 2022, Arjunan et al. 2024). These advances promise breakthroughs in noise control, architectural acoustics, medical imaging, and underwater navigation and communication, demonstrating how biomimetic insights and digital design converge to redefine sound manipulation.

- **Environmental Responsiveness**
Eco-sustainable materials that respond to pollutants—such as graphene oxide nanocomposites for heavy-metal adsorption (Liu et al. 2019) or functionalized

conductive polymers sensing VOCs (Volatile Organic Compounds)—are deployed in purification devices and environmental sensors (Jain et al. 2023). CAD/CAM systems optimize these materials by tailoring membrane porosity for oil–water separation or refining sensor lattice geometries to boost efficiency, selectivity, and dynamic control (Kumar et al. 2025). Mangrove roots filter and cleanse water (Biomimicry New Zealand 2024), while mussels and other bivalves remove microscopic particles—including heavy metals—from aquatic environments, making them both natural purifiers and bio-indicators of water quality.

- **Magnetic-Responsive Materials**
Magnetic-responsive materials, such as magneto-elastic materials or magnetic metamaterials, can be used in magnetic detection devices, magnetic actuators, and advanced sensing systems. Thanks to the use of CAD/CAM, it is possible to design and optimize micro- and nanoscale structures that specifically respond to magnetic fields, enabling the implementation of technologies such as magnetic tracking in medical devices or magnetic analysis in biotechnology. A natural example can be observed in certain fish, such as Chinook salmon, which use a magnetic system within their bodies to perceive and orient themselves in Earth's magnetic fields, similar to the artificial systems used in marine navigation. Magnetically responsive materials, such as magnetoelastic materials or magnetic metamaterials, can be used in magnetic sensing devices, magnetic actuators, and advanced sensor systems. Through the use of CAD/CAM technologies and 3D/4D printing technologies-*material extrusion (MEX), powder bed fusion (PBF), binder jetting (BJ), vat photopolymerization (VPP), material jetting (MJ), directed energy deposition (DED),* and *sheet lamination (SHL)* - it is possible to design and optimize structures and materials that respond specifically to magnetic fields, enabling the implementation of technologies such as magnetic tracking in medical devices or magnetic analysis in biotechnology (Zhai et al. 2025). A natural example can be observed in certain fish species, such as the Chinook salmon, which use a magnetic system within their bodies to perceive and orient themselves within Earth's magnetic fields—much like artificial marine navigation systems (De Stefano 2020).

- **Chemical Response**
Some chemically reactive materials respond to specific chemical stimuli, such as the presence of certain gases or contaminants. These materials are used in chemical sensors for environmental monitoring and industrial safety. Sharma et al. (2023) offer a comprehensive review of CAD/CAM technologies applied to the design of chemical sensors with highly specific shapes and properties, improving the accuracy of chemical detection:

"Compared to traditional manufacturing processes, 3D printing offers numerous advantages for miniaturized electrochemical point-of-care diagnostic devices, such as rapid prototyping, custom-shaped devices, flexible fabrication designs, low energy consumption, reduced time to market, and reduced waste generation." In nature, carnivorous plants, such as the Venus flytrap (*Dionaea muscipula*), possess mechanisms that respond to chemical signals released by prey, demonstrating how even natural materials can react to chemical changes (Hedrich and Kreuzer 2023).

- **Tunable Materials**

Tunable materials can alter their properties in response to external stimuli such as pressure, temperature, electric field, or magnetic field (Abdelraouf et al. 2022). These metamaterials, made with precise configurations and adapted to devices that require dynamic modifications of physical properties, are used in applications such as optical devices, adaptive antennas, vibration control devices, and telecommunications materials. Yuan et al. (2020) explored the use of multi-material composites in 3D printing. Some insects, such as butterflies from the *Papilionidae* family, use tunable surfaces in their scales to change color based on light, adapting their appearance in real time. Zhang et al. (2014) explore how the structure of butterfly and moth wing scales can inspire characterization, modeling, and fabrication of new material design.

- **Auxetic Materials**

Materials with a negative Poisson's ratio (also known as *auxetic materials*) expand laterally when compressed, rather than contracting as conventional materials do. This unusual property can be exploited for shock-absorbing devices, protection of delicate structures, and self-healing surfaces (Fig. 4). Lecina-Tejero et al. (2023) explore the use of auxetic constructs in skin wound healing.

The use of parametric and generative CAD/CAM is essential for designing specific geometries that maximize auxetic behavior, which is useful, for example, in materials for ballistic protection or sports footwear materials. Traditional materials have mechanical properties that are directly linked to their molecular or atomic structure. In contrast, mechanical metamaterials derive their properties from the specific spatial arrangement of their constituent building blocks. By designing these architectures, it is possible to create mechanical metamaterials with extraordinary properties, such as negative Poisson's ratio, negative mass density, negative stiffness, vanishing shear modulus, and an ultra-high strength-to-density ratio. In recent years, substantial research has focused on developing new mechanical metamaterials by either designing novel architectures or incorporating new functional materials (Zheng et al. 2021). However, there are still several challenges to overcome:

(1) The design space is severely constrained by the design methods, as most architectures are generated using computer-aided design (CAD), which may not always lead to optimal performance and requires significant computational resources, especially for complex and intricate geometries;

(2) Mechanical metamaterials often rely on zero-energy deformation modes, leading to a fixed mechanical response after fabrication;

(3) Most previous design approaches are forward-design methodologies, where an architecture is created first, and its mechanical properties are then analyzed through experimentation or simulations. Zheng (2023) presents a state-of-the-art review of the computational design and fabrication of mechanical metamaterials, offering a comprehensive overview of existing methodologies and challenges, and introduces innovative deep-learning–based frameworks for the controllable inverse design of auxetic metamaterials (Fig. 5).

Fig. 4. Auxetic 3D-printed metamaterial with re-entrant cells.

Moreover, Zheng proposes a simplified mathematical approach to model mechanical metamaterials, using MATLAB codes and implicit functions to design a 3D auxetic metamaterial with optimal performance. This material, 3D printed with a rubber-like substance, exhibits auxetic behavior dominated by buckling instability. To enhance its properties, a nickel layer was added for increased stiffness, strength, and conductivity, without sacrificing auxeticity or resilience. Data maps of key mechanical properties provide guidance for practical applications. Additionally, he introduces reprogrammability in mechanical metamaterials, achieved through light-responsive shape-memory polydimethylsiloxane (SM-PDMS), enabling the reprogramming of their mechanical responses. Three types of reprogrammable SM-PDMS metamaterials were designed, including auxetic, chiral, and buckling-induced versions, with the latter used to create a soft actuator capable of reprogrammable locomotion. This strategy allows for adaptive, flexible materials with potential uses in soft robotics, actuation, and safety equipment. Furthermore, he proposes a deep learning framework to enable inverse design of 2D and 3D mechanical metamaterials. Using 2D and 3D conditional generative adversarial networks (CGANs), this framework generates metamaterials with specific target properties, such as relative density, Young's modulus, and Poisson's ratio. This approach offers a powerful tool

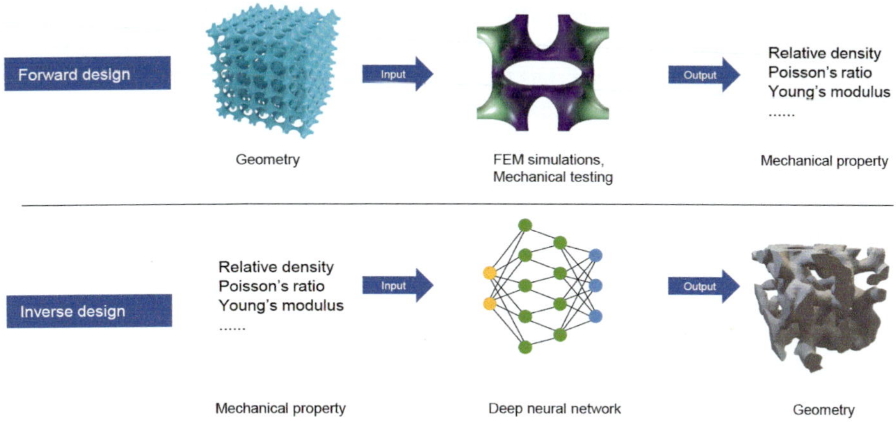

Fig. 5. Comparison between forward design and inverse design (Zheng 2023).

for applications requiring tailored architectures with desired material properties. This work introduces new computational design methods and materials for mechanical metamaterials, with potential applications including nickel-coated auxetic catalysts, reprogrammable soft robots, and CGAN-generated scaffold implants. The inverse design approach facilitated by deep learning can optimize metamaterials for superior properties, accounting for architectural designs, material choices, and interactions with external stimuli. Considering studies in biomimetics inspired by natural organisms, certain biological tissues—such as fish skin and the peels of fruits like bananas—exhibit nonlinear expansion behaviors. These natural structures could serve as models for the development of advanced synthetic materials with a negative Poisson's ratio, offering potential applications in areas such as flexible protection systems, soft robotics, and adaptive materials engineering, like advanced composite foam structures for impact absorption (Coastal Automotive 2022).

- **Shape-Memory Alloys (SMAs) with Non-Conventional Responses**

Shape memory alloys (SMAs) are smart materials with the extraordinary ability to remember a previously programmed shape (Zadafiya et al. 2021). When deformed, they can spontaneously return to their original shape when exposed to a specific stimulus, usually an increase in temperature. In more advanced models, however, these alloys do not just change shape with heat. Some can be activated at ambient temperature or respond differently depending on external conditions, such as humidity, pressure, or magnetic fields. These hybrid behaviors allow the material to exhibit multifunctional properties: for example, it can selectively stiffen, change length, or twist, making it extremely useful in fields such as soft robotics, aerospace, or regenerative medicine. SMAs are used to design smart structures with intrinsic functional properties and enhanced efficiency, especially in the aerospace industry. One of the most common SMAs is the NiTi alloy, which, however, has the significant limitation of poor machinability, making it difficult to produce complex shapes. Additive manufacturing processes allow overcoming this limitation, but optimizing the process for NiTi alloy is complex due to the effects of high temperatures, which can alter the material's properties. Shape memory alloys (SMAs) are materials used to design smart structures with intrinsic functional properties and enhanced efficiency, especially in the aerospace industry. One of the most common SMAs is the NiTi alloy, which, however, has the significant limitation of poor machinability, making it difficult to produce complex shapes. Additive manufacturing processes allow overcoming this limitation, but optimizing the process for NiTi alloy is complex due to the effects of high temperatures, which can alter the material's properties. This study focuses on optimizing the process parameters for NiTi alloy printed through Laser Powder Bed Fusion (L-PBF) to ensure optimal pseudoelastic behavior, essential for the design of structural dampers. The process was started with simple structures and then applied to the design of a damper that combines the pseudoelasticity of NiTi with load-bearing capacity. Biasutti et al. (2024) focus on optimizing the process parameters for NiTi alloy printed through Laser Powder Bed Fusion (L-PBF) to ensure optimal pseudoelastic behavior, essential for the design of structural dampers. The process was started with simple structures and then applied to the design of a damper that combines the pseudoelasticity of NiTi with load-bearing capacity. In the study by Rodinò, Curcio, and Maletta (2023), the authors discuss the use of active composite materials with shape-morphing capabilities, which are gaining popularity due to their efficiency and weight reduction. Shape memory alloys (SMAs), known for their smart properties, are used as embedded actuators in these applications. These materials, employed in fields such as aerospace and automotive, improve aerodynamic performance by enabling continuously morphing surfaces and simplify designs by reducing the number of movable parts. The study presents a design for an SMA–polymer composite plate with morphing capabilities, developed and manufactured using 3D printing processes. Additionally, the authors created a simple and effective analytical model to predict the shape-morphing behavior of the composite. The accuracy of this model was validated by comparing it with experimental data from the prototypes. In nature, the muscular systems of jellyfish and octopuses can be seen as shape memory models, where changes in the body are dictated by contractions and extensions that respond to environmental stimuli. These aspects inspire biomimetic solutions for soft-robotic designs (Villanueva et al. 2011). In this study, the authors explore the development of a robotic jellyfish, called *Robojelly*, which mimics the movement and functionality of natural jellyfish. The actuators of this robot are made of shape memory alloys, materials capable of returning to their original shape when exposed to a specific stimulus, such as heat. The paper provides insights into the use of biomimetic approaches in robotics and highlights the potential of integrating shape memory alloys to create efficient, flexible, and adaptive robotic systems.

- **Quantum Dots (QD) Materials**

Quantum dots (QD) are nanomaterials that possess unique optical and electronic properties due to their extremely small size. These materials, which are on the nanometer scale, exhibit distinctive behaviors due to the quantum effect that arises when the size of the particle is comparable to its wavelength. Essentially, the confinement of electrons and holes within such a small region alters the energy levels compared to bulk materials, creating new optical properties, such as emitting light at specific wavelengths depending on the size of the quantum dot. These characteristics make them ideal for a wide range of technological applications, from photonics, where they are used in devices such as color displays and solar cells, to biomedicine, where they are employed as probes for cellular imaging. Their ability to absorb and emit light in a highly controllable manner, combined with significant thermal and chemical stability, also offers the potential to enhance existing technologies and develop new electronic and optoelectronic devices. Additionally, quantum dots are being researched for their ability to improve solar cell efficiency through the use of tunable bandgaps and for their application in low-cost lasers and solid-state memory devices. In medicine, their biocompatibility and the ability to be functionalized with biological molecules make them suitable for use as agents in tumor treatment or advanced diagnostic analysis. The continued development of synthesis methods and a deeper understanding of their quantum behaviors are opening up new pathways for futuristic applications in various technological fields (Fan et al. 2024). In a study by Jeon et al. (2024), the authors present a simple method to create 3D structures using a special material made from perovskite quantum dots (PQDs) and a polymer. These materials are important for future electronics because they have great properties, like emitting colorful light. The method, called direct-ink writing (DIW), uses an ink made of PQDs and a polymer that is easy to handle at room temperature. By changing the mix of chemicals, the researchers can create red, green, and blue light from the material. This technique allows the creation of precise 3D shapes with special light-emitting features, which could be useful for things like anti-counterfeiting and secure information systems. In nature, bioluminescent algae, dinoflagellates, and fireflies produce light through a biochemical mechanism that, although not identical, shares the characteristic of manipulating optical properties at the molecular level (Perin et al. 2022).

- **Photovoltaic Materials with Enhanced Efficiency**

Some advanced photovoltaic materials, such as organic solar cell materials or perovskites, have significantly improved energy conversion efficiency through the integration of emerging technologies. By leveraging the power of CAD/CAM, it is possible to optimize molecular arrangements and material structures at the nanoscale, leading to enhanced performance and greater efficiency in sustainable energy applications (Dada and Popoola 2023). This approach not only aids in the design of more efficient solar cells but also contributes to the development of next-generation energy solutions (Hunde and Woldeyohannes 2023). A natural parallel in this case is photosynthesis in flowers and plants, which convert sunlight into chemical energy with remarkable efficiency through biochemical processes. Just as advanced photovoltaic materials strive to capture solar energy and convert it into usable electrical energy, plants have perfected a process called photosynthesis to convert sunlight into chemical energy. This natural phenomenon occurs in the chloroplasts of plant cells, where light energy is absorbed by chlorophyll and used to convert carbon dioxide and water into glucose, a form of chemical energy. This process is incredibly efficient, allowing plants to produce energy to sustain their growth and development, as well as to fuel the entire food chain. Similarly, modern solar technologies aim to mimic this natural ability by developing materials that can efficiently capture and convert sunlight into energy, as seen in the advancements of organic solar cells and perovskite materials. The study of photosynthesis has even inspired the design of *artificial photosynthesis* systems, which aim to mimic nature's process of energy conversion to develop more sustainable energy solutions (Zhou et al. 2020). The incredible efficiency with which plants perform photosynthesis could serve as a model for improving photovoltaic technologies, pushing the boundaries of energy conversion efficiency for future green energy applications.

In the next chapter, the concept of material smartness will be expanded and refined, exploring the emergence of group intelligence in living organisms (such as *murmuration* and *swarm* intelligence) and how these phenomena inspire fascinating developments and technological-scientific innovations. By examining how collective behaviors in nature enable adaptive and efficient problem-solving, the chapter will highlight how these biological principles are being applied to engineering and design. From distributed networks to self-organizing systems, the study of nature's collective intelligence offers groundbreaking insights into how we can leverage similar principles in materials science, robotics, and artificial intelligence. This exploration paves the way for a future where technology evolves through collaboration and shared intelligence, pushing the boundaries of what is possible in both material design and complex system optimization.

References

3Dnatives (2024). Mycelium as a material for 3D printing. 3Dnatives. Available at: https://www.3dnatives.com/en/mycelium-as-a-material-for-3d-printing-140620244/ [Accessed: 17 May 2025].

Abdelraouf, O. A. M., Wang, Z., Liu, H., Dong, Z., Wang, Q., Ye et al. (2022). Recent advances in tunable metasurfaces: Materials, design, and applications. ACS Nano, 16(9): 13339–13369.

Abugu, C. (2025). Integrating Building Information Modelling (BIM) for effective construction planning, execution, and lifecycle management. International Journal of Research Publication and Reviews, 6: 4347–4364.

Adlington, K. (2021). New fungi acoustic panel system offers high-quality sustainable design solution for work spaces. Arup. Available at: https://www.arup.com/news/new-fungi-acoustic-panel-system-offers-high-quality-sustainable-design-solution-for-work-spaces/ [Accessed: 17 May 2025].

Alima, N., Colmo, C. and Goidea, A. (2021). Mycology for Architecture: 3D Printing Mycelium Composites [Video]. YouTube. Available at: https://www.youtube.com/watch?v=971AHk4VJ4k [Accessed: 17 May 2025].

Anderson, N. (2020). 715-million-year-old fungi microfossils found. Sci.News. Available at: https://www.sci.news/paleontology/neoproterozoic-fungi-microfossils-08049.html [Accessed: 17 May 2025].

ArchDaily (2013). Hygroskin: Meteorosensitive Pavilion by Achim Menges Architect in collaboration with Oliver David Krieg and Steffen Reichert. ArchDaily. Available at: https://www.archdaily.com/424911/hygroskin-meteorosensitive-pavilion-achim-menges-architect-in-collaboration-with-oliver-david-krieg-and-steffen-reichert [Accessed: 18 May 2025].

Arjunan, A., Baroutaji, A., Robinson, J., Vance, A., and Arafat, A. (2024). Acoustic metamaterials for sound absorption and insulation in buildings. Building and Environment, 251: 111250.

AskNature (2016). Generative design software inspired by slime mold and human bones. AskNature. Retrieved May 17: 2025, from https://asknature.org/innovation/generative-design-software-inspired-by-slime-mold-and-human-bones/.

Autodesk (2023). Decarbonize architecture: Mycelium fungi as a building material. Autodesk University. Available at: https://www.autodesk.com/autodesk-university/class/Decarbonize-Architecture-Mycelium-Fungi-as-a-Building-Material-2023 [Accessed: 17 May 2025].

Autodesk Research (2025). Project Dreamcatcher: Generative Design Solutions in CAD. Autodesk Research. Available at: https://www.research.autodesk.com/projects/project-dreamcatcher/ [Accessed: 17 May 2025].

Awad, A., Pang, W., Lusseau, D., and Coghill, G. (2023). A survey on Physarum polycephalum intelligent foraging behaviour and bio-inspired applications. Artificial Intelligence Review, 56(1): 1–26.

Banerjee, H., Suhail, M., and Ren, H. (2018). Hydrogel actuators and sensors for biomedical soft robots: Brief overview with impending challenges. Biomimetics, 3(3): 15.

Bar-Cohen, Y., and Anderson, I. (2019). Electroactive polymer (EAP) actuators—background review. Mechanics of Soft Materials, 1.

Bensaude-Vincent, B. (2019). Bio-informed emerging technologies and their relation to the sustainability aims of biomimicry. Environmental Values, 28(5): 551–571.

Biasutti, T., Bettini, P., Nespoli, A., Grande, A. M., Scalia, T., Albano, M. et al. (2024). 3D printing of shape memory alloys for complex architectures of smart structures. Acta Astronautica, 221: 206–217.

Biomimetics in Botanical Gardens—Educational Trails and Guided Tours (2023). Biomimetics, 8(3): 303. Available at: https://doi.org/10.3390/biomimetics8030303 [Accessed: 18 May 2025]. License: CC BY 4.0.

Biomimicry New Zealand (2024, June 23). Mangrove-inspired seawalls: A biomimetic approach to coastal protection. Available at: https://biomimicry.org.nz/mangrove-inspired-seawalls-a-biomimetic-approach-to-coastal-protection/ [Accessed: 18 May 2025].

Bionic Surface Technologies GmbH (2025). Bionic surface technologies. Available at: https://www.bionicsurface.com [Accessed: 18 May 2025].

Bruschi, A., Donati, D. M., Choong, P., and Wallace, G. (2021). DEA working as artificial muscle. In Dielectric elastomer actuators, neuromuscular interfaces, and foreign body response in artificial neuromuscular prostheses: A review of the literature for an in vivo application. Advanced Healthcare Materials. Available at: https://www.researchgate.net/figure/DEA-working-as-artificial-muscle_fig2_352137693 [Accessed: 18 May 2025].

Case, K., Wood, I., Graham, B., and Abdul, M. S. (2003). CAD genetic algorithms for evolutionary form and function design. Journal for Manufacturing Science and Production, 5(1-2): 11–16.

Chen, D. A., Ross, B. E., and Klotz, L. (2015). Lessons from a coral reef: Biomimicry for structural engineers. Journal of Structural Engineering, 141(4): 02514002.

Chipilska, I. (2024). What are the challenges smart city initiatives are facing? Modeshift. Available at: https://www.modeshift.com/what-are-the-challenges-smart-city-initiatives-are-facing/[Accessed: 17 May 2025].

Choi, C., Bansal, S., Münzenrieder, N., and Subramanian, S. (2020). Fabricating and assembling acoustic metamaterials and phononic crystals. Advanced Engineering Materials, 23(3): 2000988.

City Vision. (2024). Benefits of CAD design for urban modeling. City Vision. Available at: https://cityvision.co.id/benefits-of-cad-design [Accessed: 17 May 2025].

Coastal Automotive (2022). Biomimicry: How human bones & banana peels inspire advanced composite foam structures for impact absorption. Coastal Automotive. Available at: https://coastal-automotive.com/news/biomimicry-how-human-bones-banana-peels-inspire-advanced-composite-foam-structures-for-impact-absorption/ [Accessed: 18 May 2025].

Dada, M., and Popoola, P. (2023). Recent advances in solar photovoltaic materials and systems for energy storage applications: A review. Beni-Suef University Journal of Basic and Applied Sciences, 12: 66.

Daneluzzo, M., Macruz, A., Tawakul, H., and Al Hashimi, M. (2023) Multispecies design: 3D-printed biomimetic structures to enhance humidity levels. Architectural Intelligence, 2, Article 9. Available at: https://link.springer.com/article/10.1007/s44223-023-00027-y [Accessed: 18 May 2025].

De Stefano, S. (2020, May 11). Salmoni di Chinook: percepiscono il campo magnetico e forse ora sappiamo come. Lega Nerd. Available at: https://leganerd.com/2020/05/11/salmoni-percepiscono-il-campo-magnetico-e-forse-ora-sappiamo-come/ [Accessed: 18 May 2025].

Dong, E., Cao, P., Zhang, J., Zhang, S., Fang, N. X., and Zhang, Y. (2022). Underwater acoustic metamaterials. National Science Review, 10(6), nwac246. Available at: https://doi.org/10.1093/nsr/nwac246 [Accessed: 18 May 2025].

Ehrlich, D., Hakimshafaei, M., and Elek, O. (2022). Scaffolding generation using a 3D Physarum polycephalum simulation. In Proceedings of the 7th Annual ACM Symposium on Computational Fabrication (pp. 30:1–30:2). Association for Computing Machinery.

Engel, K., Kilmartin, P., and Diegel, O. (2021). Recent advances in the 3D printing of ionic electroactive polymers and core ionomeric materials. Polymer Chemistry, 13.

Evolv AI. (2019). AI 101: Intro to evolutionary algorithms. Evolv AI. Available at: https://blog.evolv.ai/ai-101-intro-to-evolutionary-algorithms [Accessed: 17 May 2025].

Heckman, D. S., Geiser, D. M., Eidell, B. R., Stauffer, R. L., Kardos, N. L., and Hedges, S. B. (2001). Molecular evidence for the early colonization of land by fungi and plants. Science, 293(5532): 1129–1133.

Fan, J., Han, C., Yang, G., Song, B., Xu, R., Xiang, C. et al. (2024). Recent progress of quantum dots light-emitting diodes: Materials, device structures, and display applications. Advanced Materials, 36: 2312948.

Fan, R., and Lai, K. O. (2022). Understanding how kinesin motor proteins regulate postsynaptic function in neurons. FEBS Journal, 289(8): 2128–2144.

Félix, L. S., and Costa, L. F. (2021). Biomimetic applications of Mimosa pudica L. in the theoretical development of a pneumatic actuator. Brazilian Archives of Biology and Technology, 64, e20210058. Available at: https://doi.org/10.1590/1678-4324-2021200584 [Accessed: 18 May 2025].

Fernández, M. (2014). The myths behind the smart city technological imaginary (PhD brief notes #1). Ciudades a escala humana. Available at: https://www.ciudadesaescalahumana.org/2014/07/the-myths-behind-smart-city.html [Accessed: 17 May 2025].

Fiorello, I., Ronzan, M., Speck, T., and Mazzolai, B. (2024). A biohybrid self-dispersing miniature machine using wild oat fruit awns for reforestation and precision agriculture. Advanced Materials, 36(27): e2313906.

Gagliano, M. (2017). The mind of plants: Thinking the unthinkable. Communicative & Integrative Biology, 10: 00-00.

Gardiner, A., Domingo-Roca, R., Windmill, J. F. C., and Feeney, A. (2024). An adjustable acoustic metamaterial cell using a magnetic membrane for tunable resonance. Scientific Reports, 14: 15044.

Hamoudi, H., Berdiyorov, G., Zekri, A., Tong, Y., Mansour, S., Esaulov, V. et al. (2022). Building block 3D printing based on molecular self-assembly monolayer with self-healing properties. Scientific Reports, 12: 6806.

Hakami, A., Srinivasan, S., Biswas, P., Krishnegowda, A., Wallen, S., and Stefanakos, E. (2022). Review on thermochromic materials: Development, characterization, and applications. Journal of Coatings Technology and Research, 19.

Hedrich, R., and Kreuzer, I. (2023). Demystifying the Venus flytrap action potential. New Phytologist, 239(5): 2108–2112.

Hub for Biotechnology in the Built Environment (n.d.). Blog. Hub for Biotechnology in the Built Environment. Available at: http://bbe.ac.uk/blog-2/ [Accessed: 17 May 2025].

Hultgren, K., and Mittelstaedt, H. (2015). Color change in a marine isopod is adaptive in reducing predation. Current Zoology, 61(4): 739–748.

Hunde, B. R., and Woldeyohannes, A. D. (2023). 3D printing and solar cell fabrication methods: A review of challenges, opportunities, and future prospects. Results in Optics, 11: 100385.

Jain, T., Kumar, S., Parmar, A., Arora, V., Abdul-Zahra, D., and Nagpal, A. (2023). Sustainable materials for water treatment: A comprehensive review. E3S Web of Conferences, 430: 01103.

Jeon, H. S., Kim, J. H., Jun, M. B. G., and Jeong, Y. H. (2021) Fabrication of thermochromic membrane and its characteristics for fever detection. Materials, 14(13): 3460. Available at: https://doi.org/10.3390/ma14133460 [Accessed: 18 May 2025].

Jeon, H., Wajahat, M., Park, S., Pyo, J., Seol, S. K., Kim, N. et al. (2024). 3D Printing of Luminescent Perovskite Quantum Dot–Polymer Architectures. Advanced Functional Materials, 10 March 2024.

Kantorovitch, J. (2019). Understanding awareness in modern smart products design. Available at: https://urn.fi/URN:ISBN:978-951-39-7824-2 [Accessed: 18 May 2025].

Khalid, M. Y., Arif, Z. U., Rashid, A. A., Bukhari, S. M. Z. S., Hossain, M., and Koç, M. (2024). Shape-memory and self-healing properties of sustainable cellulosic nanofibers-based hybrid materials for novel applications. Giant, 19, 100299.

Kreis, A., Abdullah, A., and Hussein, H. M. A. (2024). Smart Connection Technology Framework – AI-based Creation of Connection Technology Elements. Computer-Aided Design and Applications, 21(4): 729–738.

Krieg, O., Christian, Z., Correa Zuluaga, D., Menges, A., Reichert, S., Rinderspacher, K. et al. (2014). HygroSkin – Meteorosensitive Pavilion. In Fabricate 2014 (pp. 97-104). UCL Press.

Kumar, A. S., Akoumeh, R., Ramanathan, A., Park, J., Thippanna, V., Patil, D. et al. (2025). Biomimetic 3D prototyping of hierarchically porous multilayered membranes for enhanced oil–water filtration. ACS Applied Materials & Interfaces, 17(5): 8285–8298.

Lecina-Tejero, Ó., Pérez, M. Á., García-Gareta, E., and Borau, C. (2023). The rise of mechanical metamaterials: Auxetic constructs for skin wound healing. Therapeutic Advances in Chronic Disease, 14, 20417314231177838.

Li, Z., Li, X., Chua, J. W., Lim, C. H., Yu, X., and Wang, Z. (2023). Architected lightweight, sound-absorbing, and mechanically efficient microlattice metamaterials by digital light processing 3D printing. Journal of Micromechanics and Microengineering, 33(2): e2166851.

Liu, L., Lei, J., Liu, T., Shang, B., Zhao, X., He, J. et al. (2023). Artificial skin with fast and robust self-healing ability for durable intelligent protection of magnesium alloys. Chemical Engineering Journal, 458: 141482.

Liu, X., Ma, R., Wang, X., Ma, Y., Yang, Y., Zhuang. et al. (2019b). Graphene oxide-based materials for efficient removal of heavy metal ions from aqueous solution: A review. Environmental Pollution, 252: 1076-1092.

Liu, X., Song, M., Fang, Y., Zhao, Y., and Cao, C. (2021a). Worm-inspired soft robots enable adaptable pipeline and tunnel inspection. Advanced Intelligent Systems, 3(1): 202100128.

Machado, I., Marques, C. F., Martins, E., Alves, A. L., Reis, R. L., and Silva, T. H. (2023). Marine gelatin-methacryloyl-based hydrogels as cell templates for cartilage tissue engineering. Polymers, 15(7): 1674.

Mancuso, S. (2023). Tree stories: How trees plant our world and connect our lives (G. Conti, Trans.). HarperOne.

Materialise (2025). Magics: 3D printing data and build preparation software. Materialise. Available at: https://www.materialise.com/en/industrial/software/magics-data-build-preparation [Accessed: 18 May 2025].

Mendes, A. C., Baran, E. T., Reis, R. L., and Azevedo, H. S. (2013). Self-assembly in nature: Using the principles of nature to create complex nanobiomaterials. Wiley Interdisciplinary Reviews: Nanomedicine and Nanobiotechnology, 5(6): 657–667.

Menges, A. (2012). HygroSkin: Meteorosensitive Pavilion. Achim Menges. Available at: https://www.achimmenges.net/?p=5612 [Accessed: 18 May 2025].

Middlesex University. (2025). London Digital Twin Research Centre. Middlesex University. Available at: https://dt.mdx.ac.uk/ [Accessed: 17 May 2025].

Mitchell Jones, A., Mautner, A., Luenco, S., Bismarck, A., and John, S. (2020). Engineered mycelium composite construction materials from fungal biorefineries: A critical review. Materials & Design, 187: 108397.

Mylo, M. D., Krüger, F., Speck, T., and Speck, O. (2020). Self-repair in cacti branches: Comparative analyses of their morphology, anatomy, and biomechanics. International Journal of Molecular Sciences, 21(13): 4630.

Mogu (2024). Radical by Nature. Mogu. Available at: https://mogu.bio/ [Accessed: 17 May 2025].

Nace, T. (2019). Scientists find one billion year old fungi, Earth's oldest. Forbes. Available at: https://www.forbes.com/sites/trevornace/2019/05/23/scientists-find-one-billion-year-old-fungi-earths-oldest/ [Accessed: 17 May 2025].

Nagy, D., Zhao, D., and Benjamin, D. (2017). Nature-based Hybrid Computational Geometry System for Optimizing Component Structure. Autodesk Research. Available at: https://www.research.autodesk.com/publications/nature-based-hybrid-computational-geometry-system-for-optimizing-component-structure/ [Accessed: 17 May 2025].

Nawade, A., Kunchanapalli, R., and Mukhopadhyay, S. (2020). Design of thermochromic materials and coatings for cool building applications. pp. 197–226. *In*: Design of thermochromic materials. Elsevier.

Neveln, I. D., Bai, Y., Snyder, J. B., Solberg, J. R., Curet, O. M., Lynch, K. M. et al. (2013). Biomimetic and bio-inspired robotics in electric fish research. Journal of Experimental Biology, 216(Pt 13), 2501–2514.

Perin, L., Moraes, G., Galeazzo, G., and Oliveira, A. (2022). Bioluminescent dinoflagellates as a bioassay for toxicity assessment. International Journal of Molecular Sciences, 23(21): 13012.

Philipsen, K. (2013). Could virtual cities make our real cities smarter? ArchDaily. Available at: https://www.archdaily.com/430041/could-virtual-cities-make-our-real-cities-smarter [Accessed: 17 May 2025].

Picon, A. (2015). Smart Cities: A Spatialised Intelligence. Wiley.

Pietroluongo, L. (2022). The golden ratio: The ultimate guide to understanding and using it. Elegant Themes. Available at: https://www.elegantthemes.com/blog/design/the-golden-ratio-the-ultimate-guide-to-understanding-and-using-it [Accessed: 17 May 2025].

Pfeifer, R., and Bongard, J. (2006). How the body shapes the way we think: A new view of intelligence. The MIT Press. Available at: https://mitpress.mit.edu/9780262537421/how-the-body-shapes-the-way-we-think/ [Accessed: 18 May 2025].

Powell, A. W., Stavrinadis, A., Christodoulou, S., Quidant, R., and Konstantatos, G. (2020). On-demand activation of photochromic nanoheaters for high color purity 3D printing. Nano Letters, 20(5): 3485–3491.

PVcase Team. (2024). Are smart cities feasible? PVcase. Available at: https://pvcase.com/blog/are-smart-cities-feasible/ [Accessed: 17 May 2025].

Qiu, Y., Wang, C., Lu, X., Wu, H., Ma, X., Hu, J. et al. (2021). A biomimetic *Drosera capensis* with adaptive decision-predation behavior based on multifunctional sensing and fast actuating capability. Advanced Functional Materials, 32: 2110296.

Rafsanjani, A., Brulé, V., Western, T., Pasini, D., and Reis, P. M. (2015). Hydro-responsive curling of the resurrection plant *Selaginella lepidophylla*. Scientific Reports, 5: 8064.

Renner, G., and Ekárt, A. (2003). Genetic algorithms in computer aided design. Computer-Aided Design, 35(8): 709–726.

Requicha, A., and Arbuckle, D. (2006). CAD/CAM for nanoscale self-assembly. IEEE Computer Graphics and Applications, 26(2): 88–91.

Reyssat, E., and Mahadevan, L. (2009). Hygromorphs: From pine cones to biomimetic bilayers. Journal of the Royal Society Interface, 6(39): 951–957.

Rodinò, S., Curcio, E. M., and Maletta, C. (2023). SMA-Polymer Composite Made by 3D Printing: Modelling and Experiments. Procedia Structural Integrity, 47: 579–588.

Sadeghipour Roudsari, M., and Pak, M. (2013). Ladybug: A parametric environmental plugin for Grasshopper to help designers create an environmentally-conscious design. Proceedings of BS 2013: 13th Conference of the International Building Performance Simulation Association, 3128–3135.

Schneider, M. S. (2025). Constructing The Universe. Constructing The Universe. Available at: https://www.constructingtheuniverse.com/ [Accessed: 17 May 2025].

ScienceDirect (2018). Electroactive polymer. In ScienceDirect Topics. Available at: https://www.sciencedirect.com/topics/chemistry/electroactive-polymer [Accessed: 18 May 2025].

Sender, J., and Grzywna, A. (2025). Influence of temperature and photoperiod on the growth and development of Lemna trisulca and Lemna minor in controlled conditions. Journal of Ecological Engineering, 26(2): 56–72.

Severi, A. (2023). MyCera | 3D printing mycelium reinforced structures. 3D WASP. Available at: https://www.3dwasp.com/en/mycera-3d-printing-mycelium-reinforced-structures/[Accessed: 17 May 2025].

Sharma, A., Faber, H., Khosla, A., and Anthopoulos, T. D. (2023). 3D printed electrochemical devices for bio-chemical sensing: A review. Materials Science and Engineering: R: Reports, 156: 100754.

Sharma, N. (2025). The role of IoT in building smart cities – 10 applications and use cases. Appinventiv. Available at: https://appinventiv.com/blog/role-of-iot-in-smart-cities-applications/[Accessed: 17 May 2025].

Shevlin, H., Vold, K., Crosby, M., and Halina, M. (2019). The limits of machine intelligence: Despite progress in machine intelligence, artificial general intelligence is still a major challenge. EMBO Rep. Oct 4;20(10):e49177.

Simard, S. W., Perry, D. A., Jones, M. D., Myrold, D. D., Durall, D. M., and Molina, R. (1997). Net transfer of carbon between ectomycorrhizal tree species in the field. Nature, 388(6642): 579–582.

Simard, S., Beiler, K., Bingham, M., Deslippe, J., Philip, L., and Teste, F. (2012). Mycorrhizal networks: Mechanisms, ecology and modelling. Fungal Biology Reviews, 26: 39–60.

Simard, S. (2018). Mycorrhizal networks facilitate tree communication, learning, and memory. pp. 145–156. *In*: Advances in the ecology and management of forest systems . Springer.

Siemens Software. (2025). NX X Manufacturing. Siemens Digital Industries Software. Available at: https://plm.sw.siemens.com/en-US/nx/products/nx-x-manufacturing/ [Accessed: 18 May 2025].

Smith, S. E., and Read, D. J. (2008). Mycorrhizal symbiosis (3ª ed.). Academic Press.

Sola, N., Trinchi, A., and Hill, A. J. (2023). Self-assembly meets additive manufacturing: Bridging the gap between nanoscale arrangement of matter and macroscale fabrication. Smart Materials in Manufacturing, 1, 100013.

Soo, S.C., and Yu, K.M. (2003). Rapid prototyping for self-similarity design. Journal of Materials Processing Technology, 139(1-3): 219–225.

Stamets, P. (2005). Mycelium running: How mushrooms can help save the world. Ten Speed Press.

Sterry, M. (2010, December 13). The Bionic City: a natural blueprint for future cities. This Big City.

Szukalski, A., Uttiya, S., D'Elia, F., Portone, A., Pisignano, D., Persano, L. et al. (2019). 3D photo-responsive optical devices manufactured by advanced printing technologies. SPIE Proceedings, 10915, 1091503.

TAAL Tech. (2025). The role of BIM in smart cities: Paving the way for a digital urban future. TAAL Tech. Available at: https://www.taaltech.com/the-role-of-bim-in-smart-cities-paving-the-way-for-a-digital-urban-future/ [Accessed: 17 May 2025].

Tech V. (2021). Intricate sound sensor inspired by bat ears. AskNature. Available at: https://asknature.org/innovation/intricate-sound-sensor-inspired-by-bat-ears/ [Accessed: 18 May 2025].

Tazwar, H. T., Antora, M. F., Nowroj, I., and Rashid, A. B. (2025). Conductive polymer composites in soft robotics, flexible sensors, and energy storage: Fabrication, applications, and challenges. Biosensors and Bioelectronics: X, 24, 100597.

Vermeulen, E. P. M. (2018). How smart technologies are transforming life & business now! HackerNoon. Available at: https://hackernoon.com/how-smart-technologies-are-transforming-life-business-now-e31641496d9b [Accessed: 17 May 2025].

VEX Robotics (2025). VEX Robotics. Available at: https://www.vexrobotics.com [Accessed: 18 May 2025]. Wei, Y., Othman, Z., Daud, K. M., Luo, Q., and Zhou, Y. (2024). Advances in slime mould algorithm: A comprehensive survey. Biomimetics, 9(1): 31.

Villanueva, A., Smith, C., and Priya, S. (2011). A biomimetic robotic jellyfish (Robojelly) actuated by shape memory alloy composite actuators. Bioinspiration & biomimetics, 6(3): 036004.

Vishnu, H., Hoffmann-Kuhnt, M., Chitre, M., Ho, A., and Matrai, E. (2022). A dolphin-inspired compact sonar for underwater acoustic imaging. Communications Engineering, 1: 10.

WeWantToLearn.net (2021). The Living Fabric Tower. WeWantToLearn.net Blog. Available at: https://wewanttolearn.wordpress.com/2021/01/20/the-living-fabric-tower/ [Accessed: 17 May 2025].

WeWantToLearn.net (2023). WeWantToLearn.net – Page 2 – Diploma Studio 10 at Westminster University School of Architecture. WeWantToLearn.net Blog. Available at: https://wewanttolearn.wordpress.com/page/2/ [Accessed: 17 May 2025].

Wright, G. and McLaughlin, E. (2023). Holistic (holistic technology). Search CIO. Available at: https://www.techtarget.com/searchcio/definition/holistic-holistic-technology [Accessed: 17 May 2025].

Wyss Institute for Biologically Inspired Engineering (2016). Octobot: A soft, autonomous robot. Harvard University. Available at: https://wyss.harvard.edu/media-post/octobot-a-soft-autonomous-robot/ [Accessed: 18 May 2025].

Wu, Y., Wang, Y., Zhang, S., and Wu, S. (2021). Artificial chameleon skin with super-sensitive thermal and mechanochromic response. ACS Nano, 15(10): 15720–15729.

Xu, C., Guo, Q., Chen, W., Xie, W., Wang, Y., Wang, M. et al. (2021). Biomimetic design of photonic materials for biomedical applications. Acta Biomaterialia, 121: 143–179.

Yang, M., Jiang, P., Zang, T., and Liu, Y. (2023). Data-driven intelligent computational design for products: Method, techniques, and applications. Journal of Computational Design and Engineering, 10.

Yahia, L. H., Bayade, G., and Cirotteau, Y. (2021). Natural coral as a biomaterial revisited. American Journal of Biomedical Science & Research, 13(6): AJBSR. MS.ID.001936.

Yuan, C., Wang, F., Qi, B., Ding, Z., Rosen, D. W., and Ge, Q. (2020). 3D printing of multi-material composites with tunable shape memory behavior. Materials & Design, 193: 108785.

Zadafiya, K., Dinbandhu, S., Kumari, S., Chatterjee, S., and Abhishek, K. (2021). Recent trends in non-traditional machining of shape memory alloys (SMAs): A review. CIRP Journal of Manufacturing Science and Technology, 32: 217–227.

Zhai, H., Li, X., Yu, S., Wang, J., Chang, Y., Li, J. et al. (2025). Review on the 3D printing technology and application of magnetic materials: Material–process–structure–application. Composites Part B: Engineering, 298: 112387.

Zhang, D., Zhang, W., Gu, J., Fan, T., Liu, Q., Su, H. et al. (2014). Inspiration from butterfly and moth wing scales: Characterization, modeling, and fabrication. Progress in Materials Science, 68: 1–32.

Zhang, W., Chen, Y., and Wang, J. (2024). A Computer-aided design (CAD) file render of the acoustic metamaterial cell components [Figure]. In Acoustic Metamaterials with Programmable Band Gaps.

Zheng, X., Chen, T.-T., Guo, X., Samitsu, S., and Watanabe, I. (2021) Controllable inverse design of auxetic metamaterials using deep learning. Materials & Design, 211: 110178. Available at: https://doi.org/10.1016/j.matdes.2021.110178 [Accessed: 18 May 2025].

Zheng, X. (2023). Computational design and fabrication of mechanical metamaterials (Doctoral dissertation, University of Tsukuba). Tsukuba Repository.

Zhou, H., Xiao, C., Yang, Z., and Du, Y. (2020). 3D structured materials and devices for artificial photosynthesis. Nanotechnology, 31: 1361–6528.

Zhu, H., Gupta, A. K., Wu, X., Goldsworthy, M., Wang, R., Mikkilineni, M. et al. (2023). A validation study for a bat-inspired sonar sensing simulator. PLOS ONE, 18(1), e0280631.

6

Make it More Connected

"I am concerned with attuning myself
to nature, far more than copying it."

Georges Braque,

6.1 The Invisible Power of Murmuration

In the season preceding winter, the skies come alive with the spectacular flight of starlings (*Sturnus vulgaris*), small passerines that move in enormous, compact flocks. Thousands, sometimes hundreds of thousands of individuals, soar in synchrony, forming shifting patterns among the clouds in an aerial dance that can reach speeds of 50 miles per hour (*Breathtaking* starling murmuration captured in Italy 2024). These collective movements, perfectly coordinated, seem to follow an invisible score, creating an authentic hypnotic spectacle. The phenomenon is known as *murmuration*, a term derived from the Latin *murmuratio*, which evokes the vibrating and incessant sound generated by the beat of the wings when the birds land. The flight of the starlings, beyond captivating observers, has inspired scientists and engineers, offering insights into the study of collective movement and innovation in the field of aerodynamics. This mastery of the air by the birds is the result of millions of years of advancement, during which nature has fine-tuned structures and flight strategies with extraordinary precision, transforming the sky into a stage of pure dynamic harmony (Rader and Hedrick 2023). Humans have long regarded the high, effortless flight of birds as a glimpse into hidden meanings and omens. From Icarus's tragic ascent to Cyrano de Bergerac's fanciful lunar voyages, the yearning for wings embodies our desire to transcend earthly limits. Among these avian spectacles, the starling murmuration stands out as both a poetic metaphor and a scientific model. Its sweeping, decentralized patterns emerge from simple local interactions, revealing principles of self-organization, non-linear dynamics, and emergent behavior. Today, researchers translate these insights into bio-inspired algorithms—such as deep-belief-network models for secure, distributed web services in smart-city infrastructures. While not every finding directly lands in aerospace engineering, studying murmuration dynamics deepens our grasp of coordinated motion and

DOI: 10.1201/9781003438946-6

adaptive networks across fields (Sheeba et al. 2024). Indeed, studying starling murmurations and other animal swarms has revealed the simple local rules that give rise to self-organization and nonlinear dynamics. Engineers, mathematicians, cyberneticists, and technologists are harnessing these insights to optimize CAD/CAM/CAE/CAO workflows and to design ever more complex systems. By decoding how thousands of agents coordinate without central control, researchers are translating these principles into computational tools—from advanced optimization algorithms to machine-learning frameworks—that can tackle high-dimensional engineering challenges (Hu et al. 2023). For example, Zamani, Nadimi-Shahraki, and Gandomi (2022) proposed the Starling Murmuration Optimizer (SMO), a novel bio-inspired algorithm designed for global and engineering optimization tasks, demonstrating competitive performance in solving complex mathematical and mechanical problems. Indicatively, starling murmurations emerge from simple, local rules: each bird adjusts its flight according to its nearest neighbors, maintaining cohesion and coordination. These interactions give rise to large-scale, emergent complexity without any central control. By translating flock dynamics into mathematical and topological models, researchers have demonstrated that self-organization principles can guide the design of metamorphic and adaptive materials—systems that autonomously reconfigure their structure and properties in response to mechanical stress, temperature shifts, or magnetic fields (Benteh 2016). The study of proximity interactions in bird flocks has become a valuable blueprint for developing composite alloys or polymers with self-healing or morphologically adaptive capabilities, enhancing resilience and durability in aerospace engineering and architecture. This phenomenon—an expression of animal collective intelligence—has inspired scientists, artists, and writers alike. In some regions, murmurations reach astonishing scales, as in the Danish *Sort Sol* (Black Sun), where thousands of starlings darken the sky over the Wadden Sea in a spectacular display (Romo-Tønder Region n.d.). Only in Vancouver, Canada, has a larger murmuration ever been recorded. Such gatherings likely evolved as collective defense mechanisms against predators like the Peregrine Falcon (*Falco peregrinus*), the starling's primary natural enemy. Originally native to the UK and continental Europe, the European starling (*Sturnus vulgaris*) has since become invasive across North America, Australia, New Zealand, Fiji, South Africa, and elsewhere, following centuries of accidental and intentional introductions (Zichello et al. 2024). As starlings prepare for winter migration, their aerial displays serve both as collective defense and as a kind of shared acrobatics, a silent choreography of collective intelligence. Without any leader, each bird tracks and matches the speed and direction of its nearest six or seven neighbors—maintaining just enough distance to avoid collisions—so that a turn by one ripples instantly through the flock (The Conversation 2016). This decentralized coordination—every individual balancing personal autonomy with group needs—mirrors social structures in other animals (e.g., wolf packs or primate troops), where simple local rules and hierarchies regulate access to resources and minimize conflict (Wilson 1975). Conflict is inevitable in any society—animal or human—but groups evolve rituals to manage and resolve it. In elephant herds, for example, disputes are often diffused by soothing vocalizations and gentle trunk-to-body contact. Among primates, post-conflict grooming restores social bonds and harmony. These behaviors, like murmuration dynamics in birds, illustrate how simple, local interactions and restorative rituals underpin resilience and cohesion across species (Duboscq et al. 2014). From Edmund Selous's evocative 1905 telepathy hypothesis—that starlings "must think collectively, all at the same time… a flash out of so many minds" (Selous 1905)—to Craig Reynolds's 1987 BOIDS algorithm, which equips each agent with three local rules (separation, alignment, cohesion) to generate fluid, realistic flocking (Reynolds 1987), the murmuration phenomenon has inspired increasingly sophisticated tools. Disney's *The Lion King* (1994) harnessed BOIDS to animate a wildebeest herd, entrusting each virtual creature with Reynolds's steering behaviors to achieve natural clustering, wave-like motion, and collision avoidance that would have been impossible to hand-animate (Allers and Minkoff 1994). In Rhino®/Grasshopper® CAD tools, two implementations bring BOIDS into parametric CAD workflows: the recursive HoopSnake component, which iterates separation, alignment, and cohesion rules over point sets to produce emergent swarm patterns (McNeel Discourse 2019), and the Boid Library plugin, offering dedicated *Separation*, *Alignment*, and *Cohesion* modules for real-time simulation, layout optimization, and collective-dynamics analysis on complex geometries (Boid Library 2025, Boid Library Demonstration 2025). Biological validation comes from Cavagna and Giardina (2014), who used computer vision and 3D imaging to show that each starling responds only to its six or seven nearest neighbors, propagating turns like a wave through a topologically connected lattice—dynamics analogous to discontinuous phase transitions in ferromagnetic materials. Although *murmuration* and *flocking* typically describe starling and bird flight, coordinated group movement occurs across many biological systems. *Swarming* often refers to insects (but can apply to any self-propelled agents), *herding* to tetrapods, and *shoaling* or *schooling* to fish; even non-motile algae form dense *blooms*. The term *swarm intelligence* was coined in 1989 by Beni and Wang to describe how simple, locally interacting agents—modeled on insect swarms—produce complex, collective behaviors without centralized control. Building on these insights, a broad family of swarm-intelligence algorithms has emerged: Particle Swarm Optimization (Eberhart and Kennedy 1995), the Artificial Fish Swarm Algorithm (Zainal et al. 2015), and more recent Marine Predators variants—IMPA with adaptive weights (Chun et al. 2024), the chaotic ACMPA (Liang et al. 2022), and APEMPA for real-time resource allocation in radar networks (Lu et al. 2025). These methods are applied in several fields, confirming distributed self-organization as a universal design paradigm:

- Computational optimization (e.g., Particle Swarm Optimization – PSO, introduced by Eberhart and Kennedy (1995);
- Collective robotics (cooperation among drones, industrial robots, etc.);
- Bioinformatics and biomedicine;
- Complex systems engineering;
- Biological system simulations;
- Finance and economics: Algorithms for trading and portfolio management based on adaptive dynamics;

- Telecommunications network management: data-traffic optimization and adaptive routing in distributed networks;
- Intelligent transportation systems: Traffic optimization and fleet management using swarm-inspired models;
- Exploration and space research: Control of satellite or drone swarms for autonomous missions;
- Advanced materials design: Structural optimization via swarm-inspired algorithms to simulate molecular-level interactions;

While Nature has, over billions of years, produced and orchestrated extraordinarily sophisticated and efficient behaviors marked by deep interconnectivity, the same cannot yet be said of our human-made artificial systems. Natural ecosystems—whether microbial mats, insect swarms, or forest canopies—rely on distributed interactions, redundant pathways, and self-organizing feedback loops that confer resilience, adaptability, and long-term stability. By contrast, engineered artifacts typically emerge from top-down design processes, depend on rigid protocols, and often lack the decentralized robustness of biological networks. As a result, man-made infrastructures and software ecosystems remain vulnerable to single-point failures, suffer from interoperability gaps, and require extensive maintenance to preserve functionality. Emulating Nature's principles—local decision-making, modular redundancy, and emergent coordination—could offer a promising route to closing this divide, guiding the next generation of design toward systems that evolve, self-heal, and thrive under variable conditions. Despite these advances biomimetics portend, a richly interconnected future, contemporary industrial-strength CAD/CAM/CAE/CAO platforms still contend with inefficiencies and interoperability gaps. As product complexity increases, seamless integration across diverse design and manufacturing protocols is critical. Recent field studies (Cheng et al. 2023) have identified fourteen major challenges—from incompatible data formats to divergent workflow conventions—and proposed targeted strategies for enhancing collaborative design processes, underscoring that the ultimate success of technological ecosystems hinges not only on algorithmic innovation, but also on standardized protocols and shared practices:

- **Challenge 1:** They recommend adopting unified modeling standards and best practices to streamline workflows: by enforcing consistent procedures, teams can cut manual effort and boost collaboration efficiency. However, extending such standardization beyond a single organization—where entrenched, company-specific methods prevail—remains a formidable obstacle, even though intra-company alignment is attainable.
- **Challenge 2:** Concerns delays in model uploads, which hinder collaboration among team members. In a distributed CAD system, team members work collaboratively but asynchronously on a centralized database. Each designer checks out a CAD model to work on it locally and then integrates the changes into the shared repository. If designers upload models infrequently or forget to update the model after making changes, others cannot proceed with their work

on the file. This issue can significantly slow down design progress, especially when there are many parts in an assembly.

- **Challenge 3:** Designers often unknowingly inherit hidden dependencies when reusing legacy models—standard parts like screws or custom components—from previous projects. Pulling in an old model to save time can leave linked assets (drawings, macros, assembly files) out of sync unless manually updated, creating more work than a fresh build. The issue worsens when one designer must temporarily take over another's work: without insight into those interlinked files, tweaks can inadvertently ripple through unrelated assemblies, complicating the entire design process.
- **Challenge 4:** When multiple designers contribute to a CAD model—whether simultaneously or in rapid succession—hidden edits and unrefreshed versions can derail workflow. Without real-time propagation of changes, individuals must manually refresh their environment or repository to stay current, lest they overwrite colleagues' updates or unknowingly duplicate effort. Conflicting versions emerge when someone uploads a new file behind another's back, forcing teammates to close and reopen models to see edits. Ultimately, this lack of visibility into collaborators' modifications breeds synchronization issues and slows every stage of design.
- **Challenge 5:** In many design-review settings, teams lack true synchronous editing in their CAD tools—so although some third-generation platforms offer real-time collaborative editing, none of the interviewed designers use it. This gap prevents multiple contributors from iterating together on minor tweaks and maintaining shared context. As a workaround, designers rely on assembly configurations—lightweight variants within a single CAD file—to experiment independently. Configurations act like personal sandboxes: each user can test new parts or tweaks without committing to a full version update, preserving the master assembly while enabling parallel exploration.
- **Challenge 6:** concerns the difficulty in presenting CAD models during synchronous work. Design review meetings, where team members explore ideas, coordinate tasks, and provide updates, are crucial. However, presenting CAD models in real-time can be problematic for several reasons:
    - **Adjusting view angles:** It takes time to optimize the model's angles for proper visualization.
    - **Manual maneuvers to show specific perspectives:** Manually moving the model to show cross-sections or hide/display components can be time-consuming.
    - **Difficulty with complex models:** Live CAD models are not ideal for communicating certain features, such as products with very thin cross-sections (e.g., a laptop cover). During meetings with a technical design team, designers might choose to show the live CAD, but modifications are slow since only one person can make changes at a time. In contrast, in meetings with non-technical stakeholders (such as clients), designers prefer to present

images of the design through presentations/renderings, as these are more technically reliable than live CAD.

- **Challenge 7** concerns the lack of support for real-time conceptual design with CAD. When starting a new project, some designers meet with clients to discuss the client's requirements and brainstorm potential designs. In these meetings, many designers find it difficult to translate the client's vision into a detailed design, as they typically use hand sketches, which may overlook some design features. The reason CAD is not used during these meetings is that creating conceptual designs directly in CAD (known as conceptual CAD) takes too much time, which is a common criticism of conceptual CAD. This issue makes it difficult to use CAD effectively for quick modifications during client discussions.
- **Challenge 8:** A major hurdle lies in the incompatibility between CAD platforms and their data-management systems. Mechanical models created in one MCAD package often cannot be edited in another—or even across different versions of the same software—and ECAD electrical designs (e.g., PCB layouts or wiring diagrams) frequently cannot be imported into MCAD assemblies. To make matters worse, MCAD and ECAD files often reside in separate PDM or version-control repositories, so synchronizing up-to-date files across domains becomes a tedious, manual task. Exporting to neutral formats such as STEP or STL can temporarily bridge these silos—but it also strips away parametric history and design intent. Recipients of these *dumb solids* can measure and reimport the geometry, or rebuild it from scratch, but they lose access to the original feature tree and editable parameters.
- **Challenge 9:** Designers frequently need to inspect a model's history or roll back to an earlier iteration—say, after evaluating a prototype—but CAD systems rarely provide clear, accessible change logs. Without automatically generated summaries of what's been added, removed, or modified between versions, pinpointing differences becomes a manual, error-prone task. This opacity makes it cumbersome to revert safely to a known good state and undermines efficiency and confidence across the entire design workflow.
- **Challenge 10:** Non-PLM (Product Lifecycle Management) users often lack version control and centralized data management. Without automated naming conventions or standardized folder structures, finding the correct model version is difficult—forcing designers to work on outdated files. When teams rely on manual tools (e.g., Dropbox) without check-in/check-out, collaborators may each download, edit, and re-upload copies, risking overwrites and merge conflicts. Adopting formal PLM systems can mitigate these issues: a conservative setup automatically checks out all linked files (parts and assemblies together), preventing unsynchronized edits. In less restrictive configurations, however, individual files can still drift out of sync, perpetuating the challenge.
- **Challenge 11:** A frequent challenge is the rigid folder structure that ties all project files to a single directory, forcing designers to copy components into each project and leading to proliferating, slightly renamed duplicates. Once duplicated, a model becomes an independent file, so updates to one copy aren't

reflected elsewhere. To prevent this bloat, teams should maintain a centralized standard-parts library—complete with fixed names (e.g., "70 mm hex head screw") and shared across locations—so that common components are created once and reused universally without variation.

- **Challenge 12:** Designers often resort to *redlining* by emailing or messaging annotated screenshots to share progress, but these markups live outside the CAD system and aren't tracked in a PLM. As a result:
  - Feedback and annotations get scattered across inboxes or chat threads, making it hard to find or link to the actual model.
  - In the rush to meet deadlines, someone might upload an updated CAD file without its corresponding notes, breaking the trace between design intent and execution.
  - Screenshots and model files drift out of sync, spawning version mismatches and quality issues.

To solve this, teams can adopt integrated digital-markup tools—many CAD platforms now let you embed comments, sketches, and review notes directly in the model. Centralizing annotations inside the CAD environment restores audit trails, enforces consistency, and keeps design history intact.

- **Challenge 13:** Controlling who can modify CAD files improves traceability but often creates bottlenecks. In PLM-managed workflows, a designer taking over a project must request file access from the original author, team lead, or system administrator—adding an extra handoff that can delay progress. External collaborators without PLM credentials must rely on generic file-share services (e.g., Dropbox or Google Drive), which lack version control and make it hard to find the latest model. Because updates there are manual and laborious, shared files may only be refreshed once a day or week, leaving stakeholders out of sync.
- **Challenge 14:** When key stakeholders lack CAD licenses—or use a different platform—they cannot directly inspect or interact with the native models. Instead, they depend on colleagues to export static screenshots, 2D drawings, or neutral-format files (e.g., STEP, 3D PDF), which introduces delays and risks misinterpretation. This is particularly acute for contractors or external partners granted only limited access (for IP or security reasons) or using a different CAD system (e.g., SolidWorks® vs. NX®). The result is a communication gap: stakeholders aren't immediately aware of design iterations and must wait for someone with the right software to generate readable outputs.

Potential remedies include:

- **Web-based or free CAD viewers** (eDrawings, Autodesk Viewer, 3D PDF) so non-licensed users can rotate, measure, and comment on the model directly.
- **Standardized review formats** such as interactive 3D PDFs or lightweight HTML5 exports that preserve some model interactivity without exposing feature history.
- **Role-based access controls** in a PLM/PDM system that issue temporary, scoped licenses or viewer rights to contractors, ensuring timely, secure access.

The tech world is looking at the evolution of AI—artificial intelligence, machine learning, and computational simulation technologies—to find new perspectives for a CAD/CAM ecosystem that is increasingly coherent, interconnected, and predictive. As highlighted, major software, such as Autodesk Fusion 360®, Siemens NX®, PTC Creo®, and SolidWorks®, offer advanced tools for parametric modeling and simulation, but they often operate in closed ecosystems, with difficulties in data exchange with other platforms. This creates inefficiencies in the transition from design to production, especially in high-complexity sectors such as aerospace, automotive, and bioengineering (Meyers 2024). Many companies are developing solutions to overcome this issue. nTopology®, for example, has introduced an approach based on computational modeling and generative design, which allows for more flexible integration with additive manufacturing technologies. Here are some practical applications of nTopology®:

- **Lightweighting of structures:** nTopology® enables weight reduction of components while maintaining structural integrity, enhancing fuel efficiency, and reducing the carbon footprint (nTop 2024).
- **Thermal management:** The software enables the design of compact, high-performance heat exchangers, increasing the efficiency and longevity of thermal systems.
- **Fluid Management:** nTopology® facilitates the creation of lighter and more compact components with improved fluid dynamic performance, optimizing flow and reducing pressure losses.
- **Engineered Materials:** The platform enables the creation of highly engineered structures with customized mechanical, thermal, electromagnetic, or biological properties, opening new possibilities in material design.
- **Mass Customization:** The platform offers automated and reusable workflows to reduce the cost and time-to-market of customized products, making large-scale production of tailored items more efficient.
- **Aesthetic Design:** The software supports the creation of unique products intended for additive manufacturing, accelerating the development process and enabling greater creative freedom in design.
- **Tooling Design:** nTopology® helps save time and costs in tooling design by enhancing performance through advanced automation tailored for additive manufacturing.
- **Design for Manufacturing:** The platform optimizes designs for manufacturability, ensuring they meet production constraints and are ready for mass production.

These applications demonstrate how the computational suite nTopology® can be used to address a wide range of design challenges, enhancing efficiency, sustainability, and innovation in engineered systems. It has already been widely noted that companies such as Materialise®, NV, headquartered in Leuven, Belgium, are among the world's largest companies in the 3D printing sector, with a strong focus on 3D printing for a wide range of industrial and healthcare applications (Materialise 2025).

It has worked to develop tools that enhance compatibility between CAD files and the specific formats used by 3D printers. Here are some practical examples of how Materialise® supports various industries:

- **3D Printing Preparation Software:** *Materialise Magics®* is a powerful tool that enables the import of various data formats, the repair and optimization of 3D models for printing, improving both the efficiency and quality of the production process.
- **Personalized Healthcare Solutions:** *Materialise®* offers 3D printing services for the creation of patient-specific anatomical models, surgical guides, and customized implants, enhancing the precision and effectiveness of surgical procedures.
- **CO-AM Platform:** Materialise® has developed a platform that connects people, processes, and technologies within the production environment, improving data traceability and automating repetitive tasks to increase operational efficiency.
- **Strategic Collaborations:** Materialise® collaborates with companies like HP to integrate advanced 3D printing solutions, making the technology more accessible and accelerating the global adoption of digital manufacturing.

However, the lack of complete standardization between design and production software still greatly hinders the efficiency of the digital workflow. In the industrial sector, Orbital Materials® is experimenting with the use of AI to generate new materials with specific properties, drastically reducing development times. Solutions such as Orbital Materials offer cloud-based tools for collaborative management of production processes (Orbital Materials 2025). For this approach to become the norm, a paradigm shift in data management and CAD/CAM protocols is required—one based on open formats, real-time interoperability, and shared standards—to ensure a greater connection across all phases of the product lifecycle. Some possible directions include:

- Development of open standards: Platforms that adopt universal file formats for greater interoperability and connection between software and hardware.
- Integration with cloud computing: Solutions like Autodesk Forge® are already exploring cloud-based environments to facilitate real-time collaboration and project data access.
- AI-based automation: The use of machine learning algorithms to automate the conversion of CAD models into CAM instructions optimized for production.

The ultimate goal of current trends in computational CAD is to develop a technological ecosystem in which design, simulation, and production are perfectly integrated, eliminating the inefficiencies and barriers between the creative phase and manufacturing steps. Only through greater integration of protocols can we fully exploit the potential of new technologies and make the CAD/CAM process more efficient, flexible, connected, and increasingly predictive.

6.2 Toward a Conscious Materiality (II)

At the heart of the Renaissance—often celebrated as the revival of classical learning and the dawn of the Modern Age—philosophers and scientists overturned the notion of a static, secular cosmos and forged a new understanding of humanity's relationship with Nature. Among them, the Italian thinker Giordano Bruno (1548–1600) championed a daring, panpsychist vision of an infinite universe infused with a universal consciousness—a view for which he paid the ultimate price. His tragic execution underscores both the transformative power and perilous reception of ideas that would reshape Western thought (Yurchenko 2024). For Bruno, every single atom contributes to the cosmic harmony of a living, thinking, and interconnected cosmos—one that does not merely exist, but actively participates in creation (Blum 2021). This vision, which blends spirituality and science, forcibly suppressed, later incarnated itself within the walls of the *Wunderkammer*—artfully constructed spaces where Nature and Artifice intertwine in a synthesis of knowledge, imagination, and spirituality (BioArt Laboratories 2013). In these rooms of art, natural objects, artifacts, curiosities, and scientific discoveries coexist in an apparently disorderly chaos, set up according to the invisible traces of symbolic symmetries and arcane meanings. Within them, man has sought to understand the world not as an entity separate from his own vision, but as a microcosm in which every element—whether natural or man-made—is part of a whole that can, in some way, be *understood* and *tamed* by the human mind. Today, more than five centuries later, the boundary between Nature and Artifice continues to blur and be redefined by progress in material science, leading us to rediscover the same tension between natural order and human ingenuity (Altamirano 2014). Current research is exploring directions that challenge the traditional separation between inanimate matter, intelligence, and consciousness. Our perception of the material world has traditionally been that of an inanimate realm, where organisms and matter are seen as passive objects, devoid of awareness or intelligent response to stimuli. However, modern theories on synthetic biology and self-organization are radically changing this view. In particular, the research of biologist Michael Levin[1] on bioelectricity and cellular awareness offers a new perspective on the consciousness of living nature. Developmental biology studies how an organism develops from a single fertilized egg cell. Traditionally, it has emphasized the role of genes, proteins, and the environment in guiding developmental processes. Levin's work has expanded this framework by highlighting bioelectricity as a key factor in morphogenesis and cellular differentiation. These electrical signals influence the formation of biological patterns and play a fundamental role in processes such as tissue and organ regeneration. By manipulating these

[1] Michael Levin is an American developmental and synthetic biologist at Tufts University, where he holds the Vannevar Bush Distinguished Professorship. He is also the director of the Allen Discovery Center at Tufts University and the Tufts Center for Regenerative and Developmental Biology. Additionally, he co-directs the Institute for Computationally Designed Organisms with Josh Bongard (2025). https://facultyprofiles.tufts.edu/michael-levin-1/publications.

signals, new perspectives open up for regenerative medicine and the development of bio-inspired technologies (Levin and Stevenson 2012). The process of organic development is extremely complex and involves multiple stages, from cell division to the formation of tissues, organs, and systems. The theory of regulation suggests that cells do not act in isolation, but interact with external signals and with other cells, influencing each other in the creation of specific patterns. Similarly, the theory of position states that a cell's position within the organism plays a crucial role in its differentiation, determining its cellular fate through chemical and bioelectrical signals (Levin 2013). Levin hypothesizes that by manipulating these signals, it is possible to guide tissue proliferation and stimulate organ regeneration (Musser 2023). Furthermore, Levin argues that cells not only respond to stimuli, but also communicate and coordinate to develop an emergent awareness that exceeds the sum of the individual parts (Segall 2024). These theories invite us to envision a future in which digitally assisted design protocols evolve organically. If, as Levin argues, growth is driven not only by genetic programs but also by dynamic bioelectrical signals, then digitally conceived materials and structures need not remain static: they can behave as adaptive systems that sense, respond, and transform over time (see Chap. 5). In parametric and generative environments—using tools like Grasshopper®, Kangaroo®, Generative Components®, Dynamo® and Blender®—we move beyond crafting fixed forms toward, at least metaphorically, instilling a *material like consciousness* capable of real-time interaction and adaptation. Within this work-in-progress, the *WunderCAD/CAMmer* emerges as more than a metaphor—it becomes the ideological core of this book. Just as the *cabinets of curiosity* once united knowledge, imagination and spirituality to reveal new worldviews, the contemporary *WunderCAD/CAMmer* acts as a digital repository of future material sciences and technosciences: a *chamber of wonders* where computational tools and bio-inspired processes co-evolve. In a not-so-distant tomorrow, such a system could underpin buildings and infrastructures that not only respond to their environment but self-regenerate and adapt, thereby redefining the boundaries of engineering, medicine, architecture and design. This concept—where Nature and technology converge—frames our exploration of design's future and its ever-deepening rapport with the living world. Tim Ingold (2013), in his book *Making*, reflects on the theory of *hylomorphism*,[2] describing the creative act as a process in which idea and matter are distinctly separated: first comes mental conception, then physical materialization

[2] *Hylomorphism* is a philosophical theory developed by Aristotle, which holds that every physical object is composed of two principles: matter (hyle), which is the raw substance, and form (morphe), which determines its structure and characteristics. In this model, matter without form is indefinite, while form without matter could not exist as a physical object. The concept of hylomorphism has also had implications in modern science, particularly in scientific theories about the formation of matter and in the discussion on the dualism between body and mind. Although contemporary science may consider these principles as separate in terms of matter and physical laws, Aristotle's idea that matter without form is indefinite and form without matter is inconceivable offers an interesting perspective on the nature of the physical universe and its laws. Aristotelian hylomorphism remains one of the deepest theories in the philosophy of nature, providing a framework to understand how physical objects emerge from the interaction between matter that has the potential to take any form and a form that actualizes that potential into a determined structure.

(Ingold 2013). This Aristotelian-derived approach proves limiting within the context of contemporary digital design, particularly in generative CAD. In the current *WunderCAD/CAMmmers*, the idea is not simply imposed upon matter, but emerges from a continuous and reciprocal interaction between thought and technology. The idea becomes embodied in the technological matter itself, in a dynamic process where design is never static, but constantly evolving. This perspective resonates with emerging views in developmental science, where form and function are seen as outcomes of complex interactions between internal and external signals—electrical, chemical, and mechanical—that shape organisms in ways not reducible to genetic information alone. Similarly, in advanced digital environments, the final form is not dictated solely by an initial concept, but arises from feedback loops between designer, code, and material system. The project, like a living structure in development, is shaped by gradients, constraints, and contextual cues—it adapts, transforms, and sometimes regenerates through iterative processes. In this view, CAD/CAM systems begin to resemble dynamic morphogenetic fields, rather than static execution tools. Thus, technology is no longer merely a passive conduit for materializing thought, but an active component of a co-evolutionary process—an interface where human cognition, algorithmic logic, and material behavior collaborate (Blok 2022).The *WunderCAD/CAMmmer* is a uniquely secretive space that serves as a futuristic cradle of form, mirroring biological development by generating novelty through interaction and emergence. It is a generative womb of possibilities—an athanor of digital alchemies that is ferrying all humanity and the Earth toward evolutive dimensions whose contours remain uncertain yet stimulating. By dissolving the divide between thought and matter, it fosters a holistic ecosystem in which design, technology, and environment continuously influence one another. In a digital realm as imagined as it is realized, human creation inevitably reflects natural complexity: digital consciousness emerges from the seamless union of mind, matter, and machine rather than existing apart from material reality. Today, the fusion of design digital tools with artificial intelligence is radically reshaping how we conceive and instantiate both virtual and physical worlds. Generative design driven by machine-learning algorithms no longer follows fixed instructions but dynamically explores solutions, adapting in real time to changing constraints and objectives. This convergence forges a new synthesis of form and substance, blurring the line between the natural and the artificial. A fundamental contemporary challenge is to reinforce this interconnected ecosystem—integrating science, technology, AI, and production into an increasingly sophisticated dialogue between conception, simulation, and realization (Gao et al. 2024) and (Rashid et al. 2024).

Design Driven Innovation is the current concept that emphasizes the importance of design as the main driver of innovation, not just in aesthetic or functional terms, but as a strategic process that can generate value and differentiation. It focuses on using design as a tool to anticipate and shape future consumer needs and to innovate in a significant way, rather than just making incremental improvements to existing products or services. This approach encourages the exploration of original ideas, the creation of solutions that address emerging needs, and the design of experiences that reflect social, cultural, and ecological changes. In this mode, innovation is not just about solving technical or production problems, but also about changing the way we perceive and interact with the world, improving our lives in a sustainable and meaningful way. Design is thus seen as a fundamental factor that not only creates products but also builds new realities, business models, and interactions between humans, technology, and Nature (Verganti 2009). Connection and integration imply the abolition of traditional divisions between mind and substance, body and soul, the digital and the physical, the scientific and the humanistic, the imagined and the real, for the creation of an increasingly conscious echotech-system of the artifact realm. In this process, digital design becomes a map that allows navigation through the interconnections, the archipelagos in which the many levels of reality have unfolded, uniting the forces of the mind, nature, technology, and matter into a single, continuous interaction mediated by digital code (Gambardella 2024). The new generation of CAD/CAM/CAE/CAO tools leverages advanced algorithms to allow synthetic structures to evolve in response to external variables—physical forces, environmental conditions, and structural changes. Mirroring Levin's theories (2014), materials are no longer passive: they *learn* to react to loads, generating self-organizing dynamics. For example, Grasshopper® simulates adaptive behaviors under gravity and tension; Kangaroo® deforms models in real time according to environmental stimuli; and Generative Components® and Dynamo® enable designs to evolve based on imposed conditions. While these systems lack true biological *consciousness*, they emulate cellular self-organization by treating materials as reactive entities that communicate, adapt, and respond to stimuli.

Terahertz biosensors (THz), for example, follow this trend and represent the products of a new metamorphosis of the contemporary *WunderCAD/CAMmer* scientific realm (Peng et al. 2024). Although entirely artificial, these structures respond to external stimuli in ways that mirror living organisms. They go beyond mere sensing to exhibit a rudimentary awareness of their material state (Kaluza et al. 2024). *Terahertz biosensors* are sophisticated nanoscale devices capable of detecting signals at highly sensitive frequencies; they are among the most fascinating developments achieved in the field of new materials, with applications in medicine, biology, and security. In the vast universe of advanced sensing technologies, terahertz biosensors are becoming increasingly central. Although they may initially appear to be simple electronic devices, they actually represent a true form of *distributed intelligence* in miniature. For this reason, we can confidently classify them as *smart objects*—intelligent not merely because they collect information, but because they also read, interpret, and transmit it within more complex systems. But what exactly are they? Biosensors that operate in the terahertz range (between microwaves and infrared, roughly from 0.1 to 10 THz) use a part of the electromagnetic spectrum that, until a few years ago, was largely unexplored. Today, however, these frequencies are proving ideal for analyzing biological materials without damaging them: tissues, cells, proteins, even individual molecules. The principle is simple yet powerful: every substance has a specific signature in the terahertz band—an electromagnetic fingerprint that can be read like a code. These biosensors can thus detect mutations, identify pathogens, and monitor physiological processes—often without direct contact, without chemical markers, and in real time. Their function is based on the interaction between electromagnetic waves and microscopic structures, which alter the sensor's response in a measurable way. But what truly makes these devices *smart* is not just their sensitivity—it's their capacity for processing and connectivity. Many terahertz sensors are now integrated with artificial intelligence or machine learning systems, enabling them to distinguish meaningful signals from background noise, to learn over time, and to deliver automated diagnostics or predictions. They can also be embedded in mobile or wearable platforms, becoming active components of a distributed monitoring network. In some cases, the very design of these sensors is inspired by nature: fractal surfaces, mesh-like structures, bio-hybrid materials that enhance sensitivity or reduce interference. This makes them not only *smart* but also bio-inspired, at the intersection of biotechnology and computational design. These sensors have long been based on established physical principles, but recent research is beginning to explore models inspired by natural behaviors—such as bird flocking and the collective dynamics of ant or termite swarms (Murugan et al. 2025). In the animal kingdom, ants build *living bridges* through a collective process of interlocking connectivity and self-assembly: each individual hooks onto its neighbor with legs and mandibles, creating beams and arches that mutually support one another without any external fasteners. This *living ant bridge* not only bears the weight of the moving colony but can also adapt, heal, and dynamically reconfigure itself—much like Leonardo's *Ponte Salvatico*, an ingenious design based purely on the reciprocal interlock of wooden beams in geometric equilibrium. Inspired by this principle, a straightforward Grasshopper® definition uses an iterative circuit—a chain of parametrized components linked in a feedback loop—to generate reciprocal modules: each element bears and constrains the others, producing self-supporting structures whose shape and scale can be adjusted with a single slider. By comparing the biological "bridge," Leonardo's timber arch, and the parametric module, we uncover the same underlying logic of reciprocity and stability born from the controlled intertwining of simple units (Figs. 1–4).

Bird's murmuration, particularly, suggests the existence of a distributed animal intelligence that emerges from the collective, and similarly, the adoption of models inspired by these collective behaviors in the design of terahertz biosensors could enable the creation of more sensitive and reactive devices, capable of adapting in real time to external stimuli, demonstrating an awareness of matter in action. These sensors, designed based on interactions between microscopic units that simulate the dynamics of swarm intelligence, could be more accurate in detecting variations in terahertz signals. Just like the individual units of a swarm, each component of the biosensor could react autonomously and contribute to the collective intelligence of the system, optimizing the overall response without centralized control. This type of nature-inspired design not only enhances performance but also opens up new scenarios for the self-awareness of matter, leading to the development of surprising hypotheses that touch on many aspects of contemporary thought and culture (Shapland 2021). Moreover, a disciplined biomimetic approach—drawn from these natural models—could enable the emergence of truly smart materials that not only respond to external stimuli but may also learn and evolve over time. In the context of terahertz biosensor development for biomedical and bioengineering applications, materials science might transform from a materialistic, reductionist discipline focused solely on the physics of matter into a holistic field where a form of proto-consciousness arises through ongoing interactions between sensor materials

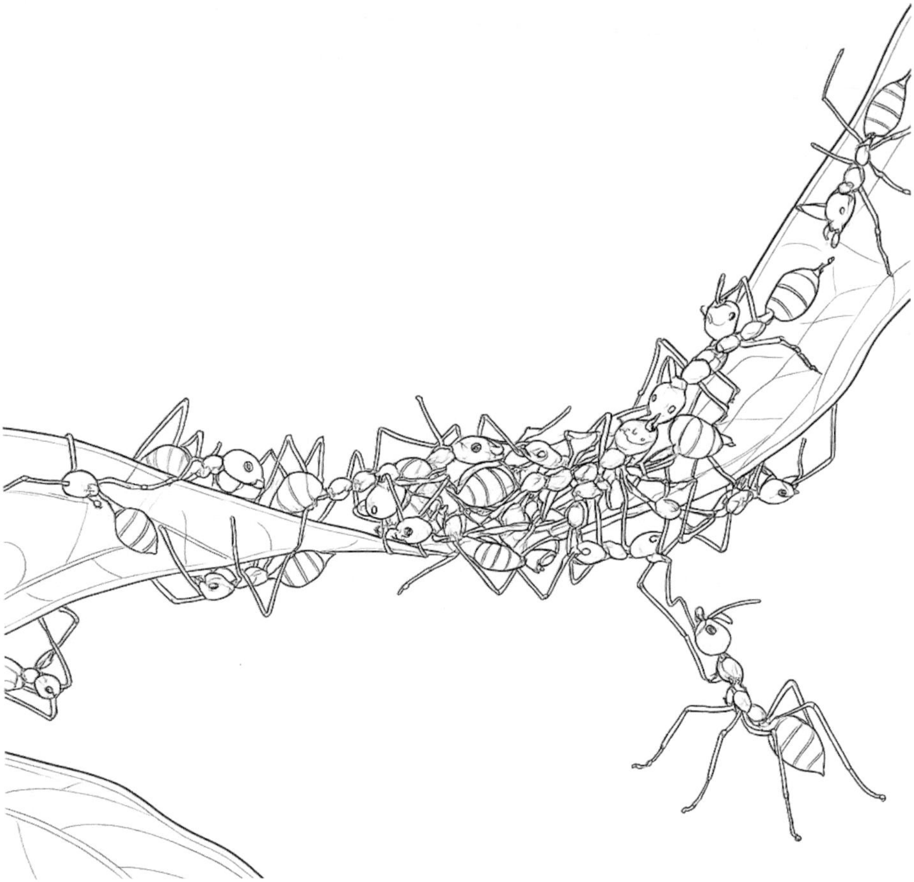

Fig. 1. The Reciprocal Ants' Living Bridge.

and their biological environment, potentially ushering in an era of predictive design and adaptive optimization.

Within this specific domain, preliminary algorithmic models for the sensor's signal-processing unit suggest a neural network that:

- Processes electromagnetic signatures from tissues quasi-continuously in real time;
- Detects and classifies spectral features across multiple layers of abstraction;
- Adapts its internal weights incrementally as new measurements arrive, improving predictive performance over time;
- Forecasts biological anomalies or mutations with probabilistic confidence scores.

While fully integrated computational CAD/CAE pipelines for terahertz biosensor production remain an emerging frontier, several recent studies illustrate how CAD-driven design and optimization are already central to THz-sensor development:

- **CAD modeling of THz imaging systems**: Zhang et al. (2024) outline a workflow in which THz-metasurface geometries are defined parametrically and imported into multiphysics simulations (CAE) to evaluate their resonance; transformative algorithms then iterate these parameters to optimize performance, and the final geometries can be exported for microfabrication processes (photolithography/CNC), although the paper does not detail CAM-ready file formats or automated protocol management.
- **Advanced THz metasurface biosensor for label-free detection:** Wekalao et al. (2025) introduce an ultrahigh-sensitivity terahertz metasurface biosensor for label-free detection of amino acids, based on an F-shaped resonator composed of graphene, gold, and silver on a silicon dioxide substrate. Finite-element simulations carried out in COMSOL Multiphysics®—a comprehensive CAE platform rather than a dedicated CAD/CAM suite, yet readily integrable with such systems—demonstrate a sensitivity of 1000 GHz/RIU and a figure of merit of 33.33 RIU^{-1} within the 0.1–0.6 THz band. By systematically varying resonator geometry, graphene's chemical potential, and the angle of wave incidence, the authors optimize performance under diverse conditions. A stacking-ensemble machine-learning framework further correlates minute resonance shifts with refractive-index changes, achieving an R^2 of 1.00 without reliance on biochemical labels. Compatible with scalable fabrication processes, this platform offers a promising route to next-generation biosensing applications in medical diagnostics, environmental monitoring, and food safety.

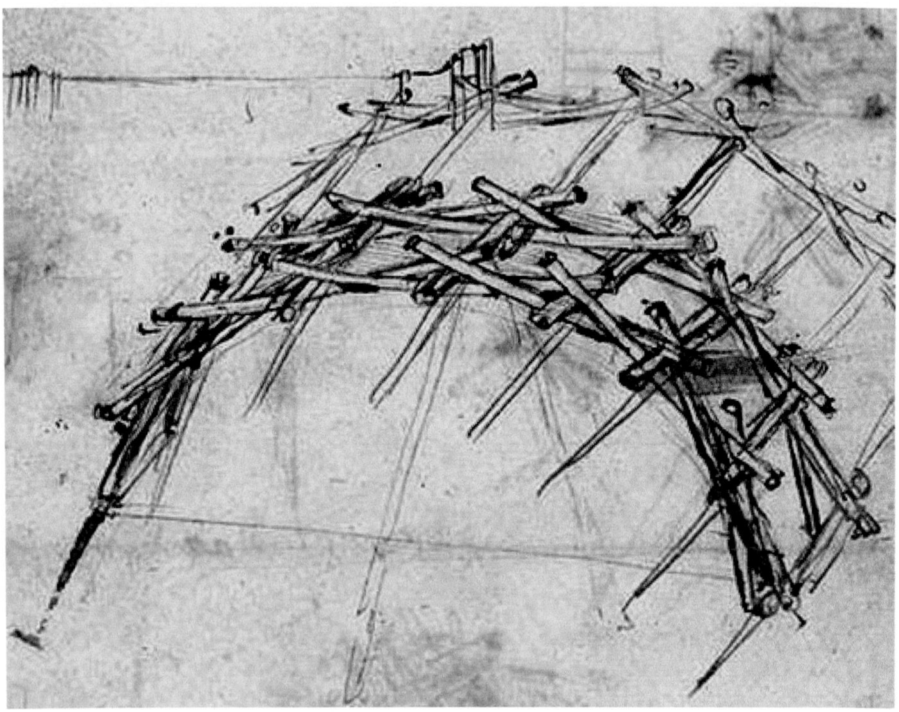

Fig. 2. 'Ponte Salvatico' – Leonardo da Vinci.

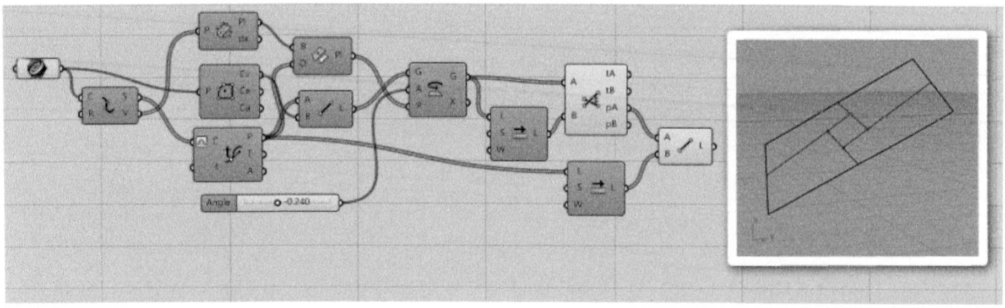

Fig. 3. Simple Grasshopper® definition for the design of reciprocal parametric modular structures.

These examples show that, although end-to-end CAD/CAM/CAE/CAO platforms tailored to terahertz biosensors are still coalescing, the core practices—parametric CAD modeling of sensor architectures and AI-driven optimization tied to CAM-compatible designs—are already in place. As these methodologies mature, we can expect true computational integration for THz biosensors, streamlining the path from digital prototype to microfabricated device.

The implications of terahertz biosensors in these technologies are particularly interesting in the context of integrating advanced design and manufacturing technologies with high-precision biological analysis. Terahertz biosensors capture real-time biological data from tissues or samples, which can then be fed into CAD/CAM workflows and integrated with multi-modal computational bio-toolsets to design and fabricate bespoke medical devices. For example, sensors measuring specific biological interactions could help create personalized implants or prosthetics that fit perfectly to the patient, based on data gathered by the sensors (Shafiee et al. 2018). With terahertz biosensors monitoring physiological engagements and material quality, CAD/CAE systems, now increasingly integrated with AI, deep learning, and quantum algorithms, could be used to engineer structures that better meet biological requirements. For instance, if a biosensor detects abnormalities in biological tissues or biomaterials, the CAD/CAE suite could automatically adapt the component design to enhance material compatibility and performance in a biological context. Terahertz biosensors could also be employed to analyze the structure and composition of biological samples or prototypes during the design phase, supporting a more accurate CAM process. For instance, if a biological prototype or medical device is being manufactured using 3D printing, terahertz sensors could monitor the material quality and cellular interactions in real time, ensuring that the production meets the necessary standards. Another area where CAD/CAM/CAE and terahertz biosensors could converge is in bioprinting—the 3D printing of biological tissues or cells. Terahertz sensors could be used to analyze the quality and effectiveness of the bioprinting process, while CAD software could optimize the design and positioning of cells or biological materials, ensuring better compatibility and functionality. In a CAM/CNC manufacturing environment, terahertz biosensors could be employed to monitor biological responses in real time during the production process. For example, when engineering tissues, the sensors could detect changes in bioelectrical properties indicating whether the biological structure is responding correctly, allowing immediate adjustments to the production process. Terahertz sensors could

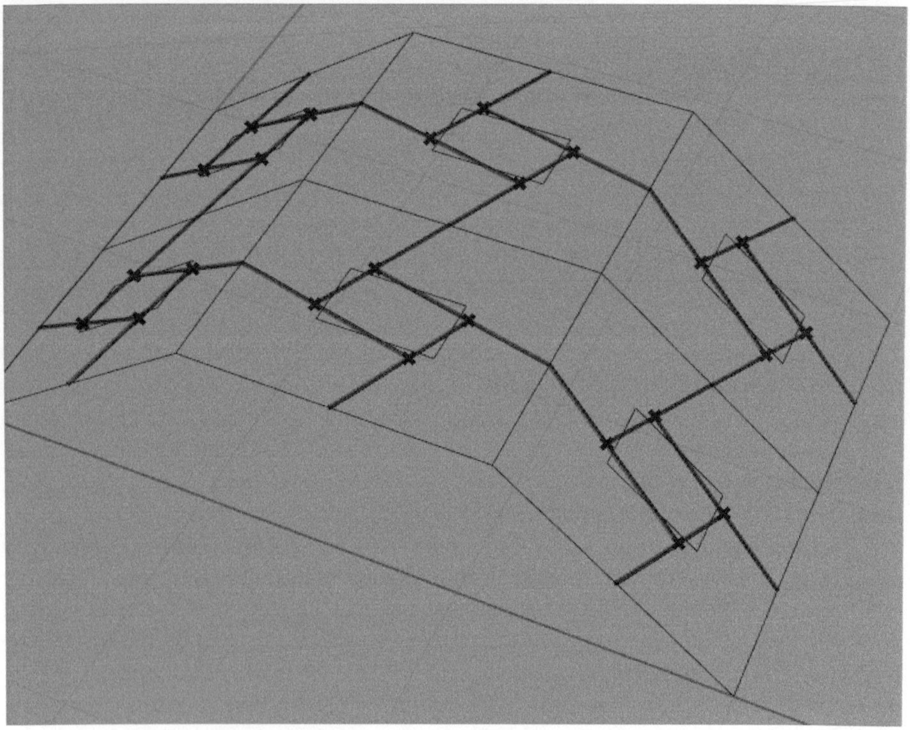

Fig. 4. Iteration of Reciprocal Geometric Modules.

also be integrated into a digital design process to simulate and test the biological responses of new medical devices or materials virtually, allowing engineers to visualize and predict how components will interact with the human body before physical production (Gezimati and Singh 2023). Integrating terahertz biosensors with CAD/CAM/CAE workflows can greatly enhance device precision and lower failure rates by enabling real-time, data-driven design adjustments. Moreover, this fusion unlocks bio-inspired innovation—allowing engineered systems to emulate natural self-organization and yielding highly personalized, efficient medical devices for cancer treatment and beyond (Hamza et al. 2024). It would seem that this new materials science aligns with Bruno's vision of a universe suffused by consciousness, where *matter* and *intelligence* are not distinct but converge in the art of design and our capacity to understand and shape the environment. Like the Renaissance *Wunderkammern*, which collected the best of science and art to give rise to integrated thinking, today the science of materials, with the aid of nature-inspired models, invites us to reconsider the boundaries between what is alive and what is artificial, suggesting that, perhaps, matter itself possesses a form of awareness capable of responding in a dynamic and intelligent way. This leads us to reflect that matter and natural life are ultimately far *more artificial* than we had imagined. It is a further step toward the idea of a world that is not only alive in a strictly adaptive and biological sense but is sentient, conscious, endowed with a form of free will that allows it to choose, program, design, take risks, and ultimately evolve continuously in relation to its local environment and the entire Universe.

In summary, murmuration and all swarm technologies serve as a metaphor for dynamic, flexible, and adaptive systems, in which decisions are made collectively and optimally, akin to the collective *behavior* of many living organisms. These advanced research areas fall within the broader scope of metamaterial creation. As already considered in Chapter 5.0, this technological class represents human-made artifacts that can have special properties not found in natural materials, and they are one of the most advanced frontiers in electromagnetism, optics, acoustics, and materials engineering.

In paragraph 5.2, *The Consciousness of the Material World (I)*, the category of so-called metamaterials is introduced within the vast field of contemporary materials. The term *metamaterial* comes from the combination of the Greek *meta* (beyond) and the Latin *materia* (material). The term was first introduced in 1967 by the Soviet physicist Victor Veselago (1929–2023), who used this term to describe a theoretical material with physical properties that do not exist in natural materials, such as a negative index of refraction. In 1999, physicist Rodger M. Walser from the University of Texas helped popularize the concept of metamaterials in the context of practical applications, especially regarding electromagnetic properties, making the term more widely known in modern physics research (Walser 2000). However, it was only in the 2000s that metamaterials became an active area of research, thanks to improvements in the design and realization of materials with extraordinary properties, such as the manipulation of electromagnetic waves.

Walser defined metamaterials as:

"macroscopic composites with a synthetic, periodic, three-dimensional cellular architecture designed to produce an optimized combination, not available in nature, of two or more responses to a specific stimulus."

His goal was to define a class of artificial materials capable of surpassing the properties of conventional materials and manipulating electromagnetic, acoustic, and mechanical waves in ways that are impossible for natural matter. Metamaterials are not simple materials, but modular engineered structures, with emergent properties derived from design rather than chemical composition, where the use of CAD/CAM systems integrated with AI and deep learning is enabling a significant increase in the design and manufacturing systems (Zheng et al. 2023). At the microscopic level, a metamaterial with optical properties, for example, has a characteristic geometry defined by its construction lattice and hierarchical scale, and is composed of repeated structural units (unit cells) smaller than the wavelength of the phenomenon it is meant to modulate. This allows for the control and manipulation of phenomena such as refraction, propagation, and absorption of waves. The structure enables the achievement of unprecedented performances such as: negative Poisson's ratio, negative mass density, negative stiffness, zero shear modulus, ultra-high strength-to-density ratio, high absorption energy, a metafluid behavior, and special optical, electrical, and acoustic *sound-absorbing properties* in 3D printing materials (Pandolfi 2021). Their applications are continuously expanding and include iconic developments in the field of optics and photonics, such as superlensing, which involves the creation of optical lenses with resolution beyond the diffraction limit, improving the resolution of optical microscopes and defining our times: *The Century of Metamaterials* (Tretyakov et al. 2017). They can be used to manipulate the phase and polarization of light to create ultra-compact optical devices, and for the creation of super-light mirrors, or for optical and acoustic cloaking, which allows light or sound waves to be bent around an object, making it literally invisible (Fleury and Alu 2014). Due to their special properties, metamaterials are increasingly being used in the development of ultraminiaturized antennas for 5G and future 6G networks, in the creation of metamaterials for interference reduction, and in devices for telecommunications such as radiation absorbers and advanced sound-absorbing materials, radar, and wireless devices. They are also used in photonic crystals that regulate light behavior in advanced optical fibers for more efficient telecommunications (Hussein et al. 2023, Kazemi 2021). In medicine, they open possibilities for the creation of bio-metamaterials for advanced imaging and innovative medical devices and for detecting biomolecules and diseases (e.g., cancer detection with THz biosensors), as well as improving safety and quality control of biomedical materials (Yang 2024). They also control the targeted release of drugs using engineered nanomaterials (Luo et al. 2025). In aerospace and defense technologies, metamaterials find applications in the creation of stealth coatings, radar-absorbing materials, and advanced thermal control systems. In robotics, they contribute significantly through auxetic metamaterials, used in soft robotics for flexible structures that expand under stress, or through the use of self-adaptive materials, which can change mechanical properties in response to external stimuli, making them ideal for prosthetics and exoskeletons (Pu et al. 2024). Regarding energy and the environment, materials are being created that can regulate heat transmission and absorption in buildings and electronic devices, increase the efficiency of solar energy conversion, and optimize water filtration through metamaterial nanostructures, also biomimetically inspired by the ultrastructure of Douglas fir wood (Wu et al. 2022, Zhang et al. 2024).

Since 1999, the concept of metamaterials has evolved rapidly, expanding its applications alongside the discovery of new properties. It has been redefined so many times that, even today, there is no single, universally accepted definition. In many ways, the structure of metamaterials mimics the hierarchical organization found in numerous natural materials. Although many applications are still in the experimental phase, metamaterials are paving the way for technologies that just a few years ago seemed impossible. Each new discovery brings us a step closer to more efficient, lightweight, and high-performance devices capable of enhancing our interaction with the world. We are only at the beginning of this revolution: with advances in computational design and advanced manufacturing, the potential of metamaterials is set to grow exponentially. Perhaps, before long, materials science will no longer be constrained by the limits imposed by nature, but only by our imagination. In the meantime, designers of metamaterials—such as *auxetic* ones—still face many challenges, as CAD/CAM software is not yet fully optimized. Over the past decade, significant efforts have been made to develop new digitally assisted design protocols that leverage recent advances in deep learning. Integrations with AI technologies are revolutionizing the process of mechanical metamaterial design, enabling property prediction and geometry generation without prior knowledge. Moreover, deep generative models can transform traditional forward design into inverse design. Several recent studies on the implementation of deep learning in mechanical metamaterials are highly specialized, and their advantages and disadvantages may not be immediately evident (Zheng 2023). Some studies argue that the main critical issues related to the current metamaterial design system include:

- The design space is heavily limited by current design methods; that is, the creation of metamaterial architectures still relies primarily on CAD (Computer-Aided Design) modeling, which does not guarantee optimal performance and requires extensive computational resources when the geometry is too complex or elaborate;
- Mechanical metamaterials always exploit zero-energy deformation modes, resulting in a fixed mechanical response after fabrication. This limits the adaptability of mechanical metamaterials compared to more advanced materials (such as smart materials), which can change their behavior in response to external stimuli (heat, electricity, magnetic fields, etc.);
- Most previous design methods follow a forward design approach, where an architecture is created and its mechanical properties are then investigated through experiments or simulations.

A peculiar class of metamaterials is represented by metasurfaces, and in particular by *Surface Plasmon Resonance (SPR)* metasurfaces— special nanofabricated structures that exploit the collective properties of free electrons in metals to enhance the interaction between light and matter at the nanoscale (Wekalao et al. 2024). Their use is particularly relevant in advanced biosensors, optical imaging, photonics, and spectroscopy. Silver-based nanostructures of this kind allow for the amplification of the optical response to tumor biomarkers. The complexity involved in designing and optimizing these devices requires the use of advanced tools for modeling, simulation, and manufacturing. SPR technology is widely employed to monitor biomolecular interactions in real time without the use of labels (Chen et al. 2024). Instruments such as the Biacore X100®, leverage this technique to study interactions between proteins, nucleic acids, and other biological molecules, providing data on the specificity, kinetics, and affinity of these interactions (Università degli Studi di Cagliari n.d.). Recent studies have shown that disordered metasurfaces can be used to control short-range surface plasmons in planar silver films. This discovery opens new possibilities for applications in imaging, sensing, and nonlinear optics (Buchmüller et al. 2023). The integration of computational technology in SPR metasurface design enables precise optimization of nanomaterial geometries and layouts to elicit targeted plasmonic responses. Because the arrangement of silver nanostructures is critical for tuning resonance at terahertz frequencies, advanced multiphysics and electromagnetic modeling software—such as COMSOL Multiphysics®, CST Microwave Studio®, Lumerical (Ansys)®, and Ansys HFSS®—is used to simulate surface-plasmon behavior under varied conditions. These platforms, though not traditional CAD authoring tools, serve as the ribs of the CAD ecosystem by providing extensive parametric control and interoperability. They support geometry import/export, adhere to common file standards, and integrate seamlessly into design-to-fabrication

workflows. By iteratively tuning parameters such as resonance wavelength, coupling efficiency, and overall sensitivity, engineers can refine metasurface geometries to meet the most demanding performance criteria.
Similarities and Compatibility:

1. **Geometric Modeling**: These tools offer advanced modeling capabilities that allow for the creation and manipulation of complex geometries, much like a CAD software. Specifically, the ability to define parametric geometries within these tools allows for dynamic modifications to the metasurface structure, similar to how it would be done in traditional CAD software.

2. **Parametric Design**: Another point of analogy is the use of parametric design. Just as in traditional CAD software, where parameters (such as dimensions or angles of a component) can be interactively changed to explore different configurations, electromagnetic simulation tools also allow for the real-time adjustment and optimization of the geometry parameters, aiming to achieve optimal metasurface performance.

3. **Design Optimization**: Much like in a traditional CAD/CAE workflow, where a designer explores different design variants to achieve the most functional or aesthetically pleasing solution, these simulation tools allow for the optimization of key parameters (such as resonance wavelength, coupling efficiency, and sensitivity). This ability to simulate different scenarios and optimize designs before fabrication significantly reduces the risk of errors during prototyping.

4. **Integration with Manufacturing**: While traditional CAD is focused on designing and producing 3D geometries for manufacturing, these simulation tools contribute to *designing for simulation* rather than just for production. This means that, although not exclusively designed for creating 3D models for manufacturing, they can be used to design geometries that will later be fabricated, providing critical information on the physical behavior of the metasurface before physical production.

5. **Interoperability**: Tools like COMSOL®, CST®, and Ansys® offer compatibility with CAD software, allowing the import of geometric models created with traditional CAD tools. Therefore, while they are not technically CAD tools, these simulation tools integrate seamlessly into the overall design workflow, which spans from geometric creation to electromagnetic simulation and design optimization. The ability to simulate complex physical phenomena, combined with parametric modeling, makes these tools an essential part of the workflow for designing innovative optical, sensing, and communication devices.

These simulations can also provide insights into how modifications to the structure affect the plasmonic response, enabling the development of metasurfaces with tailored optical and electromagnetic properties for specific applications like sensing, imaging, and communications.

FEM (Finite Element Method) and FDTD (Finite-Difference Time-Domain) simulations are essential for investigating plasmon–biomarker interactions and optimizing the device's sensitivity (Cadence System Analysis 2021). In the modern design world, talking about CAD no longer means simply drawing shapes. The term *Computer-Aided Design* is now just the starting point of a much broader process, one that involves simulation, verification, optimization, and production. This is where FEM and FDTD simulations come into play—two essential numerical methods that seamlessly integrate into the broader digital design protocols. When a designer works with CAD software, they're creating the virtual geometry of an object: a chair, a prosthetic, a micro-antenna. But for this geometry to become real and functional, it must pass another threshold—the realm of forces, waves, fields, flows, and dynamic physical behavior. This is where CAE, or Computer-Aided Engineering, steps in. It is in this phase that models are virtually tested. The FEM is perhaps the most widely used in engineering. It allows a complex object to be broken down into small parts (the *finite elements*) to analyze its behavior under mechanical, thermal, or other forms of stress. In other words, FEM lets you simulate how a beam will bend, how an electronic component will heat up, or how a frame might break under load. FDTD, on the other hand, is more specialized but equally powerful. It is used to simulate the propagation of electromagnetic waves in time. Imagine trying to understand how light behaves in an optical fiber, or how an electric field is distributed around an antenna. The FDTD method discretizes space and time into a grid and computes, moment by moment, how the waves propagate. It's widely used in optics, photonics, microwave design, radar, and biomedical applications. So, while FEM and FDTD are not CAD in the narrow sense, they live within the digital ecosystem of design (Machine Design 2024). Many modern software platforms—such as Ansys®, COMSOL®, CST Studio Suite®—combine CAD modeling and CAE simulation within a single interface, allowing smooth transitions from drawing to analysis. Finally, the last phase in the workflow is CAM: the production phase, which takes validated models and translates them into instructions for machining, 3D printing, or other manufacturing systems. We can therefore say that FEM and FDTD simulations are an integral part of the CAD *protocol*, if by CAD we mean not only drawing but a comprehensive computational design method, allowing humans to simulate, predict, refine, and finally create—before even touching the material (Hamdi et al. 2013). FEM, FDM, and FVM methods form the numerical backbone of CAE workflows by directly leveraging CAD geometry: the solid model exported from CAD (in formats such as STEP or IGES) is first discretized into a computational mesh—FEM using unstructured elements (tetrahedra, hexahedra), FDM applying a structured grid, and FVM constructing control volumes around mesh nodes—onto which boundary conditions and loads defined in the CAD environment are imposed. The CAE solver then computes field variables (stress, thermal gradients, electromagnetic fields, fluid flows) element-by-element (FEM), node-by-node (FDM), or volume-by-volume (FVM), and the resulting results are visualized and, if necessary, imported back into the CAD model for geometry refinement and material optimization, enabling an iterative design-to-simulation loop prior to physical prototyping. This design approach is now at the heart of predictive—and increasingly biomimetic, adaptive, intelligent—design. It's the key that binds form to function before matter comes into play. The main challenge in designing high-sensitivity biosensors lies in achieving a pattern that maximizes the signal amplification generated by the interaction between biomolecules and the sensor surface. For the physical fabrication of the device, manufacturing techniques must be extremely precise, as the structures are on the nanometer scale. Technologies used include *Electron Beam Lithography (EBL); Deep Ultraviolet (DUV) Photolithography; Nanoimprint Lithography (NIL)*, which allow the pattern to be imprinted with high resolution. Furthermore, the controlled deposition of metallic layers through techniques such as *PVD (Physical Vapor Deposition)* in Angstrom Engineering (2025), *CVD (Chemical Vapor Deposition)* in Sun et al. (2021), and *ALD (Atomic Layer Deposition)* in Johnson et al. (2014) ensures material quality and the stability of plasmonic properties. CAE (Computer-Aided Engineering) plays a crucial role in validating and optimizing the biosensor. Through advanced simulations and *Mie scattering models* (Microtrac Retsch GmbH 2025), it is possible to test different configurations and predict the plasmonic response to varying biomarker concentrations, thereby iteratively improving the sensor's sensitivity. Simulations also take into account environmental variables such as temperature, humidity, and optical interference, allowing for a detailed analysis of the interaction between the biosensor and its surrounding environment. CAO (Computer-Aided Optimization), integrated with AI–artificial intelligence and bio-inspired models, further enhances detection by using quantum decision networks (QDNNs) to analyze signals with superior precision. These models, inspired by quantum mechanics and supported by architectures such as *Quantum Boltzmann Machines (QBM)* and *Quantum Support Vector Machines (QSVM)*, are capable of recognizing complex patterns in data and improving the system's predictivity (Coopmans and Benedetti 2024). The use of artificial intelligence and quantum computing is revolutionizing early cancer diagnosis, enhancing the efficiency and accuracy of the devices used. Electromagnetic and mechanical simulations allow for the prediction of structural behaviors of devices, significantly reducing both development time and costs. Additionally, quantum neural networks and bio-inspired models can analyze vast datasets, accelerating the discovery of new materials with targeted applications, such as in bioengineering and medical device industries. A promising example in this field is represented by QBIC metasurfaces, which, due to their high-quality resonance factors, are opening new frontiers in biosensors, optoelectronic devices, and advanced sensor technology, providing solutions with unprecedented precision. These developments not only enhance detection capabilities but also offer a more precise and detailed understanding of molecular and biological interactions, creating opportunities for new technological applications in health and biotechnology (Joseph et al. 2021).

Connection implies *synchronization*, a property that matter exhibits at every dimensional scale. In his book *Sync: The Emerging Science of Spontaneous Order* (2003), Steven Strogatz explores the phenomenon of spontaneous synchronization, where complex systems of autonomous elements synchronize without external guidance. The text explores the phenomenon of spontaneous synchronization in nature, such as that observed in fireflies, which emit synchronized lights in groups—a phenomenon that has fascinated scientists and travelers for centuries. Bioluminescence, produced by chemical reactions between luciferin and oxygen with the enzyme luciferase, is a highly efficient process that releases light with minimal heat waste. This synchronized behavior facilitates mating among fireflies, with characteristic light patterns specific to each species. Strogatz also mentions the discovery by physicist Huygens, who in the 17th century noticed the synchronization of clock pendulums. After observing that nearby pendulums tended to synchronize, Huygens realized that the phenomenon was due to imperceptible movements of the common base on which they were mounted. This discovery led to the understanding of synchronization as a phenomenon present in numerous physical and biological systems. The concept of synchronization is also fundamental in biological systems, such as brain and heart rhythms, and in physical systems like lasers and particle accelerators (Pikovsky et al. 2001). In lasers, the synchronization of longitudinal modes allows the generation of ultrashort pulses, crucial for advanced applications in spectroscopy and medicine. Similarly, in synchrotrons, the synchronization of particles with electromagnetic fields is essential for efficient acceleration, preventing particle losses and inefficiencies in the process. Strogatz's pioneering work has played an important role in highlighting how the tendency to synchronize could be the most mysterious and pervasive force in all of nature—a theme that has intrigued some of the brightest minds of the 20th century, including Albert Einstein (1879–1955), Richard Feynman (1918–1988), Norbert Wiener (1894–1964), Brian Josephson, and Arthur Winfree (1942–2002). Elegant and fascinating, *Sync* told the story of the birth of a new science and showed that, although these phenomena may seem unrelated on the surface, at a deeper level, there exists a profound connection forged by the unifying power of mathematics. Mathematical models to describe synchronization include Kuramoto's phase equations, which describe the dynamics of a population of coupled oscillators (Strogatz 2000). These models are used to study synchronization phenomena in biological systems, neural networks, and many other contexts. Synchronization is a fundamental phenomenon in various fields of science and technology, particularly in materials and manufacturing processes. In physics, for example, the *synchronization of qubits* through entanglement and quantum coherence is crucial for the functioning of quantum computers, where the coordinated behavior of qubits enables the execution of complex calculations with unprecedented efficiency. In material manufacturing, synchronization plays an equally significant role. In semiconductor fabrication processes, the synchronization of deposition and thermal treatment phases is essential to ensure uniformity and quality of the final product. Moreover, parameters such as temperature, pressure, and chemical composition must be synchronized precisely during operations like sintering and crystallization to improve the quality and structure of materials. In the mechanical and structural properties of materials, synchronization also plays a crucial role. Synchronization of internal oscillations or vibrations can influence the dynamic behavior of the material, such as fatigue resistance or shock absorption. In composite materials, synchronization between matrix and reinforcement phases optimizes mechanical properties and improves the overall performance of the material. In smart and active materials, synchronization between physical or chemical changes is used to develop materials capable of responding to external stimuli, such as variations in temperature or electric fields. Synchronization is also essential in advanced sensors and actuators, which require materials with well-synchronized reactive properties for proper functioning. Furthermore, synchronization plays a key role in the scalability of production processes, especially on an industrial scale, where it is necessary to maintain uniformity in material properties across large volumes. This is crucial for improving the consistency of material quality, reducing variations, and increasing the reliability of the final product.

Finally, research on new materials, such as metamaterials, exploits the synchronization of properties to achieve desired behaviors not found in conventional materials. Synchronization is also relevant for the development of technologies based on quantum principles, such as materials for quantum computers and sensors. In this context, modeling and simulating atomic dynamics, based on the synchronization of molecular movements, allows optimization of material characteristics, such as strength and flexibility.

To conclude, *connection* proves to be not merely a structural constraint but the very connective tissue that holds—and propels forward—the entire materials-science ecosystem, from metamaterials to advanced alloys, as monitored in the present research. On one hand, the synergy among complex geometries, atomic structures, and functional properties demands a continuous exchange of information and simultaneous updating of CAD/CAM/CAE models and experimental data; on the other, the synchronization protocol serves as guarantor of coherence, traceability, and reliability at every stage of the process. Looking ahead, it will be precisely the ability to orchestrate in real time the multiple factors—from simulation to machining, from data versioning to predictive analyses—that ensures success not only for new generations of metamaterials but for every innovation in the materials field. And, just like a swarm of drones communicating instantaneously to adapt to their environment, swarm tech offers us the perfect metaphor: only a network of synchronized elements, capable of acting in unison, can push the boundaries of innovation and swiftly transform ever more ambitious visions into ultra–high-performance components.

References

Allers, R., and Minkoff, R. (1994). The Lion King. Walt Disney Pictures. Available at: https://walt-disney-animation-studios.fandom.com/wiki/The_Lion_King [Accessed: 16 May 2025].

Altamirano, M. (2014). Three concepts for crossing the nature-artifice divide: Technology, milieu, and machine. Foucault Studies, 11–35.

Angstrom Engineering (2025). Physical-Chemical Vapor Deposition Systems & Coated Products. Available from: https://angstromengineering.com/ (Accessed 22 May 2025).

Beni, G., and Wang, J. (1989). Swarm intelligence in cellular robotic systems. Proceedings of the NATO Advanced Workshop on Robots and Biological Systems, Tuscany, Italy, June 26–30, 1989.

Benteh (2016). A murmuration of robots, a huddle of penguins. The Visual Squirrels. Available at: https://visualsquirrels.net/science/a-murmuration-of-robots-a-huddle-of-penguins/[Accessed: 16 May 2025].

Bin Rashid, M. D., Karim Kausik, A., and Ashfakul, K. (2024). AI revolutionizing industries worldwide: A comprehensive overview of its diverse applications. Hybrid Advances, 7, 100277.

BioArt Laboratories. (2013). Video Performance. Available at: https://bioartlab.com/portfolio_page/video-performance/ [Accessed: 16 May 2025].

Blum, E. (2021). Panpsychism in Giordano Bruno and Tommaso Campanella. pp. 207–217. *In*: Traversino Di Cristo, M. (Ed.). Giordano Bruno: Law, Philosophy, and Theology in the Early Modern Era. Garnier.

Blok, V. (2022). The role of human creativity in human-technology relations. Philosophy & Technology, 35(1): 59.

Boid Library. (2025). Boid Library for Rhino/Grasshopper. Food4Rhino. Available at: https://www.food4rhino.com/en/app/boid-library [Accessed: 16 May 2025].

Boid Library Demonstration. (2025). YouTube video, 5:32. Available at: https://youtu.be/PFHY8ghBQIE?si=3FOtDepO3P8oPQRw [Accessed: 16 May 2025].

Breathtaking starling murmuration captured in Italy. (2024). YouTube. Available at: https://youtu.be/FlbWQhl87cY?si=Na1sFW_ss58lj_0l [Accessed: 16 May 2025].

Buchmüller, M., Shutsko, I., Schumacher, S. O., and Görrn, P. (2023). Harnessing short-range surface plasmons in planar silver films via disorder-engineered metasurfaces. arXiv.

Cadence System Analysis. (2021). FDTD vs. FEM vs. MoM: What are they and how are they different? Cadence. Available at: https://resources.system-analysis.cadence.com/blog/msa2021-fdtd-vs-fem-vs-mom-what-are-they-and-how-are-they-different [Accessed: 16 May 2025].

Cavagna, A., and Giardina, I. (2014). Bird Flocks as Condensed Matter. Annual Review of Condensed Matter Physics, 5: 183–207.

Chen, Y., Zhang, H., Li, R., Fan, H., Huang, J., Zhou, R. et al. (2024). Novel multifunctional meta-surface plasmon resonance chip microplate for high-throughput molecular screening. Advanced Healthcare Materials.

Cheng, K., Davis, M. K., Zhang, X., Zhou, S., and Olechowski, A. (2023). In the age of collaboration, the computer-aided design ecosystem is behind: An interview study of distributed CAD practice. Proceedings of the ACM on Human-Computer Interaction, 7(CSCW1), Article 137: 1–29.

Ye, C., Xu, H., Chen, Q., and Ye, X. Y. (2024). Improved marine predators algorithm for engineering design optimization problems. Scientific Reports, 14: 13000.

Coopmans, L., and Benedetti, M. (2024). On the sample complexity of quantum Boltzmann machine learning. Commun. Phys., 7: 274.

Duboscq, J., Agil, M., Engelhardt, A., and Thierry, B. (2014). The function of postconflict interactions: new prospects from the study of a tolerant species of primate. Animal Behaviour., 87: 107–120.

Fleury, R., and Alu, A. (2014). Cloaking and invisibility: A review. Progress in Electromagnetics Research, 147: 171–202.

Gambardella, C. (Ed.). (2024). For Nature/With Nature: New Sustainable Design Scenarios. Springer Series in Design and Innovation (SSDI, volume 38). Springer.

Gao, R. X., Krüger, J., Merklein, M., Möhring, H.-C., and Váncza, J. (2024). Artificial intelligence in manufacturing: State of the art, perspectives, and future directions. CIRP Annals, 73(2): 723–749.

Gezimati, M., and Singh, G. (2023). Terahertz imaging and sensing for healthcare: Current status and future perspectives. IEEE Access, 11: 1–11.

Hamdi, M., Aifaoui, N., Louhichi, B., Benamara, A., and Deneux, D. (2013). FEM simulation based on CAD model simplification: A comparison study between the hybrid method and the technique using a removing details (Vol. 1). Springer.

Hamza, M. N., Islam, M. T., Din, I. u., Lavadiya, S., Sanches, B. C. de S., Koziel, S. et al. (2024). Development of a terahertz metamaterial micro-biosensor for ultrasensitive multispectral detection of early stage cervical cancer. IEEE Sensors Journal, 24(20): 32065.

Hu, G., Zhong, J., Wei, G., and Chang, C.-T. (2023). DTCSMO: An efficient hybrid starling murmuration optimizer for engineering applications. Computer Methods in Applied Mechanics and Engineering, 405: 115878.

Hussein, H., Atasoy, F., and Elwi, T. A. (2023). Miniaturized antenna array-based novel metamaterial technology for reconfigurable MIMO systems. Sensors, 23(13): 5871.

Ingold, T. (2013) Making: Anthropology, archaeology, art and architecture. Routledge.

Johnson, R. W., Hultqvist, A., and Bent, S. F. (2014). A brief review of atomic layer deposition: From fundamentals to applications. Materials Today, 17(5): 236–246.

Joseph, S., Pandey, S., Sarkar, S., and Joseph, J. (2021). Bound states in the continuum in resonant nanostructures: An overview of engineered materials for tailored applications. Nanophotonics, 10(17): 4175–4207.

Kaluza, M., Walczakowski, M., and Siemion, A. (2024). Exploring the impact of 3D printing parameters on the THz optical characteristics of COC material. Materials, 17(20): 5104.

Kazemi, F. (2021). Miniaturized array antenna base on metamaterial structures. AEU - International Journal of Electronics and Communications, 138: 153832.

Kennedy, J., and Eberhart, R. C. (1995). Particle swarm optimization. pp. 1942–1948. *In*: Proceedings of the 1995 IEEE International Conference on Neural Networks (Vol. 4,). IEEE. Available at: http://doi.org/10.1109/ICNN.1995.488968 [Accessed: 16 May 2025].

Levin, M., and Stevenson, C. G. (2012). Regulation of cell behavior and tissue patterning by bioelectrical signals: Challenges and opportunities for biomedical engineering. Annual Review of Biomedical Engineering, 14: 295–323.

Levin, M. (2013). Reprogramming cells and tissue patterning via bioelectrical pathways: Molecular mechanisms and biomedical opportunities. Wiley Interdisciplinary Reviews: Systems Biology and Medicine, 5(6): 657–676.

Levin, M. (2014). Molecular bioelectricity: how endogenous voltage potentials control cell behavior and instruct pattern regulation in vivo. Molecular Biology of the Cell, 25(24): 3835–3850.

Liang, S., Pan, Y., Zhang, H., Zhang, J., Wang, F., and Chen, Z. (2022). Marine predators algorithm based on adaptive weight and chaos factor and its application. Scientific Programming, 2022, Article 4623980.

Lu, D., Cheng, S., Chen, Y., Zhang, X., Li, H., and Yang, T. (2025). An adaptive parameter evolutionary marine predators algorithm for joint resource scheduling of cooperative jamming networked radar systems. Remote Sensing, 17(8): 1325.

Luo, Y., Gao, M., Lu, C., Zhang, J., Liu, J., Tan, Y. et al. (2025). High-sensitivity tunable terahertz metasurface based on vanadium dioxide for biological protein detection. ACS Applied Optical Materials, 3(3): 789–797.

Machine Design. (2024). What's the difference between FEM, FDM and FVM? Machine Design. Available at: https://www.machinedesign.com/additive-3d-printing/fea-and-simulation/article/21832072/whats-the-difference-between-fem-fdm-and-fvm [Accessed: 16 May 2025].

Materialise. (2025). Materialise Magics | Data and Build Preparation Software. Available at: https://www.materialise.com/it/industriale/software/magics-data-build-preparation [Accessed: 16 May 2025].

McNeel Discourse. (2019). How does the HoopSnake plugin work?. Available at: https://discourse.mcneel.com/t/how-does-the hoopsnake-plugin-work/88641 [Accessed: 16 May 2025].

Meyers, J. (2024) Connected CAD and CAM software avoids multiple data transfers, improves efficiency. Siemens Digital Industries Software. Available at: https://blogs.sw.siemens.com/industrial-machinery/2024/02/01/connected-cad-and-cam-software-dpp/ [Accessed: 16 May 2025].

Microtrac Retsch GmbH. (2025). Mie scattering and Mie theory. Available from: https://www.microtrac.com/knowledge/mie-scattering-mie-theory/ [Accessed May 22, 2025].

Murugan, K., Dhivya, R., Sangeetha, C., and Alagarsamy, M. (2025). Design and analysis of a highly sensitive terahertz biosensor for early cancer detection using silver surface plasmon resonance metasurfaces and elastic reflection starling murmuration equivariant quantum decision networks. ECS Journal of Solid State Science and Technology, 14: 10.1149/2162-8777/ada4da.

Musser, G. (2023). The biologist blowing our minds. Nautilus. Available at: https://nautil.us/the-biologist-blowing-our-minds-323905/ [Accessed: 16 May 2025].

nTop. (2024). Connected CAD and CAM software: DPP. Siemens. Available at: https://blogs.sw.siemens.com/industrial-machinery/2024/02/01/connected-cad-and-cam-software-dpp/ [Accessed: 16 May 2025].

Orbital Materials. (2025). Orbital Materials. Available at: https://www.orbitalmaterials.com/[Accessed: 16 May 2025].

Pandolfi, T. M. (2021). Experimental investigations and metrological insights on the sound absorbing properties of 3-D printed space-coiled acoustic metamaterials (Master's thesis, Politecnico di Torino). Available at: https://webthesis.biblio.polito.it/17057/ [Accessed: 16 May 2025].

Peng, J., Lin, X., Yan, X., Yan, X., Hu, X., Yao, H. et al. (2024). Terahertz biosensor engineering based on quasi-BIC metasurface with ultrasensitive detection. Nanomaterials (Basel), 14(9): 799. https://doi.org/10.3390/nano14090799.

Pikovsky, A., Rosenblum, M., and Kurths, J. (2001). Synchronization: From coupled systems to complex networks. Cambridge University Press.

Pu, Y., Zheng, S., Hu, X., Tang, S., and An, N. (2024). Robotic skins inspired by auxetic metamaterials for programmable bending of soft actuators. Materials & Design, 246: 113334.

Rader, J. A., and Hedrick, T. L. (2023). Morphological evolution of bird wings follows a mechanical sensitivity gradient determined by the aerodynamics of flapping flight. Nature Communications, 14(1): 7494.

Reynolds, Craig W. (1987). Flocks, Herds and Schools: A Distributed Behavioral Model. ACM SIGGRAPH Computer Graphics, 21(4): 25–34. Available at: https://dl.acm.org/doi/10.1145/37402.37406 [Accessed: 16 May 2025].

Romo-Tønder Region (n.d.). Sådan oplever du Sort Sol. Romo-Tønder. Available at: https://romo-tonder.dk/en/saadan-oplever-du-sort-sol/ [Accessed: 16 May 2025].

Segall, M. D. (2024). Some philosophical implications of Michael Levin's new paradigm biology. Footnotes2Plato. Available at: https://footnotes2plato.com/2024/05/27/some-philosophical-implications-of-michael-levins-new-paradigm-biology/ [Accessed: 16 May 2025].

Selous, Edmund. (1905). Bird Life Glimpses. London: G. Allen. Available at: https://archive.org/details/birdlifeglimpses00selo/page/n9/mode/2up [Accessed: 16 May 2025].

Sheeba, A., Bushra, S. N., Rajarajeswari, S., and others. (2024). An efficient starling murmuration-based secure web service model for smart city application using DBN. Artificial Intelligence Review, 57(72).

Shafiee, A., Ghadiri, E., Kassis, J., Zarandi, N., and Atala, A. (2018). Biosensing technologies for medical applications, manufacturing, and regenerative medicine. Current Stem Cell Reports, 4: 105–115.

Shapland, M. (2021). Material culture and consciousness: A thought experiment. Time and Mind, 14(4): 517–535.

Strogatz, S. H. (2000). From Kuramoto to Crawford: Exploring the onset of synchronization in populations of coupled oscillators. Physica D: Nonlinear Phenomena, 143: 1–20.

Strogatz, S. H. (2003). Sync: The emerging science of spontaneous order. The Perseus Books Group.

Sun, L., Yuan, G., Gao, L., Yang, J., Chhowalla, M., and Gharahcheshmeh, M. H. (2021). Chemical vapour deposition. Nature Reviews Methods Primers, 1, 5.

The Conversation. (2016). Why do flocks of birds swoop and swirl together in the sky? A biologist explains the science of murmurations. The Conversation. Available at: https://theconversation.com/why-do-flocks-of-birds-swoop-and-swirl-together-in-the-sky-a-biologist-explains-the-science-of-murmurations-176194 [Accessed: 16 May 2025].

Tretyakov, S., Urbas, A., and Zheludev, N. (2017). The century of metamaterials. Journal of Optics, 19: 080404.

Università degli Studi di Cagliari. (n.d.). SPR Biacore X100. Available at: https://web.unica.it/unica/it/spr_biacore_x100.page [Accessed: 16 May 2025].

Verganti, R. (2009). Design Driven Innovation: Changing the Rules of Competition by Radically Innovating What Things Mean. Harvard Business Review Press.

Walser, R. M. (2000, March). Metamaterials: What are they? What are they good for? American Physical Society, Annual March Meeting, Abstract Z5.001, Minneapolis, MN.

Walser, R. M. (2001). Electromagnetic metamaterials. In Proceedings of SPIE (Vol. 4467). doi:10.1117/12.432921. Available at: https://doi.org/10.1117/12.432921 [Accessed: 16 May 2025].

Wekalao, J., Sk, M., Mallan, S., Murthy, G., Nagarajan, N., Karthikeyan, S., et al. (2024). Metasurface based surface plasmon resonance (SPR) biosensor for cervical cancer detection with behaviour prediction using machine learning optimization based on support vector regression. Plasmonics. Advance online publication.

Wekalao, J., Mehaney, A., Alarifi, N. S., Abukhadra, M. R., Elsayed, H. A., and Rajakannu, A. (2025). Advanced THz metasurface biosensor for label-free amino acid detection optimized with stacking ensemble algorithm. Physica E: Low-dimensional Systems and Nanostructures, 172: 116287.

Wilson, E. O. (1975). Sociobiology: The new synthesis. Belknap Press of Harvard University Press.

Wolfram, S. (2002). A New Kind of Science. Champaign, IL: Wolfram Media.

Wu, Q., Ling, F., Zhang, C., Zhong, Z., and Zhang, B. (2022). Water-based metamaterials absorber with broadband absorption in terahertz region. Optics Communications, 526: 128874.

Yang, Y. (2024). Overview of the current state of research on metamaterials in biomedicine. BIO Web Conf., 142: 03020.

Yurchenko, S. B. (2024). Panpsychism and dualism in the science of consciousness. Neuroscience & Biobehavioral Reviews, 165: 105845.

Zainal, N., Zain, A. M., and Sharif, S. (2015). Overview of Artificial Fish Swarm Algorithm and its Applications in Industrial Problems. In Applied Mechanics and Materials (Vol. 815), pp. 253–257. doi:10.4028/www.scientific.net/AMM.815.253. Available at: https://doi.org/10.4028/www.scientific.net/AMM.815.253 [Accessed: 22 May 2025].

Zamani, H., Nadimi-Shahraki, M. H., and Gandomi, A. H. (2022). Starling murmuration optimizer: A novel bio-inspired algorithm for global and engineering optimization. Computer Methods in Applied Mechanics and Engineering, 392: 114616.

Zhang, L., Liu, H., Song, B., Gu, J., Li, L., Shi, W. et al. (2024). Wood-inspired metamaterial catalyst for robust and high-throughput water purification. Nature Communications, 15(1): 2046.

Zheng, X., Zhang, X., Chen, T.-T., and Watanabe, I. (2023). Deep learning in mechanical metamaterials: From prediction and generation to inverse design. Advanced Materials, 35: e2302530.

Zheng, X. (2023). Computational Design and Fabrication of Mechanical Metamaterials (Doctoral dissertation, University of Tsukuba, Graduate School of Science and Technology).

Zichello, J. M., DeLiberto, S. T., and Holmes, P. (2024). Recent beak evolution in North American starlings after invasion. Scientific Reports, 14: 140.

7

Make it Smaller

"Nature is great in great things,
but it is greatest in the small ones."
Pliny the Elder

"The whole Universe is a shining pearl."
Master Gensha Shibi

7.1 The Nanotech Rabbit At Alice's Table

In the ancient *wunderkammern*, among the most coveted *artificialia* to be displayed in princely collections, a special place was reserved for strange and exotic micro-worlds: rare and precious *small wonders*, tiny objects crafted at the limits of the then-visible world. These were created by highly skilled artisans like the mysterious Adam Dircksz[1] (active 1500–1530), sometimes even by alchemists, whose working techniques—often kept secret—were shrouded in an aura of magic and mystery (The Metropolitan Museum of Art 2015). The craftsmanship of the makers of the time was such that they created miniature sculptures depicting mythological, biblical, or epic scenes, micro-carved into pits, seeds, corals, shells, and other precious materials, and enclosed in walnut shells. An example of this art can be seen today in the Kunstkammer of Archduke Ferdinand of Tyrol, at Ambras Castle, where more than 4,000 works are housed, including a rare cherry pit inlaid with tiny heads—an

[1] Adam Dircksz is a name commonly associated with a prominent Dutch sculptor, renowned for his role in the creation of around 60 of the approximately 150 surviving Gothic boxwood miniature carvings. Some scholars attribute these works to his workshop, while others suggest that a variety of artists, either individually or in groups, may have been responsible. It is possible that Dircksz was the pioneer of this particular sculptural style, inspiring subsequent works of similar nature. The British Museum suggests that he may have catered to elite patrons in the circle of Holy Roman Emperor Charles V, with strong connections to Delft. Despite his influence, very little is known about Dircksz's personal life. The majority of information comes from inscriptions and dates found on the carvings, which provide little more than his association with the works. He may have been a woodcutter, sculptor, medallist, or possibly a patron of the arts.

DOI: 10.1201/9781003438946-7

example of microtechnique highly appreciated at the time (Kunstkammer n.d.). The challenge of manual skill at the threshold of human visibility has long fascinated our ancestors, driving artificial techniques to compete with the imperceptible perfection of natural forms. In the context of miniaturization within artificial material design, we previously explored micro-scale printing technologies such as stereolithography (SLA) and digital light processing (DLP) in Chapter 4.0. Stereolithography (SLA) operates by using a 3D model created with CAD software, while the printing process is governed by CAM software, which defines the trajectories and steps for the precise, layer-by-layer polymerization of a photopolymer resin. Once the digital model is finalized, it is transferred to the stereolithographic machine, where the CAM system orchestrates the generation of the physical object. Charles W. Hull, recognized as the father of additive manufacturing, developed and patented the first functional SLA method in 1986. His invention relied on ultraviolet light to selectively cure photosensitive liquid polymers—a process detailed in U.S. Patent 4575330-A, *Apparatus for Production of Three-dimensional Objects by Stereolithography.* The method involves the creation of a three-dimensional object by sequentially generating cross-sectional patterns at the surface of a fluid medium capable of altering its physical state when exposed to targeted energy stimuli, such as radiation, particle bombardment, or chemical reaction. Each layer corresponds to a horizontal cross-section of the final object and is solidified before the next is added, resulting in a cumulative, laminar build-up that allows the object to emerge gradually from a flat surface during the fabrication process (Hull 1986). Notably, an earlier conceptual attempt at 3D printing dates back to 1967, when Wyn Kelly Swainson filed a patent for a laser holography system capable of producing three-dimensional forms. However, his work never reached the stage of a functional prototype. Although Swainson's ideas were forward-thinking, they remained largely theoretical and were ultimately eclipsed by Hull's practical and scalable stereolithographic process in the mid-1980s. Swainson, who continued to pursue research without commercial success, passed away in 1988, never witnessing the 3D printing revolution he had partially anticipated. As most experts agree, the true origin of modern stereolithography and the broader field of additive manufacturing begins with Charles W. Hull's contributions. Since then, SLA technology has undergone substantial evolution, giving rise to multiple technical variants. These can now be classified into four successive generations of stereolithographic systems, each representing a significant advance in resolution, speed, scalability, and integration with other digital design tools. It is worth considering the analysis presented by Michael Petch (2024) in his report *The Future of 3D Printing 2024: The 3D Printing Industry Executive Survey*, which offers a concise and comprehensive overview of the projections for the next decade in the CAD/CAM macro- and micro-scale 3D printing sector. Building on the insights provided in Michael Petch's article (2024), a historical overview is proposed to trace the rapid evolution of 3D printing technologies—from early macro-scale developments to the current meta-nanometric level—which represents the core focus of this chapter (Gonzalez 2025):

First Generation (1980s – 1990s):

The first generation, based on Hull's work, uses a laser beam to polymerize photosensitive resins in a layer-by-layer process. Each layer is drawn by a numerically controlled laser that solidifies the surface of the liquid material. While innovative, this approach had a low printing speed and limited ability to resolve very complex geometries.

Second Generation (1990s – 2000s):

The second generation focused on optimizing the process to increase printing speed and quality. One of the main innovations was projection stereolithography, which uses photomask technology to project the image of each layer onto the entire build plane, polymerizing each layer simultaneously rather than tracing it with a laser. Additionally, an integrated CAD module allows geometric optimization of each model before printing. For applications in DNA nanotechnology, the platform offers seamless integration with caDNAno®, enabling researchers to import design files directly, simulate structures in solution, and optimize folding parameters prior to physical synthesis (Douglas et al. 2009).

Third Generation (2015 – 2020):

In 2016, the Continuous Liquid Interface Production (CLIP) process was introduced (Janusziewicz et al. 2016). This system represented a significant leap in printing speed. Instead of building the object layer by layer, CLIP allows for continuous production of objects, drastically reducing production times. Thanks to a combination of UV light and a liquid interface that keeps resins in a semi-liquid state, parts can be printed in minutes rather than hours. This technology paved the way for high-speed, high-resolution production (Lee 2024).

Fourth Generation (2020 – Present):

The fourth generation involves volumetric stereolithography, a technique that allows 3D printing of objects in volume, using a volumetric scan area instead of separate layers. This innovation enables the production of very complex objects in seconds, using simultaneous photon projection through an entire volume of resin. Volumetric stereolithography is seen as the next step in the evolution of 3D production processes, due to its ability to manipulate structures at the volume level rather than at the individual layer level. The stereolithography process starts with a file in STL (Standard Tessellation Language), which is now the standard for every AM process. By slicing the STL file, the 3D model is translated into 2D slices containing cross-sectional information. Based on these 2D slices, the physical model can be produced layer by layer. The 3D model in STL format is represented by a polygonization of many small triangular faces. Each triangular face is described by the coordinates of three vertices and a unit vector indicating the normal direction of the face. The foundation of stereolithography is the polymerization reaction of resins, an exothermic process characterized by chemical crosslinking reactions. The reaction is initiated by UV light energy and involves two transitions during the polymerization process: gelation and vitrification. Gelation is a transition from liquid to rubber that causes a dramatic

Table 1. Comparison of Main Advanced 3D Macro Printing and Nano-Manufacturing Technologies.

Technology	Application Sector	Key Usage	Technologies Applied	Company References
Continuous Liquid Interface Production (CLIP)	Consumer Goods, Prototyping, Medical Devices	Fast 3D printing of high-resolution objects, reducing time for complex parts	Liquid resin, continuous projection, light curing	Carbon™
Digital Light Processing (DLP)	Consumer Goods, Medical Devices, Jewelry	High-speed 3D printing for fine details and intricate designs	Digital light projection, photopolymerization	EnvisionTEC® Rapid Shape™
Direct Ink Writing (DIW)	Biomedical, Aerospace, Structural Materials	3D printing with functional inks to produce complex geometries	Inkjet-like extrusion, multi-material printing	Optomec®, Xerox®
Nanoimprint Lithography (NIL)	Semiconductor Manufacturing, Optoelectronics	High-precision patterning for microchips, photonic devices	Thermal or UV nanoimprint, high-resolution molds	Nanonex® EV Group®
Stereolithography (SLA)	Biomedicine, Electronics, Automotive	Creation of highly detailed 3D models, prototyping, and custom parts	UV-curable resins, laser-based curing, CAD/CAM integration	Formlabs® 3D Systems®
Two-Photon Polymerization (TPP)	Nanomedicine, Microelectronics, MEMS	Fabrication of nanoscale 3D structures for medical implants, microelectronics	Two-photon absorption, ultrafast laser pulses, photosensitive resins	Nanoscribe® Botmingen University™
Femtosecond Laser Lithography (FLL)	Nanomaterials, Biotechnology	High-precision direct writing for nano-scale patterning and applications	Femtosecond laser pulses, multi-photon absorption	Lightwave Logic™,Nanoscribe®

increase in viscosity. During this transition, gel and sol phases coexist. Vitrification, on the other hand, is a gradual, thermoreversible process that transitions from liquid or rubbery resin to solid, glassy resin. Other known stereolithography systems include color stereolithography and thermal stereolithography.

The historical development of SLA technology laid the foundation for rapid advancements in 3D printing, initially focusing on macro-scale structures and later progressing toward micro and nano-scale fabrication. As 3D printing evolved, there was a natural progression toward more precise and smaller-scale manufacturing, fueled by advancements in nanotechnology. By the early 2000s, research into nanomaterials and molecular engineering opened the door for the development of nanoprinting techniques capable of operating at the nanoscale. Technologies like Nanoimprint Lithography (NIL) emerged in the 2000s, providing a method for creating nanoscale patterns on substrates (Tiginyanu et al. 2011). The real leap into practical nanoprinting came in the mid-2010s, when the ability to print at nanoscale resolution was significantly advanced. One of the major breakthroughs was Two-Photon Polymerization (TPP), developed in the early 2000s, which enabled the fabrication of 3D structures at the nanoscale using focused laser light (O'Halloran et al. 2022). These technological advancements made possible the creation of highly intricate and precise structures at the nanometer scale, which has since paved the way for revolutionary applications in fields such as electronics, biomedicine, and nanomaterials. These fields rely on the ability to manipulate matter at the molecular and atomic levels, unlocking new possibilities for miniaturized devices, advanced drug delivery systems, and novel materials with enhanced properties (See Table 1).

The quest to reveal the invisible realm has found new possibilities in contemporary nanotechnologies, through which we can now see and manipulate matter at incredibly tiny scales. Before the invention of the microscope, the human eye could barely penetrate the surface of reality, discerning details on the order of a tenth of a millimeter, and was necessarily plagued by blurring and optical aberrations. Where the eye could not reach, imagination filled in the gaps—often deceived by the optical and psychological laws of *pareidolia*,[2] from the Greek εἴδωλον (*eidōlon*, image) with the prefix παρά (*parà*, near): the subconscious illusion that tends to interpret random natural or artificial shapes or profiles as familiar forms (Ghose and Zimmermann 2023). In the 17th and 18th centuries, some naturalists, glimpsing vaguely anthropomorphic shadows within the ogival head of human sperm cells, supported the theory of the *homunculus (little man)*—according to which each sperm contained a miniature human being, already fully formed and ready to be nourished and developed by the vital humors of the female womb (Mental Itch 2023). The preformationist view dominant in the scientific community of the time opposed the concept of epigenesis, according to which life develops through cellular differentiation and organization (Casetta 2017). Among the proponents of this theory was Nicolaas Hartsoeker (1656–1725), a Dutch microscopist who, in his 1694 treatise *Essay de Dioptrique*, included the famous illustration of a sperm cell containing the tiny man *homunculus*, curled up inside (Hartsoeker 1694). The image appeared to reconcile the scientific perspective with certain religious interpretations of creation. Thanks to advancements in microscopic techniques and empirical observation, Caspar Friedrich Wolff (1733–1794), in his work *Theoria Generationis* (1759), disproved the existence of a *homunculus*. Through experimental studies, he demonstrated that embryonic development occurs through a process of successive organization and differentiation, placing scientific observation above philosophical

[2] *Pareidolia* is the tendency of the human mind to perceive familiar patterns, such as faces or figures, in unrelated or random stimuli, like clouds, rock formations, or abstract images. This phenomenon arises from the brain's inclination to seek meaningful shapes, often as a result of evolutionary adaptation for recognizing faces or potential threats. Pareidolia is closely related to *Gestalt theory*, which is a psychological approach that emphasizes how humans perceive patterns and wholes rather than individual components. According to Gestalt principles, the mind organizes sensory input into structured and recognizable patterns, even when the stimuli are incomplete or ambiguous. In the case of pareidolia, the brain applies Gestalt principles—such as *proximity*, *similarity*, and *closure*—to interpret random or abstract stimuli as familiar shapes or objects, like faces or animals. This tendency to find meaning in chaos is an example of how the brain seeks to make sense of the world by imposing patterns, a key concept in Gestalt psychology.

or religious assumptions (Wolff 1774). This anecdote from the history of biology highlights how the limitations of human visual perception have influenced—and often distorted—the path of scientific discovery. The well-known case of the homunculus, now regarded with a mix of wonder and irony, appears to us today as a naïve misunderstanding, typical of an era when science lacked the proper tools to explore the invisible. Yet, the fragility and limitations of human perception constantly remind us of the importance of maintaining a critical stance—even today—toward the hyper-technological images that our sophisticated digital machines provide of the nanometric world. Although the evolution of observational tools allows us to gather new evidence and continually redefine our understanding of reality, a latent risk remains within the contemporary scientific imaginary:

"The unknown is to be captured in an experiment, using instruments adequate to the task. However, we do not know whether the instrument is adequate until we are sure it gives us the correct readings. But since the phenomenon itself is unknown yet, there is no way to decide what correct readings are; we won't know if we have built a good detector until we have tried it and obtained the correct outcome. But we don't know what the correct outcome is until...and so on *ad infinitum*" (Collins 1992, as cited in Ruivenkamp and Rip 2010).

The invisible, defined as that which eludes sight due to size, distance, or concealment, is not only a physical matter but also a perceptual and conceptual one. An object can remain invisible not only because it is too small, but because our attention is not oriented to recognize it. What we do not see is not absent: it exists, awaiting revelation through new investigative tools. The French philosopher Merleau-Ponty (1908–1961) masterfully describes this paradox of perception: the world is what we see, yet, the moment we try to define it rationally, we enter a labyrinth of difficulties and contradictions (Merleau-Ponty 1968). Technological miniaturization continuously challenges our perception and representation of reality. In this context, digital scientific imaging plays a fundamental role: from advanced microscopy to molecular simulations, from volumetric reconstruction algorithms to artificial intelligence applied to scientific visualization. These tools do not merely display reality, but reconstruct it, interpret it, and sometimes even generate it anew, creating a true digital theater of wonders. The contemporary *WunderCAD/CAMmer* is reflected in the most powerful microscopes and the pixels of workCADstations, becoming an epistemic device where the visible and the invisible merge. Just as in the Renaissance Wunderkammern, where naturalia and artificialia sparked wonder and knowledge, today's graphics and CAD/CAM digital models explore and visualize what would otherwise remain hidden. Digital scientific imaging is no longer just a window to reality, but a laboratory of the visible, constantly redefining the boundary between what we can perceive and what escapes our understanding (Humanitas Research 2025). Lenses and microscopes are fundamental tools in scientific research, extending both the eye and the mind, and expanding the limits of knowledge. Thanks to the evolution of microscopy techniques and digital imaging, today we can observe previously invisible phenomena, study them, classify them, and modify them. However, it is important to remember that their use has been subject to censorship during certain historical periods (Arber 1954). Microscopy began in the

13th century with magnifying lenses, used in various civilizations to correct vision and amplify images. However, the Roman Catholic Church slowed its use until the 17th century, fearing it would desecrate divine secrets. Only with the Renaissance and the empirical approach did the use of lenses spread, leading Galileo Galilei (1564–1642) to revolutionize astronomy, despite religious persecutions, and Anton van Leeuwenhoek (1632–1723) to discover microorganisms, opening new visual frontiers for scientific research (Kutschera 2023). It has been emphasized how, over the centuries, science has relentlessly sought to penetrate the realm of the small—yet the domain of the infinitesimal has often proved elusive, ambiguous, even deceptive. The mistake has not been merely technical, but profoundly conceptual. When faced with the unknown or the undecipherable, both the human eye and mind, in their effort to complete the perceptual act and render it meaningful, tend to rely on imagination that precedes understanding. This reminds us that observing a phenomenon does not necessarily imply an *automatic* or accurate comprehension of it. To look is always to filter—and that filter reveals what we expect, or even unconsciously wish, to find. Thanks to the ongoing evolution of observation techniques, the human gaze has been able to magnify and discern what, in that mass of seminal matter—deceptively resembling miniature human forms—actually curls into the astonishing biological origami of DNA and the molecules of life. In 2012, the Italian physicist Professor Enzo di Fabrizio captured one of the first detailed and widely recognized images of a helical DNA strand, contributing to a more accurate visual rendering of reality at an otherwise invisible scale (Marini et al. 2017). The photographic capture of this iconic image was made possible because Professor Di Fabrizio and his team from the University of Genoa and the Italian Institute of Technology developed a method that preserves the molecular structure, enabling them to directly image the DNA double helix without causing any damage (Marini 2015). The technique involves the use of silicon nanostructures (wafers) to create a grid system with very small openings that support and align DNA molecules, allowing them to be examined using a transmission electron microscope (TEM) (Marini et al. 2018). The perception of *scalar dimensional factors*[3] in reality is an exceptionally intriguing aspect of human cognition (Collegio et al. 2019). As happens to *Alice in Wonderland* chasing the

[3] The perception of *scalar dimensional factors* in reality is a fundamental and fascinating aspect of human cognition, especially when we consider how the human mind responds to objects that represent smallness. Recent studies suggest that instead of merely relying on the retinal image size of an object, our brain directs attention more effectively toward objects whose inferred real-world size is small, regardless of the actual space they occupy on the retina. This phenomenon highlights how the act of *making things smaller* – whether through design, simulation, or microscopic understanding of structures – is not only a technical process but also a cognitive one. Our attention is more easily drawn to small objects, which we tend to perceive as more complex or worthy of exploration, even though they may occupy the same visual space as larger objects. This tendency becomes particularly significant in contexts where recognizing real-world size is crucial. In the medical field, for instance, radiologists must identify tumors of varying sizes, while in airport security, agents are trained to detect potential threats hidden in relatively small spaces. Both contexts benefit from understanding how the inferred size of an object can influence attention, making the analysis process more effective. In this sense, research not only highlights the complexity of visual perception but also suggests that our cognition does not simply process what we see. Instead, it actively interprets and prioritizes information based on our experience and knowledge of the world, influencing our approach to designing and manipulating the *small-scale structures*. This approach has fundamental applications in fields such as biomimetic design, where control of size and scale is essential, and in innovation processes that aim to adapt to nature's laws through miniaturization.

White Rabbit through the nanotech universe, our perception of dimensions drives us to penetrate the secrets of both the micro and the macro (Horton and Long 2010). This makes it increasingly difficult, yet stimulating, to relate them to the experiential surroundings of our natural tactile and haptic perception of the 1:1 scale. The technological theme of making things smaller can no longer be approached solely as a matter of physical scale. It now requires a thoughtful critical reflection on the quality and nature of the tools we use—not just optical ones, but cognitive, interpretative, and design tools that filter the flow of images to the brain. The fascination of penetrating and holding the keys to the realm of the ultra-invisible in order to manipulate it *from above* drives the contemporary *WunderCAD/CAMmer* to open windows and walls in order to host the wonders of nanotechnology. *There's Plenty of Room at the Bottom* is the title given to the transcript of a famous lecture delivered by physicist Richard Feynman in 1959 (Loretan and Müller 2023). With his usual humor, Feynman foresaw the visionary possibility of atomic manipulation to directly build new molecules and chemical substances with designed properties, stating that the principles of physics don't prohibit it. It is within the context of this talk that Feynman considered the possibility of directly manipulating individual atoms as a more significant form of chemical synthesis compared to the methods used in his time. Feynman explored a variety of intriguing applications of the ability to manipulate matter at the atomic scale. He was particularly fascinated by the potential to create denser computer circuits and microscopes capable of observing objects even smaller than those visible with an electron microscope. This latter possibility has been realized with the invention of the scanning tunneling microscope, followed by the atomic force microscope. These very innovative tools for exploring the invisible have proven fundamental for advancing biomimetic observations at the nanometric detail of hierarchical organization and the three-dimensional physical constitution of material design at ultra-small scales. Thanks to scanning electron microscopy (SEM) images, detailed analyses of microstructures—such as those of butterfly eggs—reveal incredible constructive details that Nature provides to living organisms. The *chorion*, for example, features a series of micropores (*micropyles*), conical structures measuring between 0.5 and 3 μm, arranged regularly across the egg's surface: these tiny openings allow gas exchange between the internal and external environments, balancing oxygen intake and carbon dioxide release, thereby ensuring a viable habitat for the developing larva (Fig. 1).

Using parametric CAD, it is possible to reconstruct virtual working models that enable design hypotheses aimed at improving, for example, food preservation, drawing inspiration from these natural mechanisms of exchange and protection (Lumini 2022) (Fig. 2).

In the wake of Feynman's insight, for more than half a century, the most exquisitely miniaturized industry, electronics, has danced around a prophecy: Moore's Law, which allowed the regular doubling of transistor density on a chip (Wood 2021). This prediction, more mythical than scientific, acted as a mandatory design horizon, almost ritualistically guiding the evolution of the technological microcosm. But today, that exponential curve has begun to flatten. Miniaturization, pushed to the limits of the nanoscale, faces severe physical, thermal, and economic constraints (Quantum News 2024). The ultraviolet rays used in classical lithography

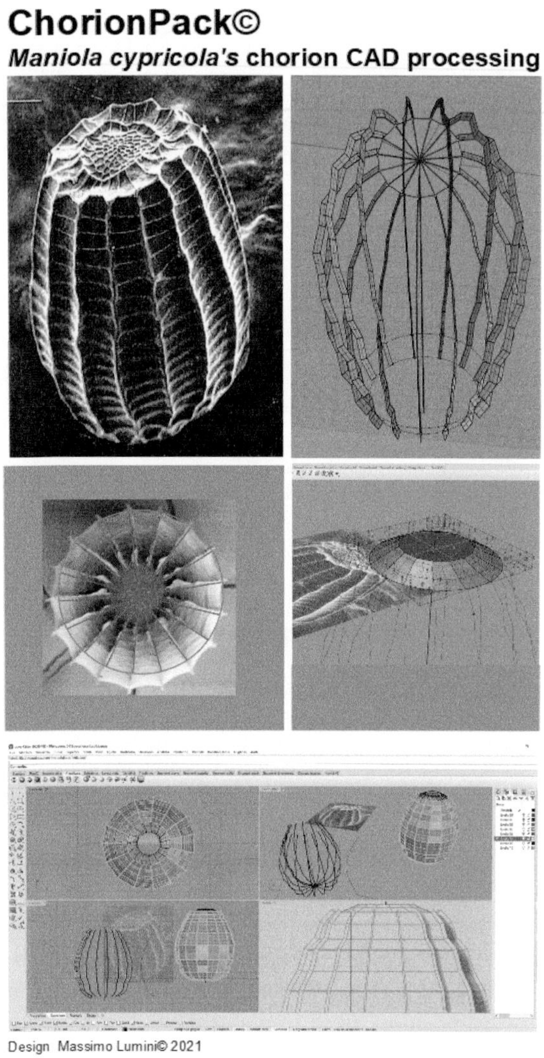

Fig. 1. *Maniola Cypricola's chorion* CAD processing (Rendered by Lumini 2021).

can no longer etch structures smaller than a certain threshold: light, as matter, has a wavelength that cannot be compressed beyond certain limits.

This is why the need for developing and introducing increasingly complex techniques arises:

- **Quadruple Patterning**, a technology that multiplies the lithographic phases to obtain increasingly finer features, but which simultaneously increases costs and the possibility of error exponentially (Shilov 2024).
- **Extreme Ultraviolet Lithography (EUV)**, which uses light with a 13.5 nm wavelength, but requires ultra-clean environments (see Chapter 4.0) and optical systems costing billions of dollars, with the final result being a minimum pitch

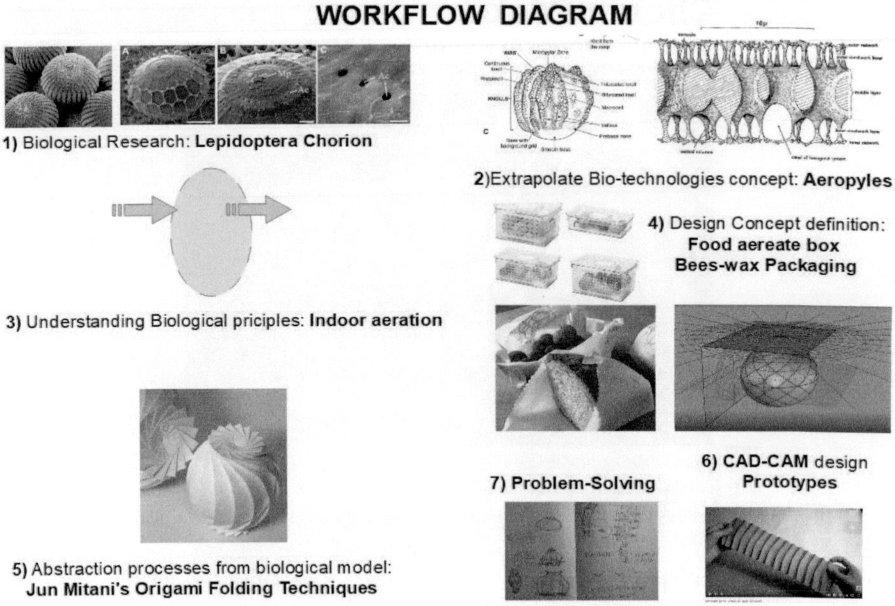

Fig. 2. Worflow Diagram.

still around 30 nm, reducible to less than 20 nm only through elaborate and expensive post-processes (Kazazis et al. 2024).

Today, miniaturization is no longer measured in microns, but in *nanometers*,[4] and it has long surpassed the visible threshold. Science, technology, and consequently, indirectly philosophy, can no longer rely on light, lenses, or eyes. Design, at this stage, becomes an increasing conflict between precision and complexity. To make things smaller, we are forced to make them more complicated. And it is for this profound conflict of our current technoscience that we increasingly turn to the miniaturization of nature, which has brilliantly solved in billions of years of research and development thousands of ways to create highly efficient, wondrous nanomaterials. The human research tension toward pushing these limits remains fertile. Thousands of studies are being published with an exponential trend. However, within this impasse lies a value: traditional technologies have taught us to control space, to draw silicon as if it were paper, through a fine digital calligraphy of extraordinary complexity and artificial efficiency. From the engraving of clay and wax tablets in antiquity to the lithographic processes used on silicon wafers, human technoscience has

[4] A nanometer (1 nm) corresponds to one billionth of a meter. To contextualize this scale, the diameter of a human hair ranges between 80,000 and 100,000 nanometers—approximately the thickness of a standard sheet of paper. In contrast, a strand of human DNA measures around 2.5 nm in diameter, and a water molecule is less than 1 nm wide. The wavelengths of visible light span from approximately 380 nm (violet) to 750 nm (red), placing many nanoscale structures below the threshold of optical resolution.Ten nanometers represent a dimension significantly smaller than one-thousandth the diameter of a human hair. At this scale, nanotechnology operates by manipulating matter at the atomic and molecular levels, enabling the engineering of materials and systems with enhanced or entirely novel properties—ranging from increased mechanical strength to targeted biochemical functionality.

progressively come to conceive of matter as a writable medium—if not directly shaped by human hands, then by the nanometric prostheses of its robotic extensions. Yet, a critical threshold has been reached: the point at which further excavation into the infinitesimal becomes unfeasible, where light can no longer reveal structural details, and where optical instruments approach the limits of shadow and resolution. At this juncture, a paradigm shift becomes necessary—one that moves beyond the act of digging into the small to one of cultivating it; beyond representing the microscopic world to programming it. In the *WunderCAD/CAMmers* of the future, the researchers of the invisible will increasingly rely on the extraordinary properties of DNA, which, as a *deus ex machina* of biology, biochemistry, and genetic engineering, emerges as the foundational element of Lilliputian-scale design (Shen et al. 2021). Nature, through DNA, offers us an extraordinary resource of precision, self-assembly, and molecular programmability that far exceeds the capabilities of traditional materials like silicon and the classical photolithography techniques used in electronics. DNA, the fundamental molecule of life, is not only an archive of genetic information but also a building material with unique chemical-physical properties. Its ability for specific and complementary recognition (thanks to base pairing) allows for the guided assembly of complex structures with atomic precision, leveraging the intrinsic mechanisms of biology. Unlike photolithography on silicon, which requires optical masks, UV light exposure, and subsequent chemical processes to create circuits, DNA-guided nanolithography allows for the three-dimensional construction of nanostructures in aqueous environments and at room temperature. This bottom-up approach, as demonstrated in the study by Shen et al. (2021), uses DNA's modular epitaxy to guide the controlled assembly of complex architectures, opening new frontiers in nanoengineering, the creation of new materials, and the fabrication of molecular devices. What makes DNA even more promising is its algorithmic programmability: thanks to synthetic biology's language, it is possible to write *molecular codes* that precisely determine the final shape of a nanometric object, just like a software that *compiles* matter. This bio-inspired logic contrasts with the rigid, top-down lithographic approach, which encounters physical limits (such as light diffraction) and material constraints (such as the scalability exhaustion of silicon) (Chen 2020). In a sense, we could say that nature is already an advanced nanofabrication laboratory. The molecular machines that operate within cells—ribosomes, polymerases, chaperones—are examples of living technologies that build with precision what our engineering still struggles to replicate. Transferring this natural intelligence into the design of new materials and devices means going far beyond the traditional limits of miniaturization. It requires us to radically rethink what we mean by design itself. In the past, design was often a matter of geometry—of shaping and arranging forms in space, usually through CAD software that allowed precise modeling and simulation. But nature doesn't work that way. Nature designs through evolution, through genetic codes, through systems that grow, adapt, and self-organize. So when we take inspiration from biology, we move from drawing lines and shapes to writing codes and programs, engaging with a kind of bio-CAD where the blueprint is not a static drawing but a dynamic script. Instead of designing a surface that repels water using geometric microstructures in a CAM-based process, we might program a living system—say, a genetically engineered cell—to produce proteins that create the same effect. Instead of machining materials layer by layer, we might instruct a population of stem cells to follow a developmental script, organizing themselves into tissues or even entire organs. The paradigm shift is profound: from pattern to program, from form to function, from static structures to dynamic, evolving systems. In this new vision, we don't just build things—we cultivate them, we grow them, we encode the conditions for their emergence. What we design is no longer the object, but the process that gives rise to it, merging the logic of CAM with the intelligence of life itself. In the contemporary *WunderCAD/CAMmer*, genetic codes are rewritten as engineering languages, self-assembling sequences are conceived, designed, and manufactured to create nanometric structures with a precision and detail that no human hand could replicate. Thus, making things smaller becomes thinking deeper, diving ever deeper into the rabbit hole. It is no longer about miniaturizing the world and etching it onto the pit of a cherry, but rather about conceiving it according to other logics, other geographies of matter. And it is in this epochal transition that the concept of artificial design hybridizes with the natural one of life, and the *WunderCAD/CAMmer* hosts new wonders, no longer made of astonishing microtechnical marvels (meaning miniaturization through traditional techniques like photolithography) or technically impossible objects, but rather entire molecular systems that build themselves (See Table 2).

Each technique has pushed miniaturization closer to its physical limits, but at increasing costs in terms of energy, complexity, and control. The final frontier is no longer just a speculative projection—it signals a conceptual shift: from *drawing on matter* to *drawing with matter*. As silicon-based miniaturization approaches its physical, thermal, and quantum boundaries, a growing consensus suggests that the future of nanomanufacturing will not lie in pushing silicon further. Although this material has driven decades of innovation—from transistors to microprocessors—it is now reaching a threshold where traditional paradigms are no longer sufficient. With Moore's Law decelerating, the industry is exploring new paths, investigating alternative materials and novel fabrication techniques in pursuit of smaller, more efficient, and more powerful devices (Ansari 2025). One of the most promising directions involves not just replacing silicon, but augmenting or transcending it through materials with superior intrinsic properties. Graphene, carbon nanotubes, and two-dimensional materials such as transition metal dichalcogenides (TMDs) have emerged as front-runners due to their exceptional conductivity, mechanical strength, and flexibility—particularly in quantum computing and the development of ultra-fast, low-energy devices. Yet, the research spotlight is increasingly shifting toward biologically derived materials. Recent studies have demonstrated that DNA, beyond its fundamental biological role, can also serve as a programmable structural material at the nanoscale (Bekkouche et al. 2023). Its remarkable capacity to self-assemble into complex architectures with atomic precision is opening new avenues in DNA origami and nanolithography (Martynenko et al. 2021). In the realm of DNA origami, the ability to design and predict the three-dimensional form of self-assembling structures is a fundamental challenge. The CanDo® framework, developed at the Massachusetts Institute of Technology (MIT) by Prof. Do-Nyun Kim in the laboratory of Prof. Mark Bathe, represents a pioneering Molecular-CAD tool specifically tailored to this task (Kim et al. 2018). Initially created to model

Table 2. Comparison of Nanolithography Techniques and Their Applications.

Technique	Wavelength/ Resolution	Advantages	Limitations	Current Status	Companies
Deep Ultraviolet (DUV)	~193 nm/down to ~ 40 nm	Well-established technique, relatively low cost	Optical limits, not suitable for sub-20 nm	Still in use in combination	ASM™ Nikon®
Multiple Patterning (Quad, Hexa)	< 30 nm (via multiple patterning)	Extends the life of DUV lithography	Complex processes, risk of cumulative errors, increased cost	Maximum reached: ~30 nm	ASML™ Canon®
Extreme Ultraviolet (EUV)	13.5 nm/ theoretically down to < 20 nm	High resolution, reduces multiple patterning steps	Very high cost, delicate masks, challenges in large-scale production	In use by TSMC, Samsung (5-3 nm nodes)	ASML™ TSMC™ Samsung®
Advanced Post-Processing	~20–22 nm (down to 16–18 nm)	Further dimension reduction	Adds complexity, uncertain in terms of structural stability	Marginal extensions	Applied Materials® Lam Research®
DNA-guided Nanolithography	Pitch 16.2 nm/ Critical Dim. 7 nm	Programmable, scalable, molecular precision	Emerging technology, still in industrial validation phase	Promising alternative frontier	Harvard University University of California

DNA assemblies built on honeycomb and square lattices via caDNAno®, CanDo® was later expanded to support off-lattice models, significantly broadening its design capabilities (Baker and Church 2009). CanDo® employs a finite element modeling approach, treating the DNA double helix as a homogeneous elastic rod with defined mechanical properties—stretching, bending, and twisting rigidities—derived from experimental data. This methodology enables the simulation of intricate 3D geometries and their flexibility, generating predictive insights that are unattainable through conventional analytical methods. One of its defining features is its seamless integration with caDNAno®, empowering researchers to directly import design files, simulate structures in solution, and optimize parameters before physical synthesis. As a result, CanDo® significantly reduces both the time and cost involved in the design of DNA-based nanostructures, functioning as a powerful CAD tool for molecular engineering. Among the most significant applications of these advances is DNA-guided nanolithography, a method that directly addresses the limitations of conventional photolithographic processes. By exploiting DNA's predictable folding behavior, this technique uses molecular strands as programmable masks or scaffolds to fabricate nanoscale patterns with unprecedented control. The ability to create features as small as 7 nm—well beyond the capabilities of current photonic methods—marks a decisive breakthrough (Basu et al. 2025). Thanks to its modularity, DNA enables the creation of tailored structures with precise geometry, spacing, and functionality. This unlocks applications across a wide range of fields: from nanoelectronics and biosensors to drug delivery and synthetic biology. Within this emerging convergence between molecular biology and digital fabrication, the use of DNA as a constructive medium constitutes a radical paradigm shift. While traditional nanolithography follows a top-down logic—etching, depositing, and patterning with external tools—DNA introduces a bottom-up, self-organizing strategy directly inspired by natural processes. This approach resonates with the principles of CAD/CAM systems, where a digital model is translated into a precise manufacturing sequence. But here, the *blueprint* is not merely geometric—it is genetic. The design file is a DNA sequence, and the manufacturing process unfolds at the molecular level (Ramezani and Dietz 2020). Such a transition challenges conventional definitions of design, leading toward a future in which CAD becomes *code-aided differentiation*: a process in which molecules, materials, and living cells self-organize in space and time according to predefined informational rules. In this context, DNA nanostructures assume a dual role: they are both the object and the operator of construction. They act simultaneously as structural templates, logic gates, and functional scaffolds—bringing us closer to a new era where design is no longer applied *to* matter, but emerges *with* it (Zhan et al. 2020). The synergy between programmable matter and digital control opens the way to a new generation of hybrid devices—biologically encoded, digitally orchestrated, and spatially printed—a fusion of genotype and geometry, scripting and shaping. As we push the boundaries of fabrication further into the molecular domain, the tools developed for DNA-guided assembly begin to align with other revolutionary frontiers—chief among them, quantum computing. The ongoing drive to create more powerful and compact devices is leading us into realms where classical methods no longer suffice. Qubits, the fundamental units of quantum information, demand fabrication approaches that operate at the very limits of material behavior. Materials like graphene and carbon nanotubes are under investigation for their potential to support quantum coherence, but the scalability of quantum architectures remains elusive (Suhail 2025). Here, the principles of nanoengineering—precise, programmable, and self-organizing—open promising avenues for rethinking the very foundations of computation. It is increasingly conceivable that the next era of computing will not be constructed on rigid, artificial substrates, but rather on biomolecular scaffolds—structures that inherently encode and process information with quantum-level precision. Within this framework, the boundaries between synthetic and organic, digital and biological, begin to dissolve, compelling a redefinition of what it means to compute, to design, and ultimately, to create. These incursions into the forbidden dimensional domains of the infinitely small confront us with the strange and counterintuitive laws that govern the fundamental nature of matter—laws that reside at the threshold of our comprehension. As a result, these advances are not only reshaping technological capabilities but are also beginning to transform the very frameworks that structure human rationality.

Human perception of space and scale has historically, and psycho-biologically, been calibrated to the physical proportions of the body anatomy and proportions. Our bodily experience provides a tacit unit of measurement for the entire universe, as famously symbolized by Leonardo da Vinci's *Vitruvian Man*—a canonical representation of the anthropometric ideal as a cosmic reference system, rooted in secret geometric correspondences (Riavis 2020). The first units of measure adopted by ancient civilizations were derived from the human anatomy—such as the cubit (forearm) or the pace—to provide a standard by which humans could measure, conquer, and represent the physical world (Tanguy and Fairbrass 2018). The first units of measurement adopted by ancient civilizations were, in fact, modeled on parts of the human body—such as the arm or the stride—in order to empower humans to represent the forms and dimensions of the world (Tanguy aand Fairbrass 2018). Even after the adoption of the metric system in 1875, Anglo-Saxon cultures continued to use human-derived measures like the inch and the foot (US Metric Association 2020). Beginning from the human body and its 1:1 perceptual space, the visual project *Powers of Ten* embarks on an imaginative journey through the dimensional continuum, unfolding within the intimate sense of scale evoked by scientific notation. Created by the American designers Charles Eames (1907–1978) and Ray Eames (1912–1988), this iconic work has profoundly impacted our perception of the true macro dimensions of the universe, as well as the micro and nano dimensions of organic life and the structure of matter (Eames Office 1977). Inspired by the book *Cosmic View* (Boeke 1957), the 1977 film navigates through exponential zooms, illuminating the relative scale of reality. The project, supported by IBM and preceded by a short film from 1968, is closely connected to the book *On Size and Life* by (McMahon and Bonner 1983), which analyzes how the size of organisms influences their physical and physiological characteristics. Volume increases with the cube of length, while surface area only increases with the square, with implications for metabolism and the energy efficiency of organisms. Organisms of different sizes face challenges related to nutrient transport, heat dissipation, and movement. Smaller organisms, with a higher surface-to-volume ratio, facilitate gas exchange but struggle to maintain a stable temperature. Size also affects the structural strength of bones, locomotion, and fluid flow. These forces influence the shape and stability of molecules, determine the physical state of a substance, and are crucial in biological interactions, such as cell adhesion or enzyme-substrate recognition. In nature, attractive and repulsive forces are exploited by many living organisms to perform essential functional tasks for their survival, such as the super-adhesiveness of the toe pads in geckos (*Gekkonidae Oppel*) (Autumn et al. 2002, Shah and Sitti 2004). In 1638, Galileo Galilei established the basic principles of biomechanics, and the Vitruvian mystical body transformed into an abstract system of invisible forces and levers. In his *Discourses and Mathematical Demonstrations Concerning Two New Sciences Relating to Mechanics and Local Motion*, he describes the mechanical limitations that organisms are subjected to (Galilei 1734). Galileo theorizes first that the bones of animals are not simply scaled-up in proportion to their size, but that their structure must be modified and adapted to biomechanical demands, with cross-sectional areas increasing more rapidly than length to support greater loads. To bear the same stresses, larger cylinders must be relatively thicker. Therefore, a large animal does not simply have larger bones, but bones whose cross-sections increase much more rapidly than their length. A giant, for example, would not have limbs with the same proportions as a human, but rather thicker limbs, suited to support its mass and the mechanical forces at play. Whether it's the thickness of an elephant's femur or the rapid beat of a hummingbird's heart, biological form is always intricately tied to scale. Larger animals require disproportionately thicker bones to support their weight, while smaller ones must navigate forces like surface tension and molecular adhesion, which become dominant at the microscale.

These observations would lay the groundwork for a new mechanistic understanding of the living body, later expanded upon by Giovanni Alfonso Borelli (1608–1679), often regarded as the father of modern biomechanics. In his seminal work *De motu animalium* (1680–1681), Borelli applied the laws of classical mechanics to the study of muscular motion and joint dynamics. Treating the human body as a system of levers, pulleys, and elastic resistances, he was the first to analyze motion not as an expression of vital spirits or humoral balance, but as the result of quantifiable mechanical forces. He showed that muscle action followed geometric and physical principles, and that bodily motion could be predicted and described mathematically. Borelli thus transformed the human organism into an object of measurable, mechanical inquiry—an idea that continues to underpin contemporary fields such as kinesiology, prosthetic design, and robotic locomotion.

Size profoundly influences design and evolutionary strategies, limiting or favoring species diversification and ecological distribution. Organisms of different sizes occupy different ecological niches, reducing competition for resources. This question has long intrigued scientists and naturalists, from Galileo to D'Arcy Thompson (1860–1948), as it subtly shapes the very fabric of life. As D'Arcy Thompson once observed in *On Growth and Form* (1917), *everything is the way it is because it got that way* (Branch 2015)—and size is often the silent architect behind that becoming (Thompson 1917). Understanding these nuances calls for an interdisciplinary lens—where biology, physics, and engineering converge to shed light on how scale governs function, behavior, and evolutionary possibility. Meanwhile, technological advances in imaging and miniaturization—such as the invention of the transistor or the electron microscope—have not only expanded our view of life's smallest architectures, but have also inspired entirely new fields like nanotechnology and biomimetic design, where nature's principles are reimagined at the atomic level. This interdisciplinary approach connects biology, physics, engineering, and other disciplines, providing a comprehensive understanding of the implications of size in *naturalia* and *artificialia*. It also highlights how the principles of scale are universal, applicable to various scientific fields, from designing biocompatible materials to understanding evolutionary nanomechanisms. Since the invention of the first microscopes, the creation of instrumentation and imaging technologies has continuously deepened the mysteries of the micro and nano worlds. At the same time, the invention of the transistor marked the beginning of nanotechnology, with extreme progress in miniaturization in structural engineering, electronics, and material science (Science 2023). In 2006, the Office of Basic Energy Sciences (BES) of the Department of Energy created the image *The Scale of Things*, comparing the sizes of natural technologies with those of human technologies at the micro and nanometer scales, synthesizing the advances made in just a few decades (Office of Basic Energy Sciences 2006). The chart serves as an educational resource to help convey the relative sizes of various structures and phenomena, from atomic dimensions to larger biological and technological systems. In materials science, the latest advances in nanofabrication are significantly enhancing the properties of alloys and composites—such as carbon nanotubes and graphene—and their biomedical, electronic, and environmental applications, allowing for more precise control over impurities and structural defects (Mishra et al. 2024). In electronics, as mentioned, these advances represent the extension of Moore's Law toward smaller and quantum-scale devices, pushing the boundaries of traditional semiconductor scaling (Ramachandran 2024). By enabling the manipulation of materials at the atomic and molecular level, nanofabrication techniques are not only permitting the continued miniaturization of components but also paving the way for entirely new paradigms in computing, such as quantum processors and neuromorphic architectures. This evolution holds the promise of drastically improved performance, energy efficiency, and functionality across a wide range of applications, from wearable technology to advanced artificial intelligence systems. In catalysis, zeolites play a crucial role due to their unique structure, consisting of a network of microporous crystals. These structures form both the models and the pores essential for various catalytic processes. Zeolites are made up of silicon, aluminum, and oxygen atoms that create a highly ordered framework. The size, shape, and distribution of these pores allow zeolites to selectively adsorb and trap molecules of specific sizes, making them highly effective in facilitating chemical reactions. Zeolites are widely used in catalytic applications, particularly in petroleum refining, where they aid in processes such as cracking, isomerization, and hydrocracking. The catalytic properties of zeolites are influenced by factors such as pore size, the presence of acidic sites, and the material's overall structure. The ability to form *models* refers to how the porous structure can be tailored to suit specific reactions, making zeolites versatile and efficient catalysts. The precise control of pore size and distribution enables highly selective reactions, improving the efficiency and yield of industrial processes. In this context, Computer-Aided Design (CAD) and Computer-Assisted Optimization (CAO) methods play a complementary role in the development of the atomic structure of the zeolites and other porous materials. While CAD enables accurate structural modeling of the crystalline framework and hierarchical pore networks, CAO—often inspired by natural systems—supports the optimization of these architectures for enhanced catalytic performance and reaction selectivity. Such integrated approaches allow researchers to simulate, evaluate, and refine materials at multiple scales, aligning structure with function (Coppens et al. 2020). The size and shape of the pores are critical in determining which molecules can enter the material and interact with active sites. CAD-CAO allows for detailed visualization and optimization of these parameters, facilitating the design of more efficient zeolites. By customizing zeolites for specific reactions, CAD-CAO contributes to the development of advanced materials for diverse catalytic applications (Kononov et al. 2022). In condensed matter physics, the nanometric scale enables the creation of perfect crystals, facilitating precise atomic and molecular arrangements. This precision minimizes imperfections, such as dislocations and defects, which are typically found in larger-scale materials.

Achieving such controlled structures at the nanoscale is crucial for a wide range of advanced materials with exceptional mechanical, optical, and electronic properties. CAD/CAM suites play a pivotal role in this process by providing tools to model, simulate, and fabricate nanostructured materials with a high degree of accuracy. Specific software packages are commonly used for these purposes. For example, SolidWorks® is frequently employed for general material design and structural simulations, while more specialized programs like Materials Studio® (by Dassault BIOVIA®) and VASP® (Vienna Ab-initio Simulation Package) are used for molecular dynamics simulations and predicting the behavior of materials at the atomic level (Dassault Systèmes 2025; VASP Software 2025). These tools allow researchers to design nanoscale structures and simulate their properties before physical production begins. Additionally, COMSOL Multiphysics® is another widely used software in the simulation of material behaviors at various scales, including the nanoscale, particularly for analyzing the mechanical, thermal, and electrical properties of nanostructures. For fabrication and manufacturing, CAM software such as Mastercam® or Fusion 360® translates the designs created in CAD into instructions for precise machine operations, ensuring the accurate realization of nanoscale structures. The combination of CAD, CAM, and advanced simulation tools empowers researchers to design perfect crystalline structures with customized properties, unlocking new possibilities in fields like electronics, nanotechnology, and quantum computing. Through these technologies, the transition from theoretical nanomaterial design to practical, defect-free production has become more feasible, leading to breakthroughs in materials science. In restorative dentistry, a similar approach is taken with materials like lithium disilicates, prized for their aesthetics, strength, and durability (Munoz et al. 2025). These materials have the potential for increased strength by modifying the crystal fraction, although such modifications typically reduce translucency. Recently, Amber® Mill lithium disilicate has been developed, promising customizable translucency through specific firing protocols without compromising flexural strength (HASS Corporation 2018). This study evaluates the effectiveness of these firing protocols by testing forty specimens for translucency under different conditions (high, medium, low translucency, and medium opacity) using the Contrast Ratio (CR) and Translucency Parameter (TP) tests. Additionally, the flexural strength of sixty specimens was assessed using a three-point bending test. The results revealed that translucency differences were most significant between the medium opacity and other translucency groups, with the flexural strength rankings being medium opacity > low translucency > medium translucency > high translucency. This highlights a trade-off between aesthetics and mechanical strength, with higher translucencies associated with reduced flexural strength. Both in condensed matter physics and materials science for dentistry, advanced CAD and CAM technologies are essential in optimizing material properties. These systems allow for the simulation of different configurations and help predict trade-offs between mechanical and aesthetic properties, ensuring that the final product meets the required standards for performance and appearance. This advancement leads to materials with optimized mechanical, thermal, and electronic properties, paving the way for more efficient and durable devices. A crucial role is played by the integration of inorganic and organic nanostructures, which enables the creation of advanced composites with hybrid properties and customizable functionalities at the molecular scale (Kaidarova et al. 2025). In molecular biology, the micro and nano scale represents the intricate level of organization where the fundamental machines of the cell, such as enzymes, ribosomes, and other molecular complexes, assemble and function. These molecular machines are responsible for essential cellular processes, including DNA replication, protein synthesis, and energy production. At this scale, interactions between biomolecules, such as proteins, nucleic acids, and lipids, occur with remarkable precision, enabling the cell to carry out complex tasks with high efficiency. Understanding these processes at the micro and nano scale is crucial for unraveling the mechanisms behind cellular function, disease development, and the design of biomolecular systems in fields like synthetic biology and nanotechnology (Ariga et al. 2024). Structures such as *histones*, *proteasomes*, *ribosomes*, and replication complexes, along with components of mitochondria and chloroplasts, operate as highly specialized nanometric systems, regulating vital processes with extraordinary efficiency. The analysis and manipulation of these structures through advanced imaging techniques and nanotechnologies are driving the development of new approaches in bioengineering and biomedicine (Tahir et al. 2021). At the same time, scalable nanofabrication is emerging as a key technological frontier, driven by the integration of advanced CAD/CAM systems with high-resolution additive manufacturing techniques. The increasing precision of nanoscale 3D printing enables the production of complex architectures with controllable features down to the atomic scale, making large-scale nanomanufacturing more feasible than ever before (Yoon et al. 2017). This paper presents a network-based CAD/CAM system designed to enable scalable nanomanufacturing through hybrid 3D printing. The system integrates multiple fabrication processes—including aerodynamic nanoparticle printing, focused ion beam milling, and spin coating—into a single coordinated platform. With a minimum resolution of 50 nm, the approach allows for the construction of complex 3D micro/nanostructures using inorganic materials. Key features include a shared process knowledge database, adaptive interpolation, and standardized process workflows. The platform promotes real-time communication between design and manufacturing stages, making it a flexible and scalable solution for high-precision nanoscale production.

To conclude, with the evolution of advanced technologies, such as scalable nanofabrication and material customization through the integration of nanometric structures, we are witnessing a new era of innovations that, by combining materials science and biotechnology, pave the way for solutions that were unthinkable just a few decades ago, transforming our approach to the design and production of devices with increasingly sophisticated and high-performance characteristics.

7.2 Nanofabrication of Metamaterials: *Artificialia* at the Edge of the Visible

Conventionally, CAD/CAM tools enable the engineered digital creation of detailed three-dimensional models by defining their geometry, material composition, and mechanical characteristics. This methodology—where parametric modeling meets the art of multiscale analysis—enables rapid modification of design variables while mastering complexity from the *macro, micro*, to the *nano* world. Achieving such mastery—once unimaginable just a few decades ago—marks one of the most astonishing milestones in the evolution of design and engineering. The tiny cherry stone, once patiently carved to astonish visitors to the artificialia collections of the ancient *Wunderkammern*, has become a reborn symbol: today, it is no longer the artisan's hand that engraves wonder into wood or ivory, but the mind itself that encodes order into the invisible substrate of matter, crafting realities at the nanoscale—beyond the thresholds of human perception. It is a silent revolution, unfolding within the deepest fabrics of existence (Van Reenen 2018). This conceptual shift is exemplified by *The Boy and His Atom* (2013), the world's smallest film, created at IBM by repositioning individual carbon monoxide molecules frame by frame using a scanning tunneling microscope (IBM Research 2013). Beyond its technical feat, the project embodies the poetic potential of nanotechnology—where storytelling itself is written atom by atom. A similar fusion of science and symbolism emerges in ALICE, a recent experiment at CERN's Large Hadron Collider, where researchers briefly achieved the age-old dream of alchemy by transmuting lead into gold. Through ultra-peripheral collisions, intense electromagnetic fields stripped protons from some lead nuclei, generating gold atoms—ephemeral and infinitesimal, yet profoundly emblematic of humanity's growing ability to manipulate matter at its most fundamental level. In this context, the cherry stone is no longer a static relic, but the living metaphor of a paradigm shift—one that redefines craftsmanship, imagination, and the boundaries of the visible. From contemporary technoscientific research, continuous confirmations emerge that the line between *naturalia* and *artificialia* is fading, challenging the very notion of what is truly natural. As we venture into uncharted territories, the boundaries of creation—once confined to Nature's design alone—are now expanding through the advent of extreme CAD/CAM technologies. Within this paradigm shift, we begin to recognize the rise of the *invisible factory*: a production model inspired by Nature's microscopic strategies, where matter is no longer shaped by visible mechanical force, but by encoded information, molecular precision, and design processes unfolding beyond the threshold of human perception (Lumini 2021). This informational approach to design demands tools capable of anticipating material behavior even before fabrication takes place. Integration with preliminary simulations, such as those based on the Finite Element Method (FEM), allows the evaluation of the structural behavior of models under stress, thermal deformation, and vibrations. In contemporary material research, conventional optical devices, for instance, are bulky and hinder miniaturization, limiting the development of ultra-compact systems. Metasurfaces, introduced by Capasso's Group (2011), provide a solution by enabling precise light manipulation while significantly reducing device thickness (Capasso Group 2025). Studies have focused on creating meta-devices for functionalities such as metalenses and meta-holograms, with CAD/CAE technologies playing a crucial role in their design and optimization. In a study by Pestourie et al. (2018), the authors present a computational framework for the efficient optimization-based *inverse design* of large-area metasurfaces, where they employ a fast approximate solver and a *locally periodic* approximation to optimize surfaces with thousands of parameters. Driven by emerging scientific and technological frontiers, conventional CAD/CAM platforms have evolved beyond traditional modeling tasks. Design now operates at the nanoscale of matter, cells, and life—where every structure embodies a fusion of geometry and function. With the integration of advanced simulation tools, including physical and chemical modeling, it becomes possible to engineer nanoparticles, porous membranes, biomimetic scaffolds, and quantum-scale electronic components. Researchers can now define molecular architectures, orient crystal domains, and control surface roughness at nanometric resolution—actions that elevate the process from mere modeling to full-fledged material engineering. As a natural consequence, the manufacturing phase has adapted accordingly. Cutting-edge techniques such as electron beam lithography, high-resolution 3D printing, laser-guided electrochemical deposition, and direct optical writing enable the fabrication of components whose physical behavior derives directly from their fine internal structure. Such manufacturing processes demand extreme precision in managing operational parameters, chemical environments, and light–matter interactions. Machine instructions, derived directly from CAD/CAM model processing, guide fabrication tools with nanometric accuracy. In fields like biomedical optics or photonic metamaterials, production workflows based on 3D models can achieve intricate forms once thought unattainable by conventional means. Yet the most disruptive change lies in the predictive capabilities now embedded within the design process. By merging CAD/CAM technologies with CAE platforms—often enhanced by artificial intelligence—engineers can simulate how nanoscale variations affect material behavior. Changes in melting point, chemical reactivity, or light responsiveness, for instance, can be anticipated long before fabrication. Similarly, the influence of structural geometry on biological compatibility or mechanical performance can be explored and optimized at the design stage. Rather than simply specifying form, designers now engage in the orchestration of emergent properties—a profound leap toward a predictive design paradigm. The convergence of digital modeling, advanced materials science, and precision nanomanufacturing is thus reshaping design as a discipline: moving from object construction to the engineering of matter's behavior. What emerges is a global-scale, *invisible factory*, capable of synthesizing the laws of quantum mechanics with human inventiveness. Such a factory redefines what can be built, opening unprecedented horizons in domains ranging from regenerative medicine to soft robotics and aerospace innovation. Within this integrated system, CAD, CAM, CAE, and AI are no longer separate tools but a seamless continuum where creativity operates not only on outward form, but on the deep structure of reality itself. Operating at the limits of what can be observed and controlled, design evolves into a method for exploring the hidden architecture of matter. As development pushes further into the nanometric domain, novel fabrication technologies have risen to meet the challenge. One notable example is ultrafast, multi-focus nanoprinting through two-photon polymerization, capable of constructing complex, unsupported forms that defy conventional manufacturing techniques—demonstrating the unprecedented potential of this new design era. In addition to additive techniques, conventional manufacturing processes, such as precision machining, have also evolved to reach the nanoscale. Technologies such as inkjet printing, ElectroHydroDynamic printing (EHD), and StereoLithogrAphy (SLA) have become standards in micro/nano manufacturing, although challenges remain regarding printing capacity and the fabrication of complex three-dimensional geometries. The expansion of the 3D printing market reflects the importance of these innovations. In 2020, the sector was valued at $13.78 billion, with a compound annual growth rate (CAGR) estimated at 21% for the period from 2021 to 2028. In 2021, 2.2 million 3D printers were shipped, with projections indicating a rise to 21.5 million units by 2030, driven primarily by developments in the medical, aerospace, and automotive sectors (Mordor Intelligence 2025). Through these 3D printing technologies, it is now possible to produce highly detailed 3D scaffolds that promote cell growth or support tissue regeneration, using a wide range of biomaterials—natural polymers, which are macromolecular compounds found in organisms, including chitosan, cellulose, alginate, and collagen (Zhang et al. 2023). With the evolution of manufacturing technologies, design and operational precision has reached increasingly extreme levels, extending down to the micro and nanometric scale. It is precisely in this regime that materials begin to behave radically differently from what is observed at macroscopic scales: properties are no longer fixed, but emergent, dependent on the size, shape, and internal organization of the structure. A notable example is the melting point depression observed in gold nanoparticles. While bulk gold melts at 1,064°C, nanoparticles with diameters below 4 nm can exhibit melting points reduced to approximately 500°C (Cortie 2004). This substantial decrease is attributed to the increased surface-to-volume ratio at the nanoscale, leading to a higher proportion of surface atoms with lower coordination numbers. These surface atoms possess reduced cohesive energy, necessitating less thermal energy to transition into the liquid phase. This seemingly counterintuitive phenomenon is now widely studied in fields such as catalysis, photonics, molecular medicine, and advanced materials engineering (Daniel and Astruc 2004). In this context, traditional engineered CAD/CAM platforms have undergone profound transformation. Computer-aided design is no longer limited to modeling macroscopic geometries, but now extends to functional structures at the nanometric scale, where form and function are intrinsically linked. Advanced software, often integrated with physical and chemical simulations, enables the modeling of nanoparticles, porous membranes, biomimetic scaffolds, and quantum-enabled electronic devices. The capacity to define molecular architectures, crystal domain orientation, or surface roughness at the nanoscale elevates design into a true act of material engineering. Consequently, the CAM phase has had to evolve in step with these new demands. Techniques such as high-resolution 3D printing, electron beam lithography, laser-guided electrochemical deposition, or direct optical writing now allow the production of objects whose physical properties arise precisely from their fine-scale structuring. These methods require stringent control over process parameters, including chemistry and light–matter interactions, and are typically driven by data directly derived from the CAD model, converted into highly detailed machine instructions. In applications such as biomedical micro-optics or photonic metamaterials, for example, it is now possible to generate fabrication sequences from 3D models with nanometric precision, producing complex architectures unattainable by conventional means.

However, the true revolution lies in prediction and simulation. The integration of CAD/CAM with CAE systems—often augmented by artificial intelligence—has made it possible to anticipate material behaviors during the very act of conception. Designers can simulate how nanoparticle size affects melting point or chemical reactivity, how geometry and porosity impact cellular adhesion, or how a nanostructured surface responds to mechanical, thermal, or photonic stimuli. This predictive capability transforms design into a proactive process—one that not only defines what is to be made but optimizes its expected function in advance. Here, form gives way to emergent behavior, marking one of the most compelling frontiers in science and design. The synergy of digital conception, advanced materials, and precision nanomanufacturing is ushering in a new paradigm: from constructing objects to constructing behavior itself. The emergence of a global, *invisible factory* realm can be observed—one that merges quantum mechanics with engineering creativity and extends the boundaries of possibility, from regenerative medicine to flexible electronics to space systems. In this integrated ecosystem, computer-aided design, guided fabrication, and predictive modeling converge into a unified creative environment—one where design operates not only on form but on the very nature of material reality. At this frontier, design becomes a near-epistemological endeavor: a means of revealing and modeling the invisible. As these demands have shifted toward the nanometric domain, a suite of advanced nanoprinting techniques has emerged. Among them is ultrafast multi-focus nano-fabrication via two-photon polymerization, capable of building intricate forms—including overhanging and self-supporting structures—that conventional approaches cannot realize. This breakthrough underscores the transformative potential of nanoscale engineering (Geng et al. 2019).

Among the most promising and widely adopted:

- **Direct Laser Writing (DWL)**

Similar to TPP but based on single-photon absorption: slightly less precise but faster. Unlike traditional photolithographic methods, it doesn't require masks or photomasks. Software directly drives the laser, making the process more flexible and efficient—especially for prototyping or low-volume production. Primarily used for surface patterning and planar microdevices (Pinheiro et al. 2024).

- **Dip-Pen Nanolithography (DPN)**

Inspired by the principle of a fountain pen, this method uses an atomic force microscope (AFM) tip to *write* molecules onto a surface with extreme precision. Ideal for biomolecular patterning and soft-matter nanotechnology (Totlani 2022).

- **Electron Beam Lithography (EBL)**

Uses a highly focused beam of electrons to directly write onto a substrate coated with an electron-sensitive resist. Offers resolutions down to below 10 nm and is extensively used in prototype development for integrated circuits, photonics, and quantum devices. Relatively slow and costly (Zhu et al. 2024).

- **Field-Assisted Electrodeposition**

Enables the localized growth of metallic nanostructures using an applied electric field. Used to fabricate microelectrodes, magnetic nanostructures, or components for next-generation energy storage devices (Hao et al. 2023).

- **Focused Ion Beam Lithography (FIB)**

Employs a finely focused beam of heavy ions (e.g., gallium) to mill, deposit, or modify materials with nanometer precision. Ideal for rapid prototyping, site-specific editing of devices, and preparing cross-sections for analysis (Asadi 2021).

- **Nanoimprint Lithography (NIL)**

A mechanical technique in which a nanostructured mold is pressed onto a soft polymer surface to transfer its pattern. Highly efficient for mass production, it offers low cost and high repeatability. Commonly used in optoelectronics, photonics, and sensor fabrication (Badiuddin 2024).

- **Two-Photon Polymerization (TPP)**

A form of 3D additive manufacturing where two photons are absorbed simultaneously at a focal point, triggering polymerization in a photosensitive resin. Enables creation of complex 3D micro- and nanostructures—such as biomimetic scaffolds, microrobots, and micro-optical components—with resolutions down to tens of nanometers. Companies like Nanoscribe® are pioneers in this domain (O'Halloran et al. 2022).

Companies like Nanoscribe® are pioneers in this domain:

"Two-Photon Polymerization, also known as Two-Photon or Multi-Photon Lithography or Direct Laser Writing, utilizes the basic physical effect of two-photon absorption. Two-photon absorption occurs when an atom or molecule absorbs two photons simultaneously, allowing it to be excited to a higher energy state. The medium is typically a liquid resin that is photosensitive and can be cured by UV light. 2PP printing uses light with lower energy, for example in the near-infrared range (NIR), which causes the printing material to solidify only if the photoresin molecules simultaneously absorb the energy of two photons. This mechanism is only likely in the focal volume of pulsed light because it requires a high light intensity inside a volume of the photosensitive material. For fast and precise printing, galvo scanners move the focal point in the focal plane at speeds of hundreds of millimeters per second. In addition, in Two-Photon Polymerization nanofabrication and microfabrication systems, a state-of-the-art stage with nanopositioning capabilities in all three dimensions moves the substrate with the photoresin to build up complex 3D microstructures (Nanoscribe 2025)."

Electron Beam Lithography (EBL) is certainly one of the most advanced nanofabrication technologies, known for its exceptional resolution capabilities of less than 10 nm. It is ideal for applications that demand extreme precision, such as photonics, advanced electronics, and quantum devices, where high-resolution patterning is crucial. However, despite its technical advantages, EBL is not the most suitable for all contexts. The process is relatively slow and expensive, particularly for large-scale production, as it involves writing structures pixel by pixel using an electron beam. This makes EBL less efficient compared to other methods like Nanoimprint Lithography (NIL), which can replicate patterns much faster and more cost-effectively. Additionally, EBL's scalability is limited, making it less practical for mass production or large-area patterning. Therefore, while EBL stands out in terms of precision and resolution, its applicability is typically reserved for high-performance, low-volume manufacturing or research settings where the utmost detail is required (Ebeam Machine 2025). Considering the scope of the technologies mentioned in Table 3, a key aspect of TPP technology is its ability to polymerize material only at specific points where two photons overlap, thus ensuring incredible definition and manufacturing precision (Zhang 2024). This technology stands out for its applicability in various fields, particularly in those where precision is critical. For example, in the microfabrication of medical devices, sensors, or complex electronic circuits, TPP proves to be an indispensable resource, as it enables the creation of highly detailed three-dimensional structures or specialized microfluidic platforms,

Table 3. Comparative Overview of Nanofabrication Techniques.[5]

Technique	Typical Resolution	Main Advantages	Key Limitations	Typical Applications
Direct Laser Writing (DWL)	~ 200 nm to 1 µm	Faster than TPP, suitable for large-area patterning	Lower resolution than TPP	Surface patterning, MEMS, optical elements
Dip-Pen Nanolithography (DPN)	~ 30 nm to 50 nm	High chemical selectivity, biomolecule-compatible	Slow, tip wear, difficult scaling	Biosensors, molecular electronics, protein patterning
Electron Beam Lithography (EBL)	< 10 nm	Extremely high resolution, direct-write flexibility	Slow, expensive, not scalable for mass production	Photonics, quantum devices, advanced electronics
Field-Assisted Electrodeposition	~ 50 nm to 500 nm	Precise metal deposition, scalable	Limited to conductive materials	Microbatteries, nanomagnetics, electrochemical sensors
Nanoimprint Lithography (NIL)	~ 20 nm to 100 nm	High throughput, cost-effective for replication	Requires master mold, less flexible design changes	Displays, nanoelectronics, metasurfaces
Two-Photon Polymerization (TPP)	~ 100 nm to 200 nm	True 3D structuring, high precision, versatile geometries	Limited throughput, expensive resins	Biomedical scaffolds, micro-optics, microrobotics

[5] These techniques are not mutually exclusive but often complementary. Selection depends on material properties, target geometry, production volume, and functional requirements.

which are perfectly suited to the demands of micro and nano-engineering, including *Lab-on-a-Chip (LOC)*[6] and *Organ-on-a-Chip (OoC)*[7] devices (Preetam et al. 2022). One of the areas where TPP shows great potential is biomedicine, particularly in the creation of scaffolds for tissue engineering and regenerative medicine. It enables the fabrication of artificially designed constructs with arbitrary geometries, mimicking the microenvironment of natural extracellular matrices, making it an ideal tool for studying cell behavior. This technique is especially valuable in cytobiology, tissue engineering, and regenerative medicine, where 3D scaffolds are essential (Jing et al. 2022b). The ability to replicate biological structures with extreme precision, enabling the engineering of artificial tissues and organs, is one of the most promising applications of TPP. In addition to biomedical applications, TPP also allows for advanced material manipulation, enabling the creation of polymer microstructures designed to have specific properties, such as conductivity or responsiveness to external stimuli (for a more detailed discussion on stimuli-responsive materials and their role in smart systems, see Chapter 5.0 on smartness). Examples of such polymers include polypyrrole, poly(N-isopropylacrylamide) (PNIPAM), and polystyrene, which can respond to factors like temperature, pH, and electrical fields. A recent study by Tariq et al. (2023) introduces *4D printing*, a technology that adds the dimension of time to traditional 3D printing. While 3D printing creates static, fixed structures, 4D printing uses stimuli-responsive materials (Chen et al. 2024). Thanks to the combination of smart materials and advanced printing techniques, 4D printing enables the creation of objects that not only perform static functions but can evolve over time, actively responding to changes in the surrounding environment (Fig. 3).

This dynamic capability allows for the creation of self-assembling, adaptive, and reconfigurable structures, making it ideal for applications in soft robotics, biomedical devices, and advanced engineering. It enables the production of objects that can change shape or properties over time in response to external stimuli such as heat, moisture, or light (Tariq et al. 2023). This ability to *change shape* or modify other physical properties, such as rigidity or conductivity, gives 4D printing extraordinary potential in a wide range of applications, including soft robotics, regenerative medicine, and biomedical devices, where adaptability and self-repair capabilities are crucial. Its application goes beyond simple prototyping: it allows the creation of devices that self-assemble, self-adapt, or even self-repair, opening new frontiers in medicine, engineering, and biotechnology. In recent advancements, several companies and institutions have been developing cutting-edge technologies in the field of 4D printing. For instance, MIT's Self-Assembly Lab has created

[6] Lab-on-a-Chip (LOC): Miniaturized devices that integrate multiple laboratory functions on a single chip, enabling rapid analysis of small fluid samples for applications in diagnostics, environmental monitoring, and biological research.

[7] Organ-on-a-Chip (OoC): Microengineered systems that replicate the physiological functions of human organs by combining living cells to replicate complex biological dynamics (such as cardiac rhythm or renal filtration) and microfluidic technology, used for drug testing, disease modeling, and personalized medicine.

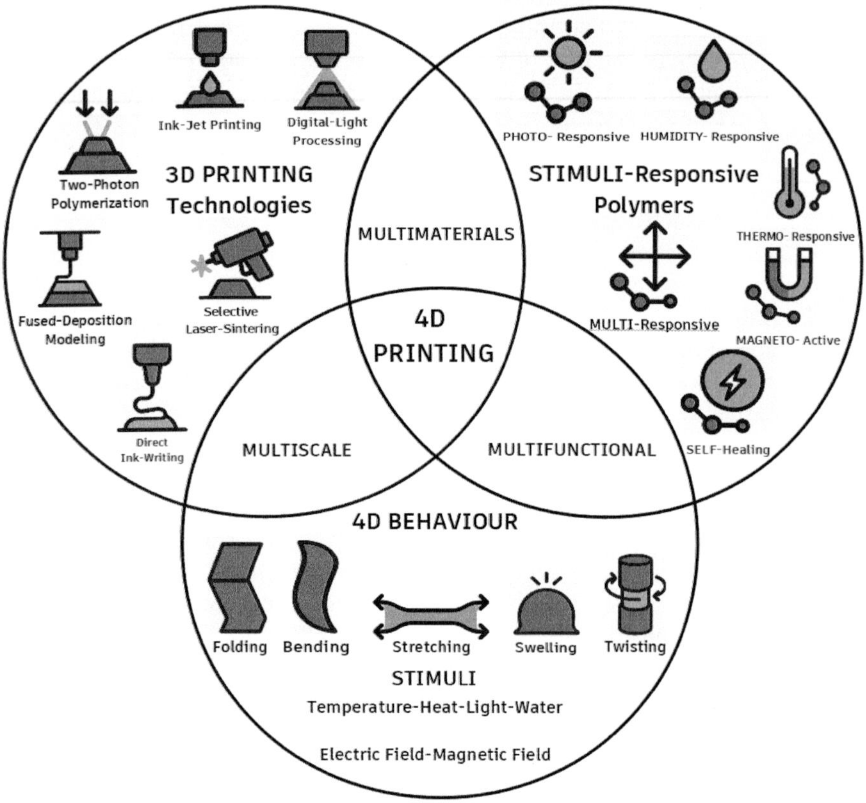

Fig. 3. 4D Printing Overview: Technologies, Materials and Behavior.

4D-printed structures that change shape when exposed to various environmental conditions, demonstrating their potential for self-assembly and adaptation:

"Self-Assembly is a process by which disordered parts build an ordered structure through only local interaction. In self-assembling systems, individual parts move towards a final state, whereas in self-organizing systems, components move between multiple states, oscillate and may never come to rest in a final configuration (Self-Assembly Lab 2025)."

A recent analysis has identified the top ten companies in the 4D printing market, highlighting their contributions to the development of this emerging technology. Notable companies include Autodesk®, Stratasys®, Materialise®, Dassault®, Organovo®, Northrop Grumman®, EnvisionTEC®, CT-Core®, Hewlett-Packard®, and Heineken®. These companies are developing advanced materials and technologies that enable the creation of objects capable of changing shape and properties over time in response to external stimuli such as heat, light, or moisture. For more details, refer to the full article available at the site of Scoop Market Us (Pangarkar 2024).

The University of Colorado Boulder has been at the forefront, particularly in developing 4D materials that mimic biological systems and can adapt to specific stimuli, pushing the boundaries of biomedical devices and soft robotics (University of Colorado Boulder 2013). In the context of 4D printing, an innovative approach to simplify the use of smart materials was proposed by Choi et al. (2021). The authors introduce three key concepts to facilitate the integration of smart materials into 3D printing. First, they present the idea of standardized kinetic components made from smart materials, which can perform basic mechanical movements such as bending and twisting, allowing for integration with rigid parts in complex mechanical assemblies. Second, they propose an open kinetic library, where users can download and share data related to kinetic components, promoting collaboration and innovation. Finally, they introduce the use of simulations based on empirical methods, using kinetic components to design and test the assembly of complex structures. The authors provide two practical examples to demonstrate the effectiveness of this approach: an icosahedron and a *mounting platform*, highlighting how these concepts can be applied in practice. The ability to customize engineered materials is crucial in advanced fields such as nanoelectronics, photonics, and communications, where there is a growing demand for devices with tailored properties and enhanced performance. This customization enables the creation of materials that meet the specific needs of emerging technologies, such as quantum computing, integrated photonic circuits, and 5G/6G communication systems, where high efficiency, miniaturization, and unique material properties are vital for advancing device capabilities. In design and rapid prototyping, Two-Photon Polymerization (TPP) stands out for its capacity to produce highly complex prototypes that would be difficult to achieve with traditional methods. The high-resolution 3D structures it creates are ideal for exploring new geometries and design concepts, influencing fields like architecture, engineering, and even art. The development of a Double-Layer Curved Compound Eye (DLCCE), as detailed by Jing et al. (2021) in their study *Design and Fabrication of Double-Layer Curved Compound Eye via Two-Photon Polymerization*, highlights the convergence of biomimicry and advanced manufacturing technologies. Inspired by the natural compound eye structure found in certain insects, this innovative optical device overcomes challenges related to defocus, miniaturization, and assembly. By leveraging Two-Photon Polymerization (TPP) technology, the researchers designed a compact, high-resolution optical system with a field of view (FOV) of 85° and the ability to integrate 33 ommatidia. These design features allow for precise refraction and focus of incident light, significantly improving optical imaging and sensing capabilities. This work exemplifies the principles of biomimetics, where nature's solutions to complex problems—such as the compound eye's ability to achieve wide-angle vision—are translated into technological advancements. The fabrication of such optical devices through TPP underscores the power of biomimetic design in pushing the boundaries of micro-optics and nanostructures. The integration of this technology opens up possibilities for miniaturized, efficient, and optimized optical systems in applications ranging from imaging and detectors to micro-optical devices (Jing et al. 2022). The combination of biomimetic inspiration and cutting-edge fabrication techniques represents a significant step forward in the evolution of optical device design. This flexibility also enables the creation of customized

Table 4. Comparison of Companies Producing Hardware and Software for Two-Photon Polymerization (TPP).

Company	Hardware	Software	Website
3D MicroPrint®	3D printers for TPP, ideal for biotechnology, medical devices, microelectronics, and the fabrication of highly detailed and customizable structures.	3D design software integrated with devices for creating models to be used with their printing systems.	www.3dmicroprint.com
Bioplotter® by EnvisionTEC	3D printers for photopolymerization technologies, including TPP, used in tissue engineering and biomedicine applications.	Design and simulation software for structures intended for printing, with specific applications in medicine and biotechnology.	www.envisiontec.com
Laser Lines®	Supplies lasers and optical components necessary for TPP systems, providing high-performance solutions for advanced 3D fabrication processes like TPP.	Control software for the precise operation of laser systems, which can be integrated into TPP applications.	www.laserlines.com
Nanoscribe®	Provides Two-Photon Polymerization solutions, including the Photonic Professional GT2 system for creating high-resolution 3D structures.	DeScribe software for designing and managing 3D models intended for TPP, highly integrated with hardware systems.	www.nanoscribe.com
Oxford Instruments®	Microscopy and photonics instruments usable for TPP, often combined with other fabrication technologies.	Software for analyzing and designing micro and nanostructures, well-integrated with TPP systems.	www.oxinst.com
Photonic Solutions®	Optical components for micro- and nano-fabrication, used in many TPP applications.	Simulation and control software that adapts to the requirements of advanced fabrication, also supporting 3D design.	www.photonic-solutions.com
Süss MicroTec®	Advanced photolithography and microfabrication systems that can be combined with TPP.	Design and simulation software to create precise structures for microfabrication.	www.suss.com
Thorlabs®	High-precision optical and laser components used in TPP applications.	Software for controlling and automating optical systems, integrated into TPP applications.	www.thorlabs.com

devices, pushing the boundaries of design possibilities. TPP also has significant applications in optoelectronics, particularly in creating microscopic devices such as lenses, waveguides, and other optical components (Andão et al. 2022). Fabricating complex optical structures at the nanometric scale is crucial for technologies like integrated photonics and computer vision, where miniaturization and precision are essential for optimal performance. Unlike traditional manufacturing processes that remove material, such as cutting or milling, TPP is an additive process, using only the material needed to create the desired structure. This minimizes waste, making TPP a more sustainable technology. Its approach enhances production efficiency, making it an environmentally friendly choice for creating advanced components. Additionally, TPP supports multimaterial processing, using various photosensitive resins. The ability to combine different materials in a single 3D structure significantly expands the technology's capabilities, allowing the creation of highly specialized components for diverse applications across industries. In summary, Two-Photon Polymerization is a transformative technology that enables the creation of high-precision 3D microstructures with applications spanning bioengineering, nanotechnology, microelectronics, and sustainability. Its ability to work with functional materials, produce complex geometries, and minimize waste makes it a vital tool for future technological and scientific advancements (See Table 4).

References

Adão, R., Alves, T., Maibohm, C., Romeira, B., and Nieder, J. (2022). Two-photon polymerization simulation and fabrication of 3D microprinted suspended waveguides for on-chip optical interconnects. Optics Express, 30(3): 9623–9642.

Ansari, A. (2025). Moore's Law and Beyond: The Future of Semiconductor Innovation. BISinfotech. Available at: https://www.bisinfotech.com/moores-law-and-beyond-the-future-of-semiconductor-innovation/ [Accessed: 14 May 2025].

Arber, A. (1954). The mind and the eye: A study of the biologist's standpoint. Cambridge University Press.

Ariga, K., Song, J., and Kawakami, K. (2024). Molecular machines working at interfaces: Physics, chemistry, evolution and nanoarchitectonics. Physical Chemistry Chemical Physics, 26: 13532.

Asadi, N. (2021). Focused Ion Beam (FIB) Principle. University of Florida. Available at: https://faculty.eng.ufl.edu/navid-asadi/wp-content/uploads/sites/84/2021/09/Lecture-13-FIB-Principle.pdf [Accessed: 15 May 2025].

Autumn, K., Sitti, M., Liang, Y. A., Peattie, A. M., Hansen, W. R., Sponberg, S. et al. (2002). Evidence for van der Waals adhesion in gecko setae. Proceedings of the National Academy of Sciences of the United States of America, 99(19): 12252–12256.

Badiuddin, A. F. (2024). Nanoimprint Lithography (NIL) – A versatile nanopatterning technology. AZoNano. Available at: https://www.azonano.com/article.aspx?ArticleID=6776 [Accessed: 26 April 2025].

Baker, T. S., and Church, G. M. (2009). caDNAno: A web-based design tool for DNA nanostructures. caDNAno. Available at: https://cadnano.org [Accessed: 27 April 2025].

Basu, P., Verma, J., Abhinav, V., Ratnesh, R. K., Singla, Y. K., and Kumar, V. (2025). Advancements in lithography techniques and emerging molecular strategies for nanostructure fabrication. International Journal of Molecular Sciences, 26(7): 3027.

Bekkouche, I., Kuznetsova, M. N., Rejepov, D. T., Vetcher, A. A., and Shishonin, A. Y. (2023). Recent advances in DNA nanomaterials. Nanomaterials, 13(17): 2449.

Boeke, K. (1957). Cosmic view: The universe in 40 jumps. John Day Company. Available at: https://www.arvindguptatoys.com/arvindgupta/cosmic-view.pdf [Accessed: 14 May 2025].

Borelli, G. A. (1680–1681). De motu animalium (Pars Prima et Pars Altera). Romae: Ex typographia Angeli Bernabo. Available at: https://archive.org/details/McGillLibrary-osl_motu-animalium_B731m1680_81_v1-20170 [Accessed: 14 May 2025].

Branch, G. (2015). "Everything is the way it is because it got that way". National Center for Science Education. Available at: https://ncse.ngo/everything-way-it [Accessed: 14 May 2025].

Capasso Group. (2025). Metasurfaces and flat optics. Harvard John A. Paulson School of Engineering and Applied Sciences. Available at: https://capasso.seas.harvard.edu/metasurfaces-and-flat-optics [Accessed: 26 April 2025].

Casetta, E. (2017). Preformation vs. epigenesis: Inspiration and haunting within and outside contemporary philosophy of biology. Rivista di Estetica, (66): 119–138.

CERN. (2025). ALICE detects the conversion of lead into gold at the LHC. Retrieved from https://home.cern/news/news/physics/alice-detects-conversion-lead-gold-lhc.

Chen, F. (2020). Lithography resolution limits – Arrayed features. SemiWiki. Available at: https://semiwiki.com/lithography/284583-lithography-resolution-limits-arrayed-features-2/ [Accessed: 14 May 2025].

Chen, J., Virrueta, C., Zhang, S., Mao, C., and Wang, J. (2024). 4D printing: The spotlight for 3D printed smart materials. Materials Today, 77: 66–91.

Choi, W., Kim, D., Lee, S., and Lee, Y.-G. (2021). New modeling approach for 4D printing by using kinetic components. Journal of Computational Design and Engineering, 8(4): 1013–1022.

Collegio, A. J., Nah, J. C., Scotti, P. S., and Shomstein, S. (2019). Attention scales according to inferred real-world object size. Nature Human Behaviour, 3: 40–47.

Coppens, M.-O., Weissenberger, T., Zhang, Q., and Ye, G. (2020). Nature-inspired, computer-assisted optimization of hierarchically structured zeolites. Advanced Materials Interfaces, 8(5): 2001409.

Cortie, M. B. (2004). The weird world of nanoscale gold. Gold Bulletin, 37(1): 12–19.

Daniel, M.-C., and Astruc, D. (2004). Gold nanoparticles: Assembly, supramolecular chemistry, quantum-size-related properties, and applications toward biology, catalysis, and nanotechnology. Chemical Reviews, 104(1): 293–346.

Dassault Systèmes. (2025). BIOVIA Materials Studio. 3DS. Available at: https://www.3ds.com/products/biovia/materials-studio [Accessed: 25 April 2025].

Douglas, S. M., Dietz, H., Liedl, T., Högberg, B., Graf, F., and Shih, W. M. (2009). Rapid prototyping of 3D DNA-origami shapes with caDNAno. Nucleic Acids Research. Available at: https://pubmed.ncbi.nlm.nih.gov/19531737/ [Accessed: 27 May 2025].

Eames Office. (1977). Powers of Ten and the Relative Size of Things in the Universe. Eames Office. Available at: https://www.eamesoffice.com/the-work/powers-of-ten/ [Accessed: 14 May 2025].

Ebeam Machine. (2025). Step-by-step electron beam lithography for beginners. Ebeam Machine. Available at: https://ebeammachine.com/step-by-step-electron-beam-lithography-for-beginners/ [Accessed: 15 May 2025].

Fu, R., Chen, K., Li, Z., Yu, S., and Zheng, G. (2022). Metasurface-based nanoprinting: Principle, design, and advances. Opto-Electronic Science, 1(10): 220011.

Galilei, G. (1734). Mathematical discourses concerning two new sciences relating to mechanicks and local motion, in four dialogues. Translated from the Italian by Thomas Weston. Internet Archive. Available at: https://archive.org/details/bim_eighteenth-century_mathematical-discourses-_galilei-galileo_1734. [Accessed: 29 September 2025].

Geng, Q., Wang, D., Chen, P., and Chen, S.-C. (2019). Ultrafast multi-focus 3-D nano-fabrication based on two-photon polymerization. Nature Communications, 10, Article 1729.

Ghose, T., and Zimmermann, K. A. (2023). What is pareidolia? Live Science. Available at: https://www.livescience.com/25448-pareidolia.html [Accessed: 14 May 2025].

Gonzalez, M. C. (2025). Infographic: The history of 3D printing. ASME. Available at: https://www.asme.org/topics-resources/content/infographic-the-history-of-3d-printing [Accessed: 28 May 2025].

Hao, X., Chen, F., Zhang, Y., Liu, Y., Gao, G., Song, Y. et al. (2023). Magnetic-field-assisted electrodeposition regulates the Ni:Fe ratio for water oxidation. Materials Today Sustainability, 24: 100556.

Hartsoeker, N. (1694). Essay de dioptrique. Internet Archive. Available at: https://archive.org/details/BIUSante_07012 [Accessed: 14 May 2025].

HASS Corporation. (2018). Amber® Mill User's Manual: New Frontier of Lithium Disilicate CAD/CAM Blocks. Memodent. Available at: https://www.memodent.nl/uploads/pdf/Amber-Mill_users-manual_0806EN_180815p.pdf [Accessed: 25 April 2025].

Horton, L., and Long, S. (2010). Alice in Nanoland. NISE Network. Available at: https://www.nisenet.org/sites/default/files/catalog/uploads/5526/aliceinnanoland_text_dec10.pdf [Accessed: 14 May 2025].

Hull, C. W. (1986). Apparatus for production of three-dimensional objects by stereolithography. Unified Patents Portal. Available at: https://portal.unifiedpatents.com/patents/patent/US-4575330-A [Accessed: 14 May 2025].

Humanitas Research. (2025). Multiscale Imaging. Humanitas Research. Available at: https://www.humanitas-research.com/tech-platforms/imaging/ [Accessed: 14 May 2025].

IBM Research. (2013). A Boy and His Atom: The World's Smallest Movie. Retrieved from https://www.youtube.com/watch?v=oSCX78-8-q0.

Janusziewicz, R., Tumbleston, J. R., Quintanilla, A. L., and DeSimone, J. M. (2016). Layerless fabrication with continuous liquid interface production. Proceedings of the National Academy of Sciences, 113(42): 11703–11708.

Jing, X., Wang, K., Zhu, R., Lin, J., Yu, B., and Lu, M. (2021). Design and Fabrication of Double-Layer Curved Compound Eye via Two-Photon Polymerization. IEEE Photonics Technology Letters, 33(5): 231–234.

Jing, X., Fu, H., Yu, B., Sun, M., and Wang, L. (2022). Two-photon polymerization for 3D biomedical scaffolds: Overview and updates. Frontiers in Bioengineering and Biotechnology, 10: 994355.

Kaidarova, A., Naenen, V., Windey, R., Goossens, N., Sinha, T., Ignatious, V. et al. (2025). Hybrid organic-inorganic functional nanocomposites: From basis to applications in stretchable sensing and energy devices. Materials Science and Engineering: R: Reports, 163: 100933.

Kazazis, D., Santaclara, J.G., van Schoot, J., Mochi, I., and Ekinci, Y. (2024). Extreme ultraviolet lithography. Nat. Rev. Methods Primers 4: 84.

Kim, D.-N., Bathe, M., and Dietz, H. (2018). About. CanDo DNA Origami. Available at: https://cando-dna-origami.org/about/[Accessed: 14 May 2025].

Kononov, P., Kononova, I., Moshnikov, V., Maraeva, E., and Trubetskaya, O. (2022). Step-By-Step modeling and demetallation experimental study on the porous structure in zeolites. Molecules, 27(23): 8156.

Kunstkammer. (n.d.). Home. Kunstkammer. Available at: https://www.kunstkammer.com/index.php/en/ [Accessed: 2 October 2025].

Kutschera, U. (2023). Antonie van Leeuwenhoek (1632–1723): Master of fleas and father of microbiology. Microorganisms, 11(8): 1994.

Lee, B. J. (2024). Continuous liquid interface production (CLIP): A new paradigm in photopolymerization-based additive manufacturing. JMST Advances, 6: 363–369.

Loretan, C., and Müller, A. (2023). Nano, bits, and Feynman's dream: There's plenty of room at the (molecular) bottom. Journal of Chemical Education, 100(3): 1366–1370.

Lumini, M. (2021). The invisible factories: Nature's technologies and design of artificial innovation. Cuadernos del Centro de Estudios en Diseño y Comunicación, (149): 145–166. Available at: https://dialnet.unirioja.es/descarga/articulo/8269192.pdf [Accessed: 15 May 2025].

Lumini, M. (2022). The Maniola, Lycaenidae, and Other Lepidoptera Eggs as an Inspiration Source for Food Storage and Packaging Design Solutions. *In:* Palombini, F.L., Muthu, S.S. (eds.). Bionics and Sustainable Design. Environmental Footprints and Eco-design of Products and Processes. Springer, Singapore.

Marini, M., Falqui, A., Moretti, M., Limongi, T., Allione, M., Genovese, A. et al. (2015). The structure of DNA by direct imaging. Science Advances, 1(7): e1500734.

Marini, M., Limongi, T., Falqui, A., Genovese, A., Allione, M., Moretti, M. et al. (2017). Imaging and structural studies of DNA-protein complexes and membrane ion channels. Nature Nanotechnology, 12(5): 429–436.

Marini, M., Allione, M., Lopatin, S., Moretti, M., Giugni, A., Torre, B. et al. (2018). Suspended DNA structural characterization by TEM diffraction. Microelectronic Engineering, 187–188, 39–42.

Martynenko, I. V., Ruider, V., Dass, M., Liedl, T., and Nickels, P. C. (2021). DNA origami meets bottom-up nanopatterning. ACS Nano, 15(7): 10769–10774.

McMahon, T. A., and Bonner, J. T. (1983). On size and life. Internet Archive. Available at: https://archive.org/details/on-size-and-life [Accessed: 14 May 2025].

Mental Itch. (2023). The strange theory of the homunculus. Mental Itch. Available at: https://mentalitch.com/the-strange-theory-of-the-homunculus/[Accessed: 14 May 2025].

Merleau-Ponty, M. (1968). The visible and the invisible: Followed by working notes (A. Lingis, Trad.). Northwestern University Pres.

Mishra, R. K., Verma, K., and Singh, D. S. (2024). Defect engineering in nanomaterials: Impact, challenges, and applications. Smart Materials in Manufacturing, 2, 100052.

Mordor Intelligence. (2025). 3D Printing Market Growth | Industry Analysis, Size & Forecast Report (2025–2030). Available at: https://www.mordorintelligence.com/industry-reports/3d-printing-market [Accessed: 15 May 2025].

Munoz, A., Louca, C., and Vichi, A. (2025). Characterization of a lithium disilicate CAD/CAM material with firing temperature-controlled translucency. Materials, 18(7): 1591.

Nanoscribe. (2025). 2PP – Two-Photon Polymerization. Nanoscribe. Available at: https://www.nanoscribe.com/en/microfabrication-technologies/2pp-two-photon-polymerization/ [Accessed: 15 May 2025].

Office of Basic Energy Sciences. (2006). The Scale of Things [Chart]. U.S. Department of Energy. Available at: https://sc-drcds.osti.gov/bes/Community-Resources/Scale-of-Things-Chart [Accessed: 14 May 2025].

O'Halloran, S., Pandit, A., Heise, A., and Kellett, A. (2022). Two-Photon Polymerization: Fundamentals, Materials, and Chemical Modification Strategies. Advanced Science, 9(1): 202204072.

Pangarkar. (2024). Top 10 4D printing companies. Available at: https://scoop.market.us/top-10-4d-printing-companies/?utm_source=chatgpt.com [Accessed: 28 April 2025].

Pestourie, R., Pérez-Arancibia, C., Lin, Z., Shin, W., Capasso, F., and Johnson, S. G. (2018). Inverse design of large-area metasurfaces. Optics Express, 26(16): 20757–20766.

Petch, M. (2024). The future of 3D printing 2024: The 3D printing industry executive survey. 3D Printing Industry. Available at: https://3dprintingindustry.com/news/the-future-of-3d-printing-2024-the-3d-printing-industry-executive-survey-228135/ [Accessed: 14 May 2025].

Pinheiro, T., Morais, M., Silvestre, S., Carlos, E., Coelho, J., Almeida, H. V. et al. (2024). Direct Laser Writing: From Materials Synthesis and Conversion to Electronic Device Processing. Advanced Materials, 36(15): 202402014.

Preetam, S., Nahak, B. K., Patra, S., Toncu, D. C., Park, S., Syväjärvi, M. et al. (2022). Emergence of microfluidics for next generation biomedical devices. Biosensors and Bioelectronics: X, 10: 100106.

Quantum News. (2024). Nanoelectronics: Pushing the limits of miniaturization. Quantum Zeitgeist. Available at: https://quantumzeitgeist.com/nanoelectronics-pushing-the-limits-of-miniaturization/ [Accessed: 14 May 2025].

Ramachandran, A. S. (2024). Pushing the boundaries: From nanoscale to interstellar. LinkedIn. Available at: https://www.linkedin.com/pulse/pushing-boundaries-from-nanoscale-interstellar-abeneth-s-ec8nf [Accessed: 15 May 2025].

Ramezani, H., and Dietz, H. (2020). Building machines with DNA molecules. Nature Reviews Genetics, 21(1): 5–26.

Riavis, V. (2020). On a Human Scale. Drawing and Proportion of the Vitruvian Figure. Diségno, (7): 43–54.

Ruivenkamp, M., and Rip, A. (2010). Visualizing the invisible nanoscale study: Visualization practices in nanotechnology community of practice. Science Studies, 23(1): 3–36.

Science. (2023). The tiny ten: Experts weigh in on the top 10 challenges remaining for nanoscience & nanotechnology. Science. Available at: https://www.science.org/content/resource/tiny-ten-experts-weigh-top-10-challenges-remaining-nanoscience-nanotechnology [Accessed: 14 May 2025].

Self-Assembly Lab. (2025). MIT's Self-Assembly Lab. Massachusetts Institute of Technology. Available at: https://selfassemblylab.mit.edu/[Accessed: 15 May 2025].

Shah, G. V., and Sitti, M. (2004). Modeling and design of biomimetic adhesives inspired by gecko foot-hairs. pp. 218–223. *In:* Proceedings of the IEEE International Conference on Robotics and Biomimetics (ROBIO 2004). IEEE.

Shen, J., Sun, W., Liu, D., Schaus, T., and Yin, P. (2021). Three-dimensional nanolithography guided by DNA modular epitaxy. Nature Materials, 20(6): 683–690.

Shilov, A. (2024). SMIC and Huawei could use quadruple patterning for China-made 5nm chips: Report. Tom's Hardware. Available at: https://www.tomshardware.com/tech-industry/semiconductors/smic-and-huawei-could-use-quadruple-patterning-for-chinese-5nm-chips-report [Accessed: 14 May 2025].

Suhail, A. (2025). The quantum potential of carbon opens the door to more sustainable computing. AZoQuantum. Available at: https://www.azoquantum.com/Article.aspx?ArticleID=594 [Accessed: 14 May 2025].

Tahir, U., Shim, B., Kamran, M., Kim, D.-I., and Jeong, M. (2021). Nanofabrication techniques: Challenges and future prospects. Journal of Nanoscience and Nanotechnology, 21: 10.1166/jnn.2021.19327.

Tanguy, D., and Fairbrass, M. (2018). The scale of things: Mind-blowing proportions, remarkable ratios and extraordinary facts. Hardie Grant. Available at: https://books.google.it/books?id=va9hswEACAAJ [Accessed: 14 May 2025].

Tariq, A., Arif, Z. U., Khalid, M. Y., Hossain, M., Rasool, P. I., Umer, R. et al. (2023). Recent advances in the additive manufacturing of stimuli-responsive soft polymers. Advanced Engineering Materials.

The Metropolitan Museum of Art. (2015). Small Wonders: Gothic Boxwood Miniatures. The Metropolitan Museum of Art. Available at: https://www.metmuseum.org/press-releases/small-wonders-2016-exhibitions [Accessed: 14 May 2025].

Thompson, D. W. (1917). On growth and form. Cambridge University Press. Available at: https://archive.org/details/ongrowthform00thom/page/n7/mode/2up [Accessed: 14 May 2025].

Tiginyanu, I., Ursaki, V., and Popa, V. (2011). Nanoimprint lithography (NIL) and related techniques for electronics applications. pp. 280–329. *In:* Makhlouf, A. S. H., and Tiginyanu, I. (eds.). Nanocoatings and Ultra-Thin Films. Woodhead Publishing.

Totlani, R. (2022). Nanolithography: A promising tool for precision nanofabrication. International Journal of Nanomaterials and Nanostructures, 8(2).

University of Colorado Boulder. (2013). (October 22) CU Boulder researchers develop 4D printing technology for composite materials. University of Colorado Boulder. Available at: https://www.colorado.edu/today/2013/10/22/cu-boulder-researchers-develop-4d-printing-technology-composite-materials [Accessed: 16 May 2025].

US Metric Association. (2020). Metric Convention of 1875. US Metric Association. Available at: https://usma.org/laws-and-bills/metric-convention-of-1875 [Accessed: 14 May 2025].

Van Reenen, C. (2018). A media history of the (digital?) Wunderkammer: A case study of Samuel Quiccheberg's 1565 proposal for an ideal Wunderkammer. MA Thesis, Concordia University. Available at: https://spectrum.library.concordia.ca/983796/1/vanReenen_MA_F2018.pdf [Accessed: 15 May 2025].

VASP Software. (2025). VASP: Vienna Ab-initio Simulation Package. Available at: https://www.vasp.at/ [Accessed: 25 April 2025].

Wolff, C. F. (1774). Theoria generationis. Halae Ad Salam: Typis et sumtu Io. Christ. Hendel. Internet Archive. Available at: https://archive.org/details/b30506256/page/n3/mode/2up [Accessed: 14 May 2025].

Woods, A. (2021). The death of Moore's law: What it means and what might fill the gap going forward. MIT CSAIL Alliances. Available at: https://cap.csail.mit.edu/death-moores-law-what-it-means-and-what-might-fill-gap-going-forward [Accessed: 14 May 2025].

Yoon, H.-S., Lee, H.-T., Jang, K.-H., Kim, C.-S., Park, H., Kim, D.-W., et al. (2017). CAD/CAM for scalable nanomanufacturing: A network-based system for hybrid 3D printing. Microsystems & Nanoengineering, 3: 17072.

Zhan, P., Peil, A., Jiang, Q., Wang, D., Mousavi, S., Xiong, Q. et al. (2023). Recent advances in DNA origami-engineered nanomaterials and applications. Chemical Reviews, 123(7): 3976–4050.

Zhang, S., Chen, X., Shan, M., Hao, Z., Zhang, X., Meng, L. et al. (2023). Convergence of 3D Bioprinting and Nanotechnology in Tissue Engineering Scaffolds. Biomimetics, 8(1): 94.

Zhang, W. (2024). Research on the application of two-photon polymerization technology in micro and nano manufacturing. Academic Journal of Science and Technology, 13: 114–117.

Zhu, C., Ekinci, H., Pan, A., Cui, B., and Zhu, X. (2024). Electron beam lithography on nonplanar and irregular surfaces. Microsyst. Nanoeng. 10: 52.

8

Make it More Organic

"Art always serves beauty, and beauty
is the joy of possessing form,
and form is the key to organic life since
no living thing can exist without it."
Boris Pasternak

8.1 Organic, All Too Organic[1]

In everyday thought, the notions of *organ*, *organism* and *organic* convey the idea of a living order, where parts and wholes are dynamically interwoven in mutual dependence. Although their meanings vary by context, these terms fundamentally describe the creation of *organismic systems* in which the components are synergistic, interconnected, and work together to maintain the overall balance, whether in *natural* or *artificial* settings (Rosslenbroich 2023). In Ancient Greek, the word Ὄργανον (*órganon*) referred to various meanings associated with the concept of a *tool*, *implement*, or *means* used to perform a specific task or function (Liddell and Scott 1996). In Aristotle's philosophy, the term also took on the meaning of *instrument of knowledge* or *method*, referring to the *intellectual tools* used for philosophical and scientific inquiry. Over time, the term evolved into the Latin *organum*, which retained the sense of *device* and *musical instrument*, and later transformed into the modern *organ* (an anatomical, physiological, and functional unit that performs

[1] The title *Organic, All Too Organic* is intended as an ironic paraphrase of Nietzsche's famous "Human, All Too Human" ("Menschliches, Allzumenschliches"), shifting the focus from the human condition to the biological realm. It is an intertextual play that redirects the axis of reflection toward the living and the organic, poised between critique and contemplation. It also serves as a reference to Nietzsche's reflection highlighting how, within the biomimetic enthusiasm of contemporary technoscience, there is often an exaggerated exaltation of the perfection of the organic—without acknowledging that Nature itself is permeated by contradictions and fragility. Even the experiment of terrestrial life does not escape the condition of a painful and tormented search for a presumed perfection, continually undermined by dramatic geo-environmental events (such as mass extinctions) that have cyclically interrupted the course of evolution in unforeseen and catastrophic ways.

DOI: 10.1201/9781003438946-8

specific tasks within an animal or plant organism, and in mechanics, any device capable of performing a specific function). Similarly, the Greek term Ὀργανισμός (*organismós*), meaning *organization* or *structure*, shares the same root as Ὄργανον (*órganon*). It passed into Latin as *organismus* and subsequently into English as *organism*, preserving the meaning of a complex, functional living structure—such as a human, animal, or plant, or, figuratively, a structured entity designed to meet specific technical or functional needs: e.g., architectural organism (Lewis and Short 1879). In Latin, the term *organicus* referred to something related to the concept of an organ, maintaining the sense of connection to living structures or functions. This evolution reflected the term's association with both physical structures and functional aspects, whether in nature or human-made environments, highlighting the dynamic and coordinated functioning of both living and artificial systems. In modern English, the derived term *organic* has expanded further to refer not only to biological life but also to natural processes, ecological systems, and products (such as organic food) made without synthetic chemicals. *Organic* can thus be considered a powerful metaphorical synonym for the *vital state of matter*, as both refer to the biochemical processes that underpin life:

"Life is a unique state of matter, a state that is highly organized and structured, a system of processes that maintain themselves in the face of external forces (Schrödinger 1944)."

However, their union has never been free from tensions. Since the advent of Darwinian theories, beneath the experimental rigor of biochemistry lie the ashes of philosophical disputes, which each new discovery about the origin of life continues to fuel. In chemistry and biology, the term refers to substances primarily composed of carbon and hydrogen, often combined with oxygen, nitrogen, sulfur, and phosphorus. Organic molecules—proteins, carbohydrates, lipids, and nucleic acids—form the foundation of essential biological processes. Biochemistry has shown that certain fundamental chemical reactions involving these molecules are key to deciphering the complex molecular interactions at the origin of life. Life, in the biological sense, is manifested through metabolism, growth, reproduction, and response to stimuli. Molecules such as pantetheine or amino acids, synthesized in the Miller-Urey experiment and subsequent studies, are essential for these processes, reinforcing the link between the organic and the vital. In philosophy, the organic has historically been associated with the structural complexity and intrinsic organization of life (Encyclopaedia Britannica 2025). Nineteenth-century vitalism proposed the existence of a vital force beyond chemistry—a view that, while now largely outdated, still leaves conceptual traces and doubts in the debate on the transition from inorganic matter to life. Contemporary biochemistry, through abiogenetic theories, has demonstrated how the primordial soup of inorganic matter gave rise to the molecular foundations of the first vital processes. Self-assembly and subsequent ecological organization, through trial-and-error phases, gave birth to the complex functional systems that characterize life on Earth. However, the question of the origin of life (both terrestrial and universal) has not yet been fully resolved, and at present, we can affirm that at least seven major theories exist and are being debated (Dutfield and Choi 2022):

- **Electric spark**: This theory suggests that lightning may have triggered chemical reactions in the primordial atmosphere, leading to the formation of complex organic molecules (Dutfield and Bailey 2024).
- **Clay molecules**: Proposed by organic chemist Alexander Graham Cairns-Smith, this hypothesis suggests that the first molecules of life formed on clay crystal surfaces, which would have organized the molecules into complex structures (Kloprogge and Hartman 2022).
- **Hydrothermal vents**: This theory proposes that life originated in the deep ocean, near hydrothermal vents, where mineral-rich water and heat would have facilitated essential chemical reactions (Georgieva et al. 2021).
- **Origin from ice**: According to this hypothesis, during periods when the Earth was covered by ice, the low temperatures would have protected organic molecules from degradation, allowing them to interact and form more complex structures (Price 2007).
- **Understanding DNA**: This theory explores how DNA and RNA molecules could have developed through natural chemical processes, leading to the formation of structures capable of self-replication (Liu 2024).
- **Simple Beginnings**: This hypothesis suggests that life began with simple molecules that, through gradual processes, acquired complexity and biological functionality (Sharov 2016).
- **Life from space**: Also known as panspermia, this theory proposes that life, or at least its fundamental components, arrived on Earth through meteorites or comets from space (Sithamparam et al. 2022).

Overall, developments in the field of abiogenesis continually offer new perspectives on the origin of life, pushing research beyond the boundaries of our planet into the realm of astrobiology. Discoveries of complex organic molecules in interstellar dust, in space clouds, and, most notably, in NASA's samples from the asteroid Bennu, have opened up the possibility that life not only emerged on Earth but could have spread throughout the universe through mechanisms like panspermia (Glavin et al. 2025). Beyond the heated debate within the scientific community, the common denominator remains that research into the origin of life, while exploring different hypotheses, continues to offer new perspectives and push the boundaries of science into the unknown:

"The most important hypothesis in all of biology, for example, is that everything an animal does, the atom does. In other words, there is nothing that living creatures do without it being traceable to the fact that they are made of atoms in accordance with the laws of physics (Feynman et al. 1963)."

Richard Feynman (1918–1988), renowned theoretical physicist and science communicator, emphasized that biological processes, despite their complexity and wonder, are ultimately the result of fundamental interactions between atoms and molecules, governed by the laws of physics and chemistry. The reductionist approach does not aim to diminish the complexity of biology but rather to reveal its deep connection to the physical world, offering new research avenues into the great mysteries that still surround our understanding of the emergence and evolution of terrestrial organic life (Wonders of Physics 2022). This concept is the foundation of structural biology, a rapidly developing field that studies the shape and function of biomolecules, with its primary goal being to understand and design biomolecular structures based on their atomic organization (Journal of Structural Biology 2025). The principle of being able to arrange atoms in any desired manner, another pioneering insight attributed to Feynman,[2] today finds application in modern tools of computational modeling and molecular engineering, which allow for the prediction and control of biomolecular interactions with unprecedented precision (Eskandar 2023). The integration of structural biology with CAD/CAM and CADD (Computer-Aided Drug Design) digital tools is transforming the way we understand and manipulate living systems, as it allows the physical and chemical laws governing biomolecules to be translated into detailed computational models, predictive simulations, and even nanometer-scale fabrication processes (Baig et al. 2018). Thanks to advanced software such as AlphaFold®, *molecular docking tools* (Lai-Yuen and Lee 2013), and *biofabrication platforms* (Addario 2024), we can now design, modify, and optimize proteins, enzymes, medicines, and bioinspired materials with engineering precision. The idea that what the animal does, the atom does, becomes, in this context, a design guide: every biological function, from enzymatic catalysis to cellular communication, can be described, simulated, and replicated through increasingly refined molecular models for biomedical engineering advances (Anyanwu et al. 2024). Biology is no longer just a field of observation, but a science of design, where the boundary between the natural and the artificial dissolves, paving the way for new forms of synthetic life, biomimetic materials, and tailor-made biotechnological solutions (Peng et al. 2024). We are facing a shift that marks a crucial turning point in the evolution of artificial technologies: from understanding biological structures *as they are*, to designing structures *as they could be*, anticipating emerging functionalities, and leveraging the very principles that govern organic life at the atomic and molecular level (Gazo et al. 2024). Research is now enabled to create incredible *ultranatural materials* for biotech use, such as in human dental health, reflecting the evolution from traditional CAD/CAM (Prause et al. 2024). It is precisely these innovations that make the expression *Organic, all too organic* particularly fitting to describe the current state of the art in contemporary biotechnological *WunderCAD/CAMmer*—a paraphrastic reference to Friedrich Wilhelm Nietzsche (1844–1900) that highlights the pervasive, and at times overemphasized, presence of organic metaphors and models within technoscientific design practices. Within the context of contemporary *digital wonder rooms*, a particularly significant example is represented by a highly specialized *biotech cabinet* managed by a team engaged in structural biology research at Vanderbilt

[2] Richard Feynman, in his famous lecture *There's Plenty of Room at the Bottom* (1959), also emphasized the potential of direct manipulation of atoms and molecules, stating:
"What I want to talk about is the problem of manipulating and controlling things on a small scale. As soon as I mention this, people tell me about miniaturization, and how far it has progressed today. But that is not what I am talking about. What I am talking about is not just miniaturization. It is a much deeper idea: that we can arrange the atoms the way we want; the very atoms, all the way down!"

University. Founded in 1873 in Nashville (Tennessee), Vanderbilt is one of the leading research universities in the United States, consistently ranked among the top twenty nationwide (Vanderbilt University 2025). The Vanderbilt University Medical Center (VUMC) serves as a reference hub for translational research, while interdisciplinary institutes such as VIIBRE promote advanced approaches in synthetic biology and molecular engineering (Vanderbilt University-VIIBRE 2025). In this highly integrated environment, the synergy between life sciences and digital tools has generated a new interpretative paradigm: a novel *WunderCAD/CAMmer*, a contemporary digital *studiolo* where, through the advances of molecular biology intertwined with simulation and visualization technologies, nature and artifice merge in a cutting-edge dialogue that blurs their boundaries and distances. These contemporary *digital wonder rooms* are therefore not configured as mere collections of extraordinary examples of Nature interpreted through sophisticated digital artifices. In these centers, cognitive systems are developed and tested, capable of generating scenarios, testing hypotheses, and anticipating alternative technological futures through the sophisticated language of digital simulations. Structural biological research, in this perspective of hybridization, emerges as an extraordinary form of design manipulation of the living world. The human-scale reconstruction of complex molecular structures—such as those that govern microbial life—is made possible by the use of ultra-high-resolution techniques, such as cryo-electron microscopy (cryo-EM) and macromolecular crystallography (Singh et al. 2023a). These tools act as perceptual amplification devices, allowing the observation, modeling, and reinterpretation of the organization of living matter. One of the most promising areas of research developed at Vanderbilt focuses on the structural analysis of the *bacterial flagellar motor*—one of the most refined and functionally efficient models of biological machinery. This rotary nanomachine, shaped by ancestral evolutionary processes, is essential for microbial locomotion, enabling movement toward nutrients, the avoidance of environmental hazards, and rapid changes in direction (*chemotaxis*). It operates by converting chemical energy into mechanical motion through a complex assembly of protein structures, including stators and rotors. The motor is powered by proton motive forces or ion gradients across the bacterial membrane—a mechanism that not only drives locomotion but also plays a critical role in cellular signal transduction. The analysis of the structural and functional dynamics of this biological system at the nanometric scale is enabling the identification of new molecular targets in the CADD (Computer-Aided Drug Design) sector, to develop competitive, targeted, and customized therapeutic interventions (Singh et al. 2023b). The inhibition of bacterial motility mechanisms is indeed emerging as a key strategy for drug design, enabling the development of biomolecular protocols capable of reducing, if not completely suppressing, bacterial virulence and hindering the spread of infections—particularly in the context of rising antibiotic resistance (Singh 2024a). As previously stated, *chemotaxis* is the process by which living cells move in response to chemical signals in their environment. It plays a crucial role in many biological functions, from immune defense to embryonic development. For example, bacteria use *chemotaxis* to navigate toward nutrients or away from toxic substances. Similarly, immune cells such as neutrophils rely on chemotactic signals to reach sites of infection or inflammation. Depending on the nature of the stimulus, the movement can be either toward higher concentrations of a chemical (*positive chemotaxis*) or away from them (*negative chemotaxis*). This ability to detect and respond to chemical gradients is essential for maintaining homeostasis and coordinating cellular behavior across complex biological systems. This specialized research line is conducted within the Department of Pharmacology and Biochemistry, under the direction of Prof. Tina M. Iverson, founder of the Vanderbilt High-throughput Biomolecular Crystallization Facility. The structural analysis of the flagellar motor—and more broadly, of the protein architecture composing it—is coordinated by Dr. Preshant K. Singh, a senior researcher specializing in cryo-EM and structural biology, who continues the historic research line on the bacterial flagellar motor (Singh et al. 2024b). His work confirms that studying this molecular configuration represents one of the most significant avenues for catalyzing technological innovation at the intersection of biology, physics, and digital design. For decades, the architectural complexity and mechanical precision of this molecular system have made it one of the most fascinating and articulated objects of study in structural biology. Particular attention is paid to the study and modeling of the cytoplasmic C-ring (also known as the *switch complex*, that can spin in both CCW and CW directions), whose architecture is still not fully understood and defined, thus limiting comprehension of the mechanisms of directional control and biomechanical torque transmission (Fig. 1).

Studies focus on *Salmonella enterica*, a model organism that shares important structural features with *Escherichia coli*, offering experimentally transferable results. A precise understanding of this molecular mechanism paves the way for new strategies in the design of targeted drugs. These interventions could selectively interrupt the mechanical and molecular functionality of the bacterial flagellar motor, representing a promising route in the fight against infections (Xu et al. 2024).

Main challenges in bacterial flagellar motor research are:

- **Motility and survival**: The bacterial flagellum enables movement in response to environmental stimuli (e.g., chemotaxis, phototaxis). This behavior allows bacteria to seek nutrients and avoid harmful conditions, making motility a key to survival.
- **Natural nanomachine**: The flagellum operates as a rotary engine, with a rotor, stator, shaft, and regulatory components, echoing the design of human electric motors. It achieves high energy efficiency by converting electrochemical gradients (typically of protons or sodium ions) into mechanical rotation — offering a fascinating model for biomimetic nanotechnology (Robinson 2024).
- **Evolutionary significance**: The structural and functional complexity of the flagellum is often discussed in relation to the concept of irreducible complexity — a controversial theory suggesting that certain biological systems cannot be simplified further without losing function. This places the flagellum at the center of debates between evolutionary and synthetic biology (CSIRO 2022).
- **Bioengineering and nanorobotics**: The architecture and mechanics of the flagellum are influencing the development of nanoscale technologies (Chikere

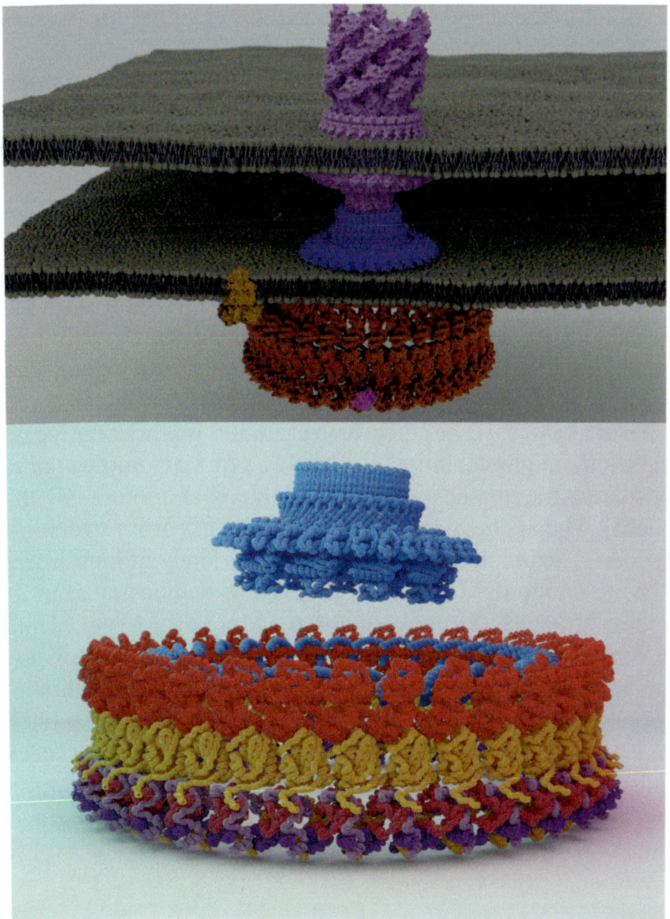

Fig. 1. Panoramic view of the bacterial flagellar motor through the double cell membrane (UP). High-resolution (~4 Å) cryo-EM reconstruction of the bacterial flagellar motor MSC-ring in wireframe, with polypeptide chains colored by functional subunit to reveal its precise symmetry and torque-coupling architecture (DOWN).

et al. 2024). Its biological design inspires the creation of micromachines or medical nanorobots capable of navigating bodily environments for targeted drug delivery or diagnostic purposes from quantum computing perspective (Kumar et al. 2024) (Fig. 2).

- **Clinical relevance**: Pathogens such as *Salmonella* and *Escherichia coli* use the flagellum to colonize and infiltrate host tissues. Understanding the flagellum's structure and function may thus support the development of innovative therapies that inhibit these invasive capabilities without relying on conventional antibiotics.

To transform cryo-electron microscopy (cryo-EM) data into meaningful models, Dr. Singh and his research team rely on an extensive ecosystem of computational tools

(Singh et al. 2024b). These software platforms are essential for the workflow from data acquisition to digital simulation (See Table 1).

In recent years, within the techno-scientific landscape, in structural biology there has been an exponential proliferation of computational applications undergoing rapid evolution and reciprocal hybridization. This dynamic environment can easily disorient researchers and requires constant updates and comprehensive usage guidelines. The standard workflow described by Singh at Vanderbilt typically begins with the acquisition of biological sample images using Cryo-Electron Microscopy (CryoEM) (Fig. 3).

The core principle of CryoEM lies in the ultra-rapid freezing of samples, a process that prevents the formation of ice crystals and thus preserves the native ultrastructure. This technique, known as *vitrification*, transforms water into a glass-like amorphous state, stabilizing delicate molecular assemblies without the need for chemical fixation or staining (Wu et al. 2020). By immobilizing samples in a near-native conformation, CryoEM enables the high-resolution visualization of macromolecular complexes—such as ribosomes, membrane proteins, and flagellar motors—with exceptional structural fidelity. The applications of CryoEM are vast: from structural elucidation of proteins, viruses, and molecular complexes to the study of their interactions. This technology has revolutionized structural biology, as demonstrated in the resolution of the Zika virus structure (Sirohi et al. 2016). Moreover, CryoEM plays a crucial role in drug development by enabling a detailed understanding of how biological molecules interact with each other and with pharmaceuticals, facilitating the design of more targeted therapies. The incorporation of advanced artificial intelligence (AI) and machine learning (ML) algorithms is expected to significantly accelerate the 3D reconstruction process, paving the way for more accurate simulations and real-time molecular visualizations. The CryoEM process involves several stages. Once the sample is prepared, it is inserted into the electron microscope and exposed to a high-

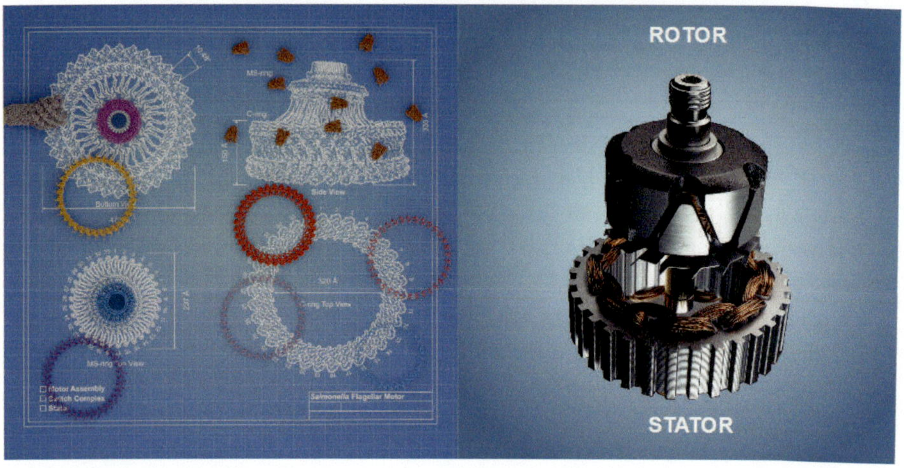

Fig. 2. Orthogonal projection of the MSC-ring with 3D-printed component details that highlight the remarkable analogy between the molecular motor's rotor–stator system and its human-engineered counterpart.

Table 1. Integrated Platforms for Structural Biology, CAD/CAM Design, and Scientific Simulation.

Category	**Software**	**Main Function**
Atomic and Structural Modeling	AlphaFold®, CCP4®, Coot®, DynDom®, HKL2000®, MolProbity®, REFMAC®, Rosetta®	Modeling, refinement, and structural validation of atomic structures.
Molecular Visualization	Blender®, Chimera®, ChimeraX®, MolecularNodes®, PyMOL®, VMD®	Visualization and 3D modeling of molecular structures and dynamics.
Molecular Dynamics	AMBER®, GROMACS®, Tinker®	Molecular dynamics simulations.
CAD/CAM and 3D Prototyping	Autodesk Fusion 360®, Materialise Mimics®, Materialise 3-matic®, PreForm®, Rhinoceros® + Grasshopper®, SolidWorks®	Parametric modeling, biological prototyping, and 3D printing optimization.
Cryo-EM and Reconstruction	CryoDRGN®, CryoSPARC®, DeepEMhancer®, RELION®, Scipion®	Cryo-EM data processing and 3D structure reconstruction.
Docking and Molecular Interactions	AutoDock®, LIGPLOT®, LightDock®	Prediction and analysis of molecular docking and interactions.
Scientific Computing & Scripting	Google Colab®, Jupyter Notebook®, MATLAB®, Python®	Data analysis, modeling, and scientific scripting.
Image Processing & Illustration	Adobe Illustrator®, Fiji®, ImageJ®, Shapr3D®	Image editing, scientific illustration, and 3D design.
Genomics and Chemistry	ChemDraw®, CLC Genomics®	Genomic analysis and chemical modeling.
Data Mining & Visualization	Orange®, Python®	Data analysis, pattern recognition, and visualization.
Collaborative Simulation	NVIDIA Omniverse®	Real-time collaborative simulations and digital twin environments.

energy electron beam. The process of using molecular CAD software such as Coot® to build protein structures from 3D volumes starts with thousands of pictures of a *bacterium*, called *projections*, which are captured from different angles, allowing for the collection of three-dimensional information about the sample. The resulting 2D images are then aligned and recombined using sophisticated mathematical algorithms to create a 3D volumetric map of the sample. This 3D reconstruction process is essential for achieving a detailed visualization of the molecular structure. Once the reconstruction is complete, the 3D model is further refined to reach atomic-level resolution. CryoEM data analysis goes beyond mere observation of molecular structures and includes modeling and data interpretation. Programs such as Coot® and ChimeraX® are used to build and visualize molecular models, while tools like Blender® can be employed to animate the dynamics of macromolecules, providing an interactive view of their function. This multidisciplinary approach has opened new possibilities in structural biology research, especially for molecules that cannot be

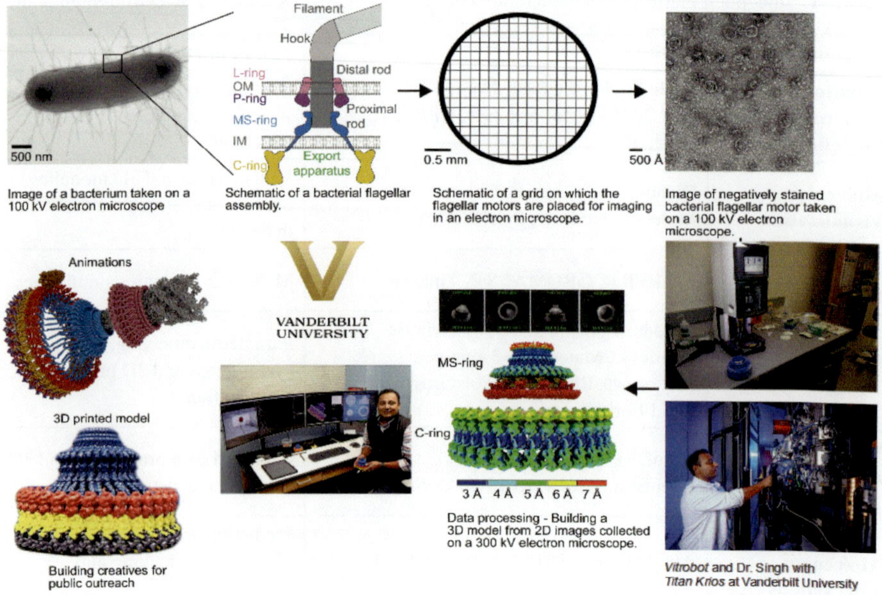

Fig. 3. This diagram summarizes the multi-step workflow employed by Dr. Singh's lab at Vanderbilt University to unravel the structural biology of the bacterial flagellar motor.

crystallized, such as membrane proteins or large protein complexes. The applications of CryoEM are vast and include the structuring of proteins, viruses, and molecular complexes, as well as the study of their interactions, and it has revolutionized the field of structural biology, enabling high-resolution visualization of complex biological molecules. Moreover, this technique is essential for drug development, as it allows for a detailed understanding of how biological molecules interact with each other and with pharmaceuticals, facilitating the design of more targeted treatments. Despite its numerous advantages, CryoEM also presents challenges. One of the main difficulties is image resolution, which depends on sample quality and the amount of acquired data. This volume is populated with amino acids using Coot®, although the 3D volume itself is not created in Coot®, but rather by other software. The protein structure is then transferred to ChimeraX® for further analysis. Finally, the workflow includes the creation of a structural model through 3D printing, using the protocol: *map to model – model to STL – STL to 3D printing*. The 2D images obtained are processed to create a 3D map or volume, which is then used in Coot® to build a molecular model of the flagellar motor. This model is further refined and animated in Blender®, allowing for dynamic visualizations of the structure and function of the motor.

One of the main issues is the disconnect between molecular modeling tools and animation or simulation software, which can hinder data transfer while risking loss of structural accuracy. Maintaining consistent amino acid bond lengths and angles throughout the modeling pipeline is essential to ensuring biological relevance and model reliability. Despite the atomic resolution achieved by CryoEM, data acquisition and processing continue to demand significant computational resources and analysis time. Moreover, interpreting the complex structural data requires expert knowledge in molecular modeling and computational biology. Image resolution also depends heavily on sample quality and data volume. Still, ongoing advances in CryoEM instrumentation and algorithmic support—particularly through AI and ML—are continuously improving image resolution and reducing processing time. These developments promise to further enhance the precision and accessibility of structural modeling in life science research. One of the key aspects emerging from Singh's workflow description is the forward-looking objective of developing 3D-printed models of the bacterial flagellar motor that incorporate moving components. The goal is to simulate molecular motion with increasing realism, transforming abstract molecular dynamics into concrete and tangible representations. By integrating advanced CAD software with high-precision 3D printing techniques, it may become possible to fabricate intricate structures in which each bond and atom moves in relation to others—faithfully mimicking the actual behavior of molecular assemblies. This approach holds significant potential for enhancing our understanding of biomolecular mechanics through physical, interactive models. Although various optimization strategies have been implemented to enhance model fidelity, these improvements are generally the result of broader research methodologies rather than intrinsic features of the CAD/CAM software itself. Nevertheless, the potential of these digital suites for this type of research remains substantial. These tools serve as powerful platforms for educational and research purposes, enabling the detailed exploration and visualization of molecular structures. However, current limitations—particularly the need for expanded molecular libraries and resources—suggest significant room for improvement Looking ahead, the integration of machine learning (ML) and artificial intelligence (AI) is expected to transform the field of molecular simulation. These technologies promise to enhance real-time modeling capabilities, offering more dynamic and precise visualizations of molecular systems. Dr. Singh's commitment also extends to using social media platforms for the public dissemination of these highly specialized topics, providing a clear example of how outreach and interdisciplinary collaboration can strengthen structural biology research (Singh 2025). The way Singh shares his *esoteric knowledge* and work reveals a deliberate intention to move beyond the secluded walls of their digital *studiolo*, making available to the public not only data and results, but also images that evoke fascination and wonder. This gesture recalls the ethos of the *Wunderkammern* of the early modern period—spaces in which *naturalia* and *artificialia* were collected and exhibited not only to amaze, but to provoke inquiry and deepen understanding. In this light, contemporary scientific dissemination becomes a digital reimagining of those *cabinets of curiosity*: an open, networked environment where complex molecular structures are no longer confined to the closed circuits of academic research, but are made visible and appreciable to broader audiences. As noted by Daston and Park in *Wonders and the Order of Nature* (1998), wonder was not opposed to knowledge in the pre-modern world—it was its necessary starting point, the emotional catalyst that made observation and explanation urgent. In the same way, the visual language of today's molecular research—animated, interactive, and accessible—restores to science its capacity to astonish, transforming the molecular world into a shared space of discovery, beauty, and intellectual engagement. Such an approach could not only enhance understanding in educational contexts but also prove valuable for researchers studying protein folding, enzymatic mechanisms, and drug interactions. The ability to physically manipulate these models would allow scientists to explore complex conformational changes and binding sites, bridging the gap between virtual simulations and experimental intuition. As for future research, the focus is shifting towards drug discovery, particularly in the treatment of bacterial infections. However, the extraordinary precision with which biological molecules fold and interact—often compared to highly sophisticated engineering design—raises questions that go beyond simple evolutionary explanations. While natural selection may account for the progressive adaptation of biological functions, the irreducible complexity of many molecular structures—such as the bacterial flagellar motor or ribosomal complexes—suggests the possibility that such systems may result from a process that appears to be guided by a kind of design logic implemented by the cellular network, thus reconnecting the theme to new theories being developed by evolutionary biologists (McPherson 2024). This does not necessarily imply a rejection of evolution theories, but rather invites us to consider that evolutionary mechanisms themselves may be part of a deeper design, in which nature follows predefined patterns reminiscent of intelligent design logic. In this context, the ability to simulate and understand molecular motion through physical models not only clarifies how such mechanisms work, but also reopens the debate on why these structures exist in such optimized forms, directing the focus of new theories toward the concept of collective intelligence (McMillen and Levin 2024).

8.2 Nulla Ars Imitari Sollertiam Naturae Potest[3]

Throughout previous examinations, it has been repeatedly emphasized that, since antiquity, humans have sought to imitate Nature, recognizing it as the most skilled engineer, architect, and designer. Cicero's statement—"*Nulla ars imitari sollertiam naturae potest*"—encapsulates a fundamental philosophical tension that continues to resonate in contemporary debates on biomimicry, synthetic biology, and artificial intelligence (Cicero 1881). With this assertion, Cicero expresses a classical worldview in which human creativity, though admirable, remains subordinate to the wisdom and complexity of the natural world. Nature is here understood not merely as a model, but as an unsurpassable artisan—capable of producing forms, systems, and functions with a level of subtlety, efficiency, and integration that human technology can only attempt to approximate. This idea points to an ontological asymmetry: while *ars* (art, or technique) can take inspiration from *natura*, it can never fully embody its generative power. In modern terms, Cicero's view aligns with the awareness that every act of design—be it computational, architectural, or molecular—faces the challenge of emulating systems shaped by millions of years of evolution. Despite the increasingly sophisticated tools available to science and engineering, from CAD/CAM to AI-driven simulations, the creative intelligence of nature often

[3] Transl. "No human art can imitate the skillfulness of Nature."

surpasses our ability to fully decode or replicate it. Far from being a defeatist statement, it is rather an invitation to humility and wonder. It suggests that, even in our most advanced attempts to simulate or recreate biological processes, there remains an irreducible mystery—one that reminds us that nature is not only to be mastered, but also to be listened to. For thousands of years, this consideration has held true and, undoubtedly, despite today's prodigious advancements, our technologies and materials—no matter how advanced—remain relatively imperfect artifacts, limited when compared to the ingenious biological creativity revealed by the evolution of life on Earth. Nevertheless, the gap between artifice and nature is progressively narrowing, thanks to ongoing progress in fields such as synthetic biology, materials science, and artificial intelligence—domains that increasingly decipher, emulate, and even collaborate and coevolve with the deep strategies inherent in nature. The functional precision of contemporary technological processes goes now beyond visible mechanical-anatomical form, extending to microscopic functionality, the dynamics of molecular interactions, and the adaptive, evolutionary, and ecosystemic behavior in its continuous becoming. Current evolutionary theories, although refined and enriched by recent scientific discoveries, continue to depict Nature as an entity operating according to a principle of *serendipitous design*, where complex forms and functions emerge not from deliberate planning, but through an ongoing process of trial, error, and adaptation. In this paradigm, Nature is not seen as a conscious designer, but rather as an extraordinarily complex biological system that, through natural selection, evolves continuously, refining solutions through a cycle of design approach tentatives. This process, devoid of intentionality, unfolds blindly: favorable genetic traits are selected for their environmental adaptability, without a long-term goal. Time becomes the key element in the success of this unconscious design, allowing for errors and transformations over millions—if not billions—of years. Each organism, structure, or vital function emerges as the outcome of a millennia-long evolutionary process in which efficiency is measured not only in terms of immediate performance but also in sustainability and resilience over time. An extraordinary example of evolutionary resilience is *Lomatia tasmanica*, an ancient and sterile shrub endemic to Tasmania, discovered in 1937 (Kean 2023). This plant has been cloning itself for over 43,000 years. Fossilized remains of leaves identical to those of the present-day *Lomatia tasmanica*, found approximately 8.5 kilometers from the living population, have been radiocarbon-dated to around 43,600 years ago. Some estimates suggest it could be as old as 135,000 years, making it potentially the oldest known plant clone. Despite its inability to reproduce sexually and thus introduce genetic variation, it has survived in a stable ecosystem, maintaining its genetic code unchanged for millennia. New plants are created when underground rhizomes spread, producing new shoots, or when a branch breaks off, forms its own roots, and starts to grow independently. These new individuals are genetically identical, as no sexual reproduction occurs to mix their genetics or generate new seeds. This case challenges the idea that evolution always requires continuous genetic innovation, showing instead that, under specific ecological conditions, longevity and sustainability can be achieved through conservative strategies. *Lomatia tasmanica* exists as a single known colony in nature, comprising approximately 500–600 genetically identical individuals, extending about 1.2 kilometers in a secret location in the southwest of Tasmania. It represents an extreme form of *unintentional natural design*, where efficiency does not come from the ability to change quickly or innovate genetically, but from the ability to endure over time, maintaining stability and functionality in a favorable environment without active genetic evolution. Classified as critically endangered, this mysterious plant faces threats such as bushfires and fungal diseases. Conservation efforts are ongoing, including the preservation of clonal specimens at the Royal Tasmanian Botanical Gardens. Biological *supermaterials* such as the abalone shell (*Haliotis*), dragonfly wing (*Anisoptera*), or shark skin (*Selachimorpha*) not only display aesthetically optimal forms but also integrate mechanical, aerodynamic, and antibacterial properties—results of highly precise molecular configurations. The vast and refined catalog of Nature's patents, incessantly interacting within the Earth's complex ecosystem and physical phenomenological context, justifies the sharp Latin term *sollertia* (*skillful* or *ingenious*), which Cicero attributed to life itself. In a more futuristic context, this dogmatic sentence could also be seen as a challenge to human potential to surpass Nature, as is evident in current discussions on AGI (Artificial General Intelligence), bioengineering, and advanced material design. These fields aim to push the boundaries of our ability to imitate and even *improve* upon Nature. Undoubtedly, within the evolutionary framework, natural selection acts as a fundamental and tyrannical agent, determining which traits will be inherited by future generations through random genetic variation and environmental pressures. However, in a striking way, Nature, in its intricate life-design processes, often seems to generate perfectly optimized technological solutions and highly functional problem-solving capabilities, as though emerging from nothing. This creativity, however, does not appear to be the result of conscious design or deliberate planning, which would lead theoretical discussions back to forms of determinism that have been completely excluded from mainstream contemporary scientific debate, although numerous examples in the natural sciences seem to increasingly contradict this long-standing dogma. Rather, it seems to arise from Nature's ability to harness and manage chaos in an apparently organized way, operating through a process that can be described as *blindly guided by an organized management of chaos*. At the core of this concept is the principle of emergence—the idea that complex systems and patterns arise from the interaction of simpler elements, often without a central command or intelligent design. These emergent properties arise from what could initially appear to be random, chaotic events. In the natural world, the countless variables—genetic mutations, environmental pressures, and unpredictable changes—combine in a way that, over time, generates solutions that are functional, adaptive, and optimized for survival. The biodiversity of life on Earth, from the complex design of wings in birds to the intricate shapes of flowers that attract specific pollinators, emerges from this process in a non-linear and iterative manner. In parallel, self-organizing systems in nature, such as the flocking behavior of birds or the intricate patterns found in coral reefs, are examples of how order can arise from chaos. In these systems, each individual element follows simple rules or behaviors that, when combined, create a coherent, complex structure. This is reminiscent of algorithms in artificial intelligence, where each agent follows basic principles that ultimately lead to the emergence of complex, optimized outcomes. This idea also has parallels in bioinspired design and biomimetics, where scientists and engineers look to nature's problem-solving methods to inspire technological innovations. In summary, nature's ability to generate problem-solving solutions is not, at least in the strictly human sense of the term, the result of conscious or planned decisions, but rather arises from a more complex and emergent process—which, moreover, still remains largely mysterious and undecipherable. Like a tightrope walker in constant balance between vital organic order and the abyss of chaotic nothingness, life performs its magnificent dance. Organism communities orchestrate life across the planet, shaping ecological systems in which adaptive solutions emerge through a brilliant combination of continuous creativity, trial-and-error routines, self-organizing mechanisms, and responses to intense evolutionary pressures. This is a form of brilliant design—*sollertia*—that remains, for the most part, inaccessible, obscure, and incomprehensibly refined. It is a process that, though blind and unintentional, leads to highly functional and efficient outcomes—solutions that can often inspire innovation in human design, technology, and understanding of the world around us. While traditionally considered devoid of awareness or intentionality, recent theories in developmental and synthetic biology suggest that certain forms of distributed or embodied intelligence may play a role in guiding these adaptive responses (Levin 2024). Simultaneously, contemporary science is attempting to replicate this natural process through bioinspired design. Designers and engineers look to Nature for inspiration, not to replicate biological processes verbatim, but to creatively adapt and extrapolate the principles that drive natural efficiency for technological and industrial applications. Contemporary science—through genetic engineering, artificial intelligence, and other emerging technologies—seeks to simulate and apply processes that emulate this ongoing refinement, addressing both biological and engineering complexities in an interdisciplinary manner. CAD/CAM technologies are enabling increasingly accurate simulations of biological structures, but these tools, while informed by input data and algorithms, still leave room for creative exploration. Even advanced software such as AlphaFold® for protein folding or Orbital Materials® for material design highlight the limitations of prediction when faced with biological complexity, including environmental factors, random mutations, and nonlinear dynamics. Generative design, inspired by biological evolution, strives to emulate nature's logic through algorithms that simulate selection and adaptation. Systems like Grasshopper® for Rhinoceros® and Autodesk Generative Design® allow for the exploration of infinite design variations within predefined constraints and objectives. However, unlike Nature, these processes lack real adaptive capacity: while they simulate adaptation to a degree, they cannot truly evolve autonomously in response to environmental feedback, nor can they integrate modifications in real time as a living organism would. A significant example is the research by Orbital Materials®, which employs artificial intelligence to design custom materials for specific applications, with a strong focus on environmental sustainability and climate change—such as more efficient batteries or advanced filtering membranes—while seeking to bypass traditional trial-and-error laboratory phases:

“Developing new advanced materials takes years of trial and error in the lab. Orbital uses its AI to replace trial and error with rational design, taking technologies from discovery to deployment faster than with human input alone. When prompted with a desired set of properties, our AI reasons through the universe of possible chemistries to design a material from scratch. Our researchers harness this ability to reason over multiple constraints, leveraging our proprietary dataset that includes process, wet lab data, manufacturing, and performance factors. This enables us to design, synthesize, characterize, and test new materials that power our technologies for data centers.” (Orbital Materials 2025).

Although the results are impressive, the process remains confined to the predictive sphere, lacking the plasticity typical of biological systems. What is missing is the intrinsic capacity for self-organization and adaptive learning that characterizes living organisms—systems capable of modifying their structures and functions in real time in response to environmental *stimuli*. Current technologies, no matter how advanced, still rely on computational logic that requires external input and lacks genuine biological agency. Nature, on the other hand, does not simply predict—it responds and adapts. It evolves through a network of local, often non-linear interactions that are continuously renegotiated with the environment, in a process that increasingly appears less mechanical and more akin to a form of distributed evaluation or embodied intelligence. These ideas within contemporary neo-evolutionary theories need to be reconciled with emerging theories of developmental biology, in order to move beyond the purely genetic mechanism-based view that has traditionally defined the evolutionary process. This underscores that bioinspired design is not simply about copying nature, but about understanding and interpreting its underlying principles, which may align with developmental processes and adaptive mechanisms. In this way, the creative potential of bioinspired design resides in the gap between the evolutionary logic of nature and its innovative reinterpretation within technological frameworks, offering new possibilities for design and engineering. Similarly, companies like Biomason® and Colorifix® are pioneering sustainable solutions by harnessing biological processes. Biomason® is transforming the construction industry by producing building materials inspired by coral formation, drastically reducing CO_2 emissions through cement production without high-temperature processes (Biomason 2025). At the same time, Colorifix® employs microorganisms to dye fabrics, eliminating harmful chemicals and significantly reducing water usage (Colorfix 2025). These innovations exemplify how bioinspired design can lead to environmentally responsible technologies across diverse industries. Synthetic biology is opening up new frontiers, shifting the focus from imitation to active participation in biological processes. CAD platforms for biology, such as Benchling®, enable the design of genetic sequences to be inserted into living organisms, creating entirely new biological functions (Benchling 2025). Benchling® is a platform that can be considered a type of molecular CAD. It is a widely used digital tool in biotechnology management and design, primarily in the fields of molecular biology and genomics. Although it is not a traditional CAD software for designing physical objects, Benchling® operates in a similar way by enabling the design, visualization, and simulation of 3D biological structures, such as DNA sequences, proteins, and other molecular components. Benchling® allows scientists to work with genetic sequences in a way similar to how engineers use CAD software to design mechanical components. It offers tools for modeling genetic sequences, managing experiments, and simulating biological processes, enabling the virtual design and optimization of genetic modifications or protein compositions. Thanks to computational design and genetic editing, researchers are creating synthetic metabolic pathways that replicate biological functions with greater efficiency and control, opening new possibilities for pharmaceutical research, materials engineering, and environmental sustainability. One of the most authoritative sources in the field of Computer-Aided Molecular Design has been collecting research contributions in bioinformatics and Computer-Aided Molecular Design since 1987 (Springer 2025). In this context, companies like Ginkgo Bioworks® and Zymergen® leverage synthetic biology to design custom microorganisms capable of producing bioplastics, perfumes, food, and pharmaceuticals. Ginkgo Bioworks® 2025, in particular, uses a native biological CAD platform to program cells as living circuits, optimizing the production of specific compounds on an industrial scale, such as amino acid sequences and DNA optimized for codons for target organisms, ready to be built and tested (Ginko Bioworks®). These companies are at the forefront of the biotechnological revolution, transforming genetic design into a highly scalable industrial process. By programming microorganisms to produce valuable compounds, they are not only advancing biotechnology but also contributing to sustainable manufacturing practices by replacing traditional chemical processes with biological ones. Ginkgo Bioworks®, for example, allows for the precise customization of living organisms to manufacture everything from specialty chemicals to biofuels. Their approach embodies a fusion of biology and engineering, where living cells are treated as programmable systems, akin to digital code, enabling a range of applications in various industries, from healthcare to agriculture. The company offers, among other services, enzyme variant library design with custom solutions tailored to customer needs. Starting from the sequence of a target enzyme, an *in silico* library of enhanced variants can be created, specifically optimized for application requirements and engineering objectives. By utilizing in-house CAD-developed computational design tools and AI, validated through experimentation, Ginkgo®'s team generates high-potential sequence variants, providing a diverse library to explore for improvements in stability, activity, specificity, and more (Ginko Bioworks 2025). This service demonstrates how synthetic biology can be applied not only to entire microorganisms but also to individual proteins, enabling precise customization for specific industrial applications. Through AI and computational design, Ginkgo® creates enzyme solutions that enhance the efficiency and sustainability of biotechnological processes, ranging from pharmaceuticals to biofuels and food production. The ability to design and optimize enzyme libraries holds significant potential for industries reliant on enzymes for catalysis, bioremediation, and bio-based chemical production, offering innovative pathways to improve performance and reduce environmental impacts. In this context, predictive medicine, material design, and drug development are benefiting from tools that, while not yet matching the complexity of life, explore its potential in increasingly organic ways. The tension between what we can predict and what remains unknown drives innovation, reminding us that, no matter how advanced, no human endeavor can ever fully match the ingenuity of nature. Yet, it is precisely in this awareness that the possibility for a new creativity lies, one based on humility and collaboration between human ingenuity and biological intelligence. The computer system, in essence, is a kind of artificial digital ecosystem created by humans in just a few decades, akin to a highly flexible and mobile network structure. Through an exchange of inputs and outputs, it becomes possible to reaggregate informational cores, hierarchize them into a myriad of relationships, and create models (Nativi et al. 2021). By altering a single atom, one can observe the ripple effect that changes the entire system. Alternatively, by modifying the direction, order, or interconnection of relationships within the system, entirely new worlds can be created. This concept parallels the idea that small, seemingly insignificant changes at a fundamental level can lead to profound transformations, illustrating the dynamic and interconnected nature of complex systems. The interplay of these elements reveals how the manipulation of the most basic components can give rise to vastly different outcomes, highlighting the complexity and potential for creativity in both natural and artificial systems (Huang et al. 2022). The word *model* plays a central role in this way of thinking, as it acts as a bridge between abstract data and tangible outcomes. In the context of computational design, a model is not just a static 3D object, but a dynamic representation of a system that mirrors the complexity of the real world. For example, a computational model of an artificial object is not only a visual depiction; it also encapsulates the principles and relationships that define how the object behaves within a given context. In this regard, the scientific definition of a model becomes crucial. Whether it's a mathematical, physical, financial, or statistical model, the concept refers to a simplified yet comprehensive representation of reality, used to predict, simulate, and understand systems. In computational design, particularly within CAD/CAM workflows, these models are built using interconnected data points—each piece of information influences the others. When one parameter in the model is altered, it triggers a ripple effect that causes changes in other related variables. For example, if we change the size of a structural component in a CAD model, it could affect the weight distribution, the forces applied to other parts of the system, or even the choice of materials, all depending on the interconnected relationships defined in the model. Object-Oriented Programming (OOP) adds an extra layer of complexity to this dynamic modeling environment. The roots of object-oriented programming date back to the 1960s, with the programming language Simula, created by Ole-Johan Dahl and Kristen Nygaard. It was the first language to introduce the concept of objects, originally developed for the simulation of real-world systems. However, it was with the development of Smalltalk in the 1970s, under the guidance of Alan Kay, that OOP established itself as an autonomous programming paradigm (Kay 1993). In OOP, objects represent distinct entities within a system, each having its own data (attributes) and behavior (methods). OOP extends beyond the traditional role of static representation, taking on qualities similar to *quasi-organic* systems by allowing these objects to interact, evolve, adapt, optimize, and respond to changes in ways similar to living organisms in an interconnected manner. Altering a single parameter in the design can trigger a cascade of changes across the system, much like how a single mutation in a biological organism can have widespread effects. The system is not fixed, but is in a constant state of flux, adapting and evolving based on the inputs it receives, mirroring processes of adaptation and optimization found in nature. This quasi-organic programmatic characteristic enables the system to simulate a form of *digital vitality*, where the model appears to act as though it possesses a life of its own, in an innovative process deeply aligned with the principles of natural evolution (Dilshani 2023). The philosophy behind these languages is to model entities from the real world as software objects, and it is closely tied to the complexity of digital design, where engineers must represent physical entities in computational models. With the increasing complexity of CAD systems and the need to design large-scale structures and objects, OOP has emerged as an ideal solution for organizing and managing such processes. Traditional or procedural programming, as seen in languages like C or Pascal, is based on breaking the program into sequential blocks of various classes of functions or procedures. Each function represents a routine of instructions that operates on data passed as parameters. This approach is effective for simpler programs but can become complex and difficult to maintain in large-scale applications. OOP, on the other hand, models software as a collection of objects that interact with each other. This allows for greater modularity and facilitates the expansion and maintenance of the code. While procedural programming focuses on actions (functions), OOP focuses on the entities that perform those actions (objects), offering a more natural representation of complex problems. Object-oriented programming can be seen as an analogy to biology and neural networks. Objects in an OOP program are similar to cells or organs in an organism: they are self-contained units that possess specific characteristics and interact with one another to perform complex tasks:

"Even though all of the cells in your human body descend from a HumanCell super-class, and look largely the same at the very beginning of their existence, when they're first *instantiated*, they quickly diverge into specialties — muscle cells, skin cells, red blood cells, to name a few. It's almost as if each type of cell has a *constructor method* specific to its *class*, that partially overrides whatever *initialization* instructions it inherits from its Cell super-class, and that is triggered when the Cell.new *command* is given — an example of nature's *polymorphism* at work!" (Dzwonchyk 2017).

This reflects a modular organization, where the individual parts collaborate for the functioning of the entire system. Similarly, artificial neural networks, used in artificial intelligence, consist of nodes (neurons) that, like OOP objects, process inputs and produce outputs, cooperating to achieve a common goal. Each node, like an object, has a specific responsibility and communicates with other nodes to complete a task. This decentralized and modular structure enables both object-oriented programming (OOP) and neural networks to address complex problems by decomposing them into simpler, more manageable units. Early CAD suites were based on procedural programming paradigms, in which geometric manipulation was carried out through a sequence of mathematical operations on points, vectors, and surfaces. For instance, drawing a complex shape required the programmer to manually implement all geometric transformations. The advent of OOP radically changed this approach: geometric entities became autonomous objects endowed with properties and behaviors. A cylinder, for example, is no longer just a set of coordinates and equations—it becomes an object that can modify its own height, radius, or spatial position through dedicated methods, sparing the designer from handling the low-level geometric logic. This clear separation between data and behavior significantly simplified the work of CAD developers by reducing errors and improving code maintainability and scalability. Contemporary CAD platforms—such as AutoCAD®, SolidWorks®, CATIA®, Fusion 360®, Rhinoceros®, and Grasshopper®—leverage the OOP paradigm to model geometry and manage the increasing complexity of engineering design (Food4Rhino n.d.). The foundational OOP principles applied to CAD can be summarized as follows (Kim et al. 1990):

- **Encapsulation**: Each object bundles together data (attributes) and functions (methods) that operate on that data. This encapsulation conceals internal object details, facilitating complexity management. In CAD systems, objects represent physical components—such as screws, bearings, or beams—each managing its own geometry and material properties. For example, a window in a CAD model can be represented by an object that contains attributes such as size, material, color, and simulation behavior (e.g., wind resistance).
- **Inheritance**: Classes can inherit characteristics from other classes, promoting code reusability and the extension of functionalities without duplication. Inheritance allows for the creation of new CAD components based on pre-existing objects. For example, a standard mechanical component like a pin could be extended to create variants with specific modifications, without having to redefine the entire object. This approach reduces code duplication and simplifies the management of families of objects with common characteristics.
- **Polymorphism**: A single method can behave differently depending on the object that implements it. This provides greater flexibility in code, improving adaptability to varying contexts. Polymorphism in CAD allows for defining generic behaviors for geometric objects that can be implemented in different ways. For example, a *rendering* function can be defined generically for all objects, but each type of object (cubic, cylindrical, spherical) can implement its specific version of rendering, respecting the geometric context.
- **Abstraction**: Allows for creating simplified representations of complex realities, focusing only on the relevant aspects for the specific application. The ability to abstract complexity is particularly relevant in CAD suites. An engineer can create an abstract representation of a complex machine, focusing on functional aspects and ignoring details that are not relevant for the current design phase. This reduces the amount of information that must be managed simultaneously, making the design more manageable.

In particular:

- **AutoCAD®** introduced a system for managing geometric entities through objects that represent polylines, surfaces, solids, etc., improving interoperability between various components of the design.
- **SolidWorks®** uses objects to represent each physical component of the CAD model, which can be manipulated through a visual interface or programmable

APIs. Each component has properties such as mass, strength, or behavior under stress, encapsulated within the object.

- **CATIA®**, used primarily in the aerospace and automotive industries, heavily employs OOP principles to enable modular management of projects, where each part or subset can be defined as an autonomous object interacting with other elements of the system.
- **Rhinoceros®** and its plugin **Grasshopper®** apply the OOP paradigm to model geometry and manage the complexity of engineering designs. This approach is based on the idea that each design element is an object with specific characteristics and methods that define its behavior. In other words, just as in an OOP programming language objects represent entities with properties and functions, in Rhinoceros® and Grasshopper®, each geometric entity—whether a point, curve, surface, or volume—is treated as an object with its own attributes and associated operations.

Rhinoceros® is a NURBS-based modeling software structured around the concept of geometric objects. Every element within a scene is treated as an object with specific properties. For example, a curve carries information such as degree, control points, and domain, while a surface has attributes like normal vectors and curvature values. Users interact with these entities through a set of operations that modify their shape and position, analogous to methods in a programming class. OOP is also reflected in the way Rhinoceros® manages hierarchies: it enables users to organize objects into layers, assign attributes, and define nested structures, thereby facilitating the management of complex projects. Each object retains its identity while interacting with others in a coherent system. If Rhinoceros® embodies the principles of traditional OOP modeling, Grasshopper® elevates this approach by introducing a visual programming paradigm based on components. In Grasshopper®, each node functions as an object with clearly defined inputs and outputs. A *Point* node, for instance, represents a spatial coordinate, while a *Loft Surface* node generates a surface from a set of input curves. This architecture supports modularity and inheritance—two core concepts in OOP. For example, a complex parametric structure can be constructed from simpler geometric modules, which are replicated and transformed according to predefined rules. This enables scalable and flexible design processes, where each model element remains autonomous yet integral to the whole system. Another key concept is polymorphism, which in classical OOP describes an object's capacity to assume different forms or behaviors in varying contexts. While Grasshopper® doesn't implement polymorphism in the strict programming sense, it supports a similar principle: a single object can function as input across multiple operations. For instance, a single curve in Grasshopper® can serve simultaneously as:

- an extrusion path for generating solids,
- a guide for distributing structural lattice elements,
- a subdivision framework for placing points or generating derivative geometries.

This functional versatility reflects the polymorphic behavior typical of OOP, enabling geometric objects to participate in multiple design strategies without losing coherence. Equally fundamental is the reactive and dynamic nature of Grasshopper®: any modification to a parameter propagates instantaneously through the network of nodes, updating the entire model in real time. This responsive workflow allows designers to quickly explore multiple design alternatives, simulate performance scenarios, and optimize geometries for functional or aesthetic purposes. The influence of OOP extends beyond visual programming: both Rhinoceros® and Grasshopper® support scripting through languages such as Python and C#, empowering users to define custom classes, implement specific methods, and build advanced algorithms for manipulating geometric data. This capability opens pathways to sophisticated simulations, structural optimizations, and the integration of artificial intelligence in design workflows. Thanks to their object-based architectures, modern CAD platforms offer robust tools for managing the inherent complexity of contemporary design projects. These platforms ensure modularity, scalability, and dynamic interaction—qualities that are essential in fields such as architecture, materials science, computational design, and bioengineering, where the ability to model adaptable and complex geometries is increasingly vital. Ultimately, the integration of geometric modeling with object-oriented programming (OOP) has fostered a more structured, flexible, and innovative design methodology, offering unprecedented control over variables across a project's lifecycle. In OOP-based CAD environments, the goal has been to encode into software a structural and functional analogy drawn directly from biological principles—particularly those observable in neural networks and living systems. Within this paradigm, each CAD component operates like an autonomous biological cell, embedding all necessary information to perform a specific function. These components interact and collaborate, forming increasingly complex assemblies—just as cells aggregate to form tissues, organs, and entire organisms. Likewise, neural networks are composed of simple yet powerful units that transmit information and contribute collectively to complex decisions. Similarly, in CAD environments, geometric objects interact within a web of relationships that support the simulation of entire systems—mechanical, structural, or architectural The capacity to abstract, organize, and control complexity through software mirrors how nature orchestrates biological order: through modularity, adaptability, and emergent cooperation. Object-oriented programming has thus not only revolutionized traditional software development but also empowered CAD systems to meet the demands of advanced engineering and digital fabrication. Its conceptual alignment with natural systems makes OOP not just an efficient computational framework, but also a metaphorical bridge between the digital and the organic. As emphasized throughout this publication, the convergence of biological inspiration and digital execution reflects one of the core features of Industry 5.0: the gradual erosion of the boundary between what has historically been defined as natural—shaped by evolutionary processes—and what is artificial—engineered by human ingenuity. This boundary is becoming increasingly fluid. Looking ahead, the future of Earth may see the emergence of what could be called an *augmented bio-WunderCAD/*

CAMmer: a planetary scenario suspended between the real and the virtual, where nature and artifice are so deeply interwoven as to be indistinguishable—giving rise to a new kind of hybrid *bio-techno-organism*. Everything seems to converge toward the realization of a technologically advanced artificial civilization, centered on a total green economy and dedicated to overcoming the systemic imbalances and environmental crises that define the current anthropocentric phase. This vision anticipates an era regenerated by the industrial-scale application of biologically hyper-technological knowledge—fusing the elegance of nature's design with the creative excellence of human invention. We are increasingly entering the age of neo-biological hybridization, where techno-scientific development is marked by an affinity for *biophilia* and *biomimicry*. Yet, in its more radical philosophical trajectories, this trend dissolves into the speculative utopias of the post-organic, best exemplified by the ideologies of transhumanism. In a seminal and prophetic 2005 article, Swedish philosopher Nick Bostrom—widely recognized for his contributions to the study of existential risk and the anthropic principle—reconstructed the intellectual origins of transhumanist thought. He even drew upon the ancient Sumerian *Epic of Gilgamesh* to demonstrate how, from the earliest myths, humanity has been animated by the desire to transcend the inherent limitations of its organic condition. At the heart of this aspiration lies the existential fragility associated with mortality—a threshold that, across centuries and civilizations, we have continuously strived to overcome (Bostrom 2005). From a neo-transhumanist perspective, this fragility is not only something to escape, but a critical passage toward the emergence of a new form of life: one in which the biological and the technological no longer coexist as separate domains, but fuse into a unified and continuous form of post-human existence.

References

Addario, G. (2024). Biofabrication platforms: Advanced 3D (bio)printed models for the study of kidney fibrosis [Doctoral thesis, Maastricht University]. Maastricht University. Available at: https://doi.org/10.26481/dis.20241106ga [Accessed: 13 May 2025].

Anyanwu, E., Osasona, F., Akomolafe, O., Ogugua, J., Olorunsogo, T., and Daraojimba, E. (2024). Biomedical engineering advances: A review of innovations in healthcare and patient outcomes. International Journal of Science and Research Archive, 11: 870–882. https://doi.org/10.30574/ijsra.2024.11.1.0139.

Baig, M. H., Ahmad, K., Rabbani, G., Danishuddin, M., and Choi, I. (2018). Computer aided drug design and its application to the development of potential drugs for neurodegenerative disorders. Current Pharmaceutical Design, 24(11): 1251–1263. Available at: https://doi.org/10.2174/1570159X15666171016163510 [Accessed: 13 May 2025].

Benchling. (2025). Benchling: Cloud-based platform for biotech R&D. Available at: https://www.benchling.com/[Accessed: 14 May 2025].

Biomason. (2025). Biomason: Revolutionizing Cement with Biotechnology. Available at: https://biomason.com/[Accessed: 13 May 2025].

Bostrom, N. (2005). In defense of posthuman dignity. Bioethics, 19(3): 202–214.

Chikere, N. C., Lozano Voticky, S., Tran, Q. D., and Ozkan-Aydin, Y. (2024). Flagellar swimming at low Reynolds numbers: Zoospore-inspired robotic swimmers with dual flagella for high-speed locomotion. arXiv. Available at: https://arxiv.org/abs/2412.05712 [Accessed: 13 May 2025].

Colorifix. (2025). Colorifix. Available at: https://www.colorifix.com/ [Accessed: 13 May 2025].

CSIRO. (2022). Evolution or intelligent design? The story of the bacterial flagellar motor. CSIRO Synthetic Biology Future Science Platform. Available at: https://research.csiro.au/synthetic-biology-fsp/evolution-or-intelligent-design-the-story-of-the-bacterial-flagellar-motor/ [Accessed: 13 May 2025].

Daston, L., and Park, K. (1998). Wonders and the order of nature, 1150–1750. Zone Books.

Dilshani, I. (2023). Unraveling object-oriented programming: A journey through evolution. Medium. Available at: https://medium.com/@imeshadilshani212/unraveling-object-oriented-programming-a-journey-through-evolution-4b9c25cd14de [Accessed: 14 May 2025].

Dutfield, S., and Choi, C. Q. (2022). 7 theories on the origin of life. Live Science. Available at: https://www.livescience.com/13363-7-theories-origin-life [Accessed: 13 May 2025].

Dzwonchyk, A. (2017). Object Oriented Biology. Launch School. Available at: https://medium.com/launch-school/object-oriented-biology-6ed991d9e82a [Accessed: 14 May 2025].

Encyclopaedia Britannica. (2025). Miller–Urey experiment. Encyclopaedia Britannica. Available at: https://www.britannica.com/science/Miller-Urey-experiment [Accessed: 13 May 2025].

Eskandar, K. (2023). Revolutionizing biotechnology and bioengineering: Unleashing the power of innovation. Journal of Applied Biotechnology & Bioengineering, 10: 81–88. Available at: https://doi.org/10.15406/jabb.2023.10.00332 [Accessed: 13 May 2025].

Feynman, R. P., Leighton, R. B., and Sands, M. (1963). The Feynman Lectures on Physics (Vol. 1, p. 1–1). Addison-Wesley. Available at: https://www.feynmanlectures.caltech.edu/I_01.html [Accessed: 13 May 2025].

Food4Rhino. (n.d.). Objectivism. Food4Rhino. Available at: https://www.food4rhino.com/en/app/objectivism [Accessed: 14 May 2025].

Gazo, H. E., Younes, K., Roufayel, R., Khazaal, M., and Fajloun, Z. (2024). Engineering innovations in medicine and biology: Revolutionizing patient care through mechanical solutions. Heliyon, 10(4): e26154.

Georgieva, M. N., Little, C. T. S., Maslennikov, V. V., Glover, A. G., Ayupova, N. R., and Herrington, R. J. (2021). The history of life at hydrothermal vents. Earth-Science Reviews, 217: 103602.

Ginkgo Bioworks. (2025a). Ginkgo Bioworks. Available at: https://www.ginkgo.bio/ [Accessed: 14 May 2025].

Ginkgo Bioworks. (2025b). Design Systems at Ginkgo. Available at: https://www.ginkgo.bio/resources/blog/design-systems-at-ginkgo [Accessed: 14 May 2025].

Glavin, D. P., Dworkin, J. P., Alexander, C. M. O'D., Aponte, J. C., Baczynski, A. A., Barnes, J. J. et al. (2025). Abundant ammonia and nitrogen-rich soluble organic matter in samples from asteroid (101955) Bennu. Nature Astronomy, 9(2): 199–210.

Huang, S., Waeijen, L., and Corporaal, H. (2022). How flexible is your computing system? ACM Transactions on Embedded Computing Systems (TECS), 21(4), Article 37: 1–41.

Journal of Structural Biology. (2025). Journal of Structural Biology. ScienceDirect. Available at: https://www.sciencedirect.com/journal/journal-of-structural-biology [Accessed: 13 May 2025].

Kay, A. (1993). The early history of Smalltalk. SIGPLAN Notices, 28(3): 69–95.

Kean, Z. (2023). Ancient clonal tree, King's lomatia, excites scientists in Tasmania's remote south west. ABC News. Available at: https://www.abc.net.au/news/2023-07-29/kings-lomatia-tasmanica-ancient-clonal-plant-worlds-oldest/102240860 [Accessed: 13 May 2025].

Kim, W., Banerjee, J., Chou, H.-T., and Garza, J. F. (1990). Object-oriented database support for CAD. Computer-Aided Design, 22(8): 469–479. https://doi.org/10.1016/0010-4485(90)90063-I.

Kloprogge, J. T., and Hartman, H. (2022). Clays and the origin of life: The experiments. Life (Basel), 12(2): 259. Available at: https://doi.org/10.3390/life12020259 [Accessed: 13 May 2025].

Kumar, G., Yadav, S., Mukherjee, A., Hassija, V., and Guizani, M. (2024). Recent advances in quantum computing for drug discovery and development. IEEE Access.

Lai-Yuen, S., and Lee, Y.-S. (2013). Interactive computer-aided design for molecular docking and assembly. Computer-Aided Design and Applications, 3. Available at: https://doi.org/10.1080/16864360.2006.10738423 [Accessed: 13 May 2025].

Levin, M. (2024, December 2). The multiscale wisdom of the body: Collective intelligence as a tractable interface for next-generation biomedicine. BioEssays, 46(2), Article 202400196.

Lewis, C. T., and Short, C. (1879). A Latin dictionary. Oxford University Press.

Liddell, H. G., and Scott, R. 1996. A Greek-English lexicon (9th ed.). Oxford University Press.

Liu, Z. (2024, September 20). Life should be redefined: Any molecule with the ability to self-replicate should be considered life. F1000Research, 13: 736.

McMillen, P., and Levin, M. (2024). Collective intelligence: A unifying concept for integrating biology across scales and substrates. Communications Biology, 7: 378. Available at: https://doi.org/10.1038/s42003-024-06037-4 [Accessed: 13 May 2025].

McPherson, K. (2024). Understanding evolution at the cellular level. ASU News. Available at: https://news.asu.edu/b/20241120-understanding-evolution-cellular-level [Accessed: 13 May 2025].

Nativi, S., Mazzetti, P., and Craglia, M. (2021). Digital ecosystems for developing digital twins of the Earth: The Destination Earth case. Remote Sensing, 13(11): 2119. https://doi.org/10.3390/rs13112119.

Orbital Materials. (2025). Orbital Materials. Available at: https://www.orbitalmaterials.com/ [Accessed: 13 May 2025].

Peng, H., Zhao, M., Liu, X., Tong, T., Zhang, W., Gong, C. et al. (2024). Biomimetic materials to fabricate artificial cells. Chemical Reviews, 124(23): 13178–13215. Available at: https://doi.org/10.1021/acs.chemrev.4c00241 [Accessed: 13 May 2025].

Prause, E., Hey, J., Beuer, F., Yassine, J., Hesse, B., Weitkamp, T. et al. (2024). Microstructural investigation of hybrid CAD/CAM restorative dental materials by micro-CT and SEM. Dental Materials, 40(6): 930–940.

Price, P. B. (2007). Microbial life in glacial ice and implications for a cold origin of life. FEMS Microbiology Ecology, 59(2): 217–231. Available at: https://doi.org/10.1111/j.1574-6941.2006.00234.x [Accessed: 13 May 2025].

Robinson, J. (2024). Ultracold snapshots reveal in exquisite detail how a bacterial flagellum rotates. Chemistry World. Available at: https://www.chemistryworld.com/news/ultracold-snapshots-reveals-in-exquisite-detail-how-a-bacterial-flagellum-rotates/4019366.article [Accessed: 13 May 2025].

Rosslenbroich, B. 2023. Properties of life: Toward a theory of organismic biology. MITPress.

Schrödinger, E. (1944). What Is Life? The Physical Aspect of the Living Cell. Cambridge University Press. Available at: https://archive.org/details/WhatIsLife-EdwardSchrodinger [Accessed: 13 May 2025].

Sharov, A. A. (2016). Coenzyme world model of the origin of life. Biosystems, 144: 8–17.

Singh, P. K., Cecchini, G., Nakagawa, T., and Iverson, T. M. (2023a). CryoEM structure of a post-assembly MS-ring reveals plasticity in stoichiometry and conformation. PLoS ONE, 18(5): e0285343.

Singh, K., Bhushan, B., and Singh, B. (2023b). Advances in drug discovery and design using computer-aided molecular modeling. Current Computer-Aided Drug Design, 20.

Singh, P. (2025) Tweets [X profile]. X. Available at: https://x.com/prash_singh [Accessed: 13 May 2025].

Singh, P. (2024a). A leap from bioenergetics to chemotaxis. Research Communities by Springer Nature. Available at: https://communities.springernature.com/posts/a-leap-from-bioenergetics-to-chemotaxis [Accessed: 13 May 2025].

Singh, P., Sharma, P., Afanzar, O., Goldfarb, M., Maklashina, E., Eisenbach, M., et al. (2024b). CryoEM structures reveal how the bacterial flagellum rotates and switches direction. Nature Microbiology, 9(7): 1271–1281. https://doi.org/10.1038/s41564-024-01674-1.

Sirohi, D., Chen, Z., Sun, L., Klose, T., Pierson, T. C., Rossmann, M. G. et al. (2016). The 3.8 Å resolution cryo-EM structure of Zika virus. Science, 352(6284): 467–470. https://doi.org/10.1126/science.aaf5316.

Sithamparam, M., Satthiyasilan, N., Chen, C., Jia, T. Z., and Chandru, K. (2022). A material-based panspermia hypothesis: The potential of polymer gels and membraneless droplets. Biopolymer, 113(5), e23486.

Springer. (2025). Journal of Computer-Aided Molecular Design. Springer. Available at: https://link.springer.com/journal/10822 [Accessed: 14 May 2025].

Vanderbilt University-VIIBRE. (2025). Vanderbilt Institute for Integrative Biosystems Research and Education (VIIBRE). Available at: https://www.vanderbilt.edu/viibre/ [Accessed: 13 May 2025].

Vanderbilt University. (2025). Vanderbilt University. Available at: https://www.vanderbilt.edu/ [Accessed: 29 April 2025].

Xu, Q., Ali, S., Afzal, M., Nizami, A.-S., Han, S., Dar, M. A. et al. (2024). Advancements in bacterial chemotaxis: Utilizing the navigational intelligence of bacteria and its practical applications. Science of The Total Environment, 931: 172967.

Wonders of Physics. (2022). Richard Feynman explains the circle of life. Wonders of Physics. Available at: https://www.wondersofphysics.com/2022/12/richard-feynman-explains-circle-of-life.html [Accessed: 13 May 2025].

Wu, G.-H., Mitchell, P. G., Galaz-Montoya, J. G., Hecksel, C. W., Sontag, E. M., Gangadharan, V. et al. (2020). Multi-scale 3D Cryo-Correlative Microscopy for Vitrified Cells. Structure, 28(11): 1231–1237.e3.

9

Make it More Predictable

"More than anything,
I remember the future."
Salvador Dalì

9.1 Beyond CAD/CAM: The Protein's Folding, the *HolyGrail* of Contemporary Biology

In a 2020 video by DeepMind, CEO and co-founder Demis Hassabis referred to the problem of predicting protein folding as a kind of *Holy Grail* of contemporary biochemistry (Hassabis 2022b) (Fig. 1).

The history of DeepMind®, a UK startup founded in 2010 in London by Demis Hassabis, Shane Legg, and Mustafa Suleyman—three brilliant minds with very different academic backgrounds—represents a sensational case in the field of artificial implementation of human predictive capabilities assisted by digital technologies and AI (Google DeepMind 2025). This is clearly demonstrated by the groundbreaking awarding of the 2024 Nobel Prize in Chemistry to Demis Hassabis and John M. Jumper, both researchers at DeepMind®, for their pioneering contributions to the study and prediction of protein structure, as well as to David Baker, professor at the University of Washington and director of the Institute for Protein Design, awarded for his work in computational protein design. His work includes the development of the molecular software *RoseTTA Fold*®, a key tool for predicting protein structures and designing new proteins with specific functionalities (Prabhune 2024). RoseTTA Fold® was one of the first and most powerful tools for modeling protein folding and designing artificial proteins with applications in medicine, biotechnology, and nanomaterials, taking a decisive step towards the personalization of treatments and the care of complex diseases (Choudhry 2024). Within the core business of DeepMind®, the most advanced developments and potential applications of AI converge. One can even trace, through the name chosen for the company, a connection with the *Deep Thought* of Douglas Adams' 1979 book *The Hitchhiker's Guide to the Galaxy* (Bundell 2019). The book begins ironically with the following passage:

DOI: 10.1201/9781003438946-9

Fig. 1. The Holy Grail of Protein Folding. A stylized chalice image symbolizing the scientific quest to decode protein folding as a form-function mystery.

"Far, far away, in the forgotten spaces marked on no map at the extreme limit of the Western Spiral Arm of the Galaxy, there is a small and insignificant yellow sun. Orbiting around it, at a distance of one hundred and forty-nine million kilometers, is a small, rather insignificant planet" (Adam 1993).

The common thread that connects DeepMind® to *Deep Thought* could lie in their shared ambition to answer grand existential questions through advanced intelligences, whether real or imaginary. *Deep Thought* is the supercomputer devised by Douglas Adams in this book. It was created with the purpose of "finding the answer to the Ultimate Question of Life, the Universe, and Everything." After millions of years of calculations, the supercomputer provides the answer: 42.

An enigmatic result that, however, does not clarify what the original question was (Nissinen 2022). Thus, although *Deep Thought* and DeepMind® operate on different planes, their core idea revolves around solving problems through advanced intelligence—whether through a futuristic imaginary supercomputer or the AGI systems that are becoming increasingly central in the contemporary world. Beyond any literary inspiration, DeepMind® represents one of the most advanced AI companies today, founded with the goal of developing systems capable of tackling and solving problems of enormous complexity, often surpassing traditional human intelligence. DeepMind® CEO, Demis Hassabis, in particular, was a true *enfant prodige*: at just 13 years old, he was already a chess champion, and later, he obtained a PhD in cognitive neuroscience to understand the mechanisms behind memory and decision-making. This combination of skills had a profound impact on his vision. Drawing directly from neuroscience studies, he aimed to build a machine capable of learning based on human brain cognitive functions. Indeed, DeepMind® focus from the very beginning was to develop an Artificial General Intelligence (AGI)—an AI capable of autonomously learning from any context, just like a human mind, rather than merely solving specific tasks. This approach, radically different from traditional AIs designed to execute very limited and predefined functions, immediately yielded significant results. In its early years, DeepMind® quickly gained prominence for its advances in machine learning and reinforcement learning—an approach inspired by how humans and animals learn through trial and error (Dean et al. 2023). Founded in 2010, it was acquired by Google® in 2014 for approximately $500 million, marking a strategic move into AI that would influence the entire tech industry. Operating under Alphabet®, DeepMind® applied its technology to practical challenges such as reducing energy consumption in data centers. However, its global breakthrough came in 2016 with AlphaGo®, an AI system built to master the ancient and complex game of Go—known for having more possible board positions than atoms in the universe (Latson 2015). The 2016 AlphaGo® vs. Lee Sedol match marked a turning point in the evolving relationship between human intuition and artificial intelligence. Go, long considered a final frontier for AI due to its strategic depth and near-infinite move combinations, had resisted algorithmic mastery—until AlphaGo®, trained to replicate and surpass human reasoning, overcame the reigning world champion. In the second game, AlphaGo® made Move 37, an unexpected and creative strategy that no human would have played in that spot. The move demonstrated the AI's ability to act beyond human intuition, while Sedol managed to win only one game out of five. It became a symbol of creative artificial intelligence (Fig. 2).

Yet, as Benjamin Labatut reflects in *The Maniac* (Scott 2023), the confrontation revealed an unsettling truth: machines can now absorb, simulate, and even enhance the uniquely human qualities of foresight and improvisation. AlphaGo® triumph was more than a technical milestone; it destabilized traditional notions of intelligence, suggesting that AI's capacity to emulate psychological nuance and mental strategies is not only advancing—it is redefining what constitutes strategic thought itself. In the wake of this event, the meaning and role of human intuition must be re-evaluated within a landscape increasingly shaped by deep learning and cognitive computation (Gignac and Szodorai 2024). The evolution of AI through game-based learning, exemplified by AlphaGo®, reveals a fundamental shift in the hierarchy between human intuition and machine cognition. While systems trained on human data retain minor flaws—exposed by Sedol's rare and intuitive move—their dominance in structured environments highlights AI's capacity for superior calculation and pattern recognition. However, in domains governed by ambiguity, emotion, and ill-defined goals, human intuition still offers a decisive edge. The true challenge now lies not in preserving human uniqueness, but in orchestrating a synergy where intuition guides and AI amplifies, particularly in contexts like scientific innovation and strategic design. AlphaGo® victory also catalyzed a redirection toward one of the greatest scientific frontiers: *protein folding*. Misfolded proteins are at the root of numerous diseases, and for decades, traditional methods failed to resolve their dynamics. DeepMind's application of AI to this molecular enigma has initiated a new paradigm in medicine—where prediction becomes a tool for precision therapy. Yet this predictive turn brings risks: in a globalized world, algorithmic misjudgments can have systemic impacts. The pursuit of prediction, like the mythic *Holy Grail*, promises profound knowledge but confronts us with the limits of both computation and comprehension. The capacity to predict confers the power to shape; thus, predictive technologies place scientists and policymakers at an ethical crossroads. As Demis Hassabis has emphasized, the governance of artificial intelligence demands not only technical excellence but moral clarity. Anticipating

Fig. 2. Move 37.

the future becomes an act of authorship—one that carries profound responsibility for what is set in motion (Perrigo 2023). As AI accelerates, Demis Hassabis, in a lecture held on May 17, 2022, at the Sheldonian Theatre in Oxford, stresses the urgent need for responsible development to ensure its power serves the common good and safeguards the future. The open-source model fosters innovation but also heightens the risk of misuse, underscoring the need for constant ethical oversight—an urgency reflected in initiatives like Oxford's Institute for Ethics in AI and global frameworks such as UNESCO's recommendations (University of Oxford – Institute for Ethics in AI 2025). DeepMind's breakthrough at the 2020 CASP14 competition marked a transformative moment in computational biology. Its AlphaFold® system, using advanced deep learning techniques, achieved over 90% accuracy in predicting the three-dimensional structure of proteins from their amino acid sequences—an achievement long considered unattainable. This milestone demonstrated that AI could solve one of biology's most complex problems, outperforming all previous methods and generating global scientific enthusiasm. In the aftermath, DeepMind® partnered with the European Bioinformatics Institute to launch the *Google AlphaFold Protein Structure Database*, a free and open-access repository initially containing 350,000 protein structures, often described as the *Google of proteins* (AlphaFold Protein Structure Database 2025). This initiative provided researchers worldwide with an unprecedented tool to study diseases, design new drugs, and accelerate biomedical discoveries. Widely praised as one of the greatest achievements in computational biology, AlphaFold® success has been likened to the invention of the microscope for its potential to revolutionize our understanding of life at the molecular level. Since joining Alphabet®, DeepMind® has pursued advances in AI and healthcare, culminating in AlphaFold 3®—a system that shifts biology toward determinism by accurately predicting the 3D structures and interactions of biomolecules, including proteins, DNA, RNA, and drug-relevant ligands, as well as their regulatory chemical modifications (Schenker 2020). AlphaFold 3® achieves its predictive power through an advanced architecture called Pairformer, combined with a diffusion network that refines a cloud of atoms into accurate molecular structures. This AI-driven process surpasses traditional methods in precision and efficiency, marking a shift from discovering biological forms by chance to designing them intentionally (Abramson et al. 2024). This evolution marks a true transition towards the customization of life, a techno-scientific system integrating evolutive CAD/CAM technologies with synthetic biology and bio-inspired design (Frappier and Keating 2021). AlphaFold® predictions, while powerful, are not infallible—AI models can *hallucinate*,[1] producing inaccurate or implausible structures, particularly with rare sequences or limited data, thus requiring experimental validation for reliability. Depending on the type of interaction being analyzed, AlphaFold3® accuracy can range from 40% to over 80%, and the system indicates its level of confidence in each prediction. When the confidence is low, the model should be treated as a preliminary guide, with further analysis needed through complementary methods. Nonetheless, for exploratory questions—such as identifying enzymes that might degrade plastic—AlphaFold offers a significantly faster and more practical alternative to traditional techniques like X-ray crystallography (Workman 2021).

AlphaFold3® is not just a research tool but a catalyst for design innovation, enabling the decoding and creation of proteins—paving the way for designing biological structures from scratch and merging scientific insight with creative engineering (The Scientist 2024) and generating novel protein structural designs (MIT news 2023). The era of programmable biology has begun, with enormous potential and opportunities, but also with risks and ethical issues (Ramadurai 2023). Compared to past efforts, which included traditional experimental and computational methods, the technology developed by DeepMind® has drastically reduced the time and costs associated with determining protein structures, eliminating many laboratory experimental processes and shifting the focus of research and experimentation to a virtual and fully computerized plane. The big innovation lies not in the goal itself but in the ways this ancient problem is addressed. The result achieved by AI application in protein design therefore has profound implications for the future of biomedical science (Abudayyeh and Gootenberg 2024). This is evidenced by the awarding of the 2024 Nobel Prize in Chemistry, which underscores the scientific and societal importance of AlphaFold®'s contribution to the future of molecular biology and AI-driven discovery. With the digital ability to accurately predict the structure of proteins, researchers can accelerate drug development, better understand genetic diseases, and design enzymes for new industrial or environmental uses. This technology is particularly useful for the structural prediction of proteins that would otherwise be difficult or impossible to analyze through experimental methods, saving years of work and significant economic resources for the international scientific community. In conclusion, although AlphaFold®, due to its specific functional characteristics, is not exactly comparable to a traditional engineering CAD standard—or molecular—it shares many similar functionalities and, in some respects, represents an evolution

[1] AI hallucinations occur when models produce outputs that are not based on reality or the available input data, leading to errors that may appear coherent but are scientifically incorrect. In applications like protein folding, such hallucinations can mislead researchers by presenting speculative structures as credible. These risks emphasize the necessity of using AI as a complementary tool, supported by domain expertise and experimental confirmation, particularly when predictions drive high-impact decisions in research or therapeutics.

of the CAD species. It is worth remembering that, in fact, the CAD protocol has specialized over time into CADD (Computer Aided Drug Design) to assist in molecular design in the specific pharmaceutical sector, optimizing therapeutic molecules and predicting the interaction between drugs and biological targets such as proteins or enzymes (Niazi and Mariam 2023).

9.2 CADD, the Oracle of New Health

AlphaFold3® uses *diffusion algorithms*—adapted from AI image generation—to model molecular interactions, simulating how particles naturally spread from areas of high to low concentration, as seen in physical diffusion processes (Desai et al. 2024). A *diffusion algorithm* is a type of codification used in various fields of science and engineering, where information, a value, or a substance (such as a probability, a physical property, or a signal) propagates through a network or space. The term diffusion comes from the idea that the information gradually spreads from one point to another, following certain rules or laws of propagation. In AI, for example, diffusion algorithms are used to generate images or to enhance images, where a model, such as those used in GANs (Generative Adversarial Networks), spreads an image from a condition of random noise to a defined form, progressively improving the image quality. It is extremely interesting to learn how GANs are increasingly integrated to expand, automate, and improve operational processes related to CAD design systems (uMake 2025). Composed of a generator and a discriminator, GANs create design variants, explore design spaces, and optimize solutions according to specific criteria. Their main applications include conceptual design, design variation, and reverse engineering. Despite challenges such as instability during training, GANs offer significant potential to innovate the CAD sector. Kendre et al. (2023) introduce *SketchCADGAN*, a generative model based on GANs designed to automatically complete partially drawn sketches of engineering shapes. The aim is to enhance the performance of retrieval systems within CAD databases, enabling more effective recognition even when user inputs are incomplete or roughly drawn. The system transforms these sketches into geometrically plausible and complete forms, fostering the integration of manual creativity with AI-assisted search in the context of engineering design. Traditional CAD, originally created to assist design in fields such as mechanics, engineering, architecture, and industrial design, has evolved into CADD (Computer-Aided Drug Design), exploring methods and solutions for biochemical design at the molecular scale (Ajjarapu 2023). Molecular engineering still effectively considers the atomic domain of organic and inorganic matter as a sophisticated biochemical LEGO®, where the fundamental building blocks of life can be treated as a manageable and malleable abacus of elements to be skillfully assembled, in order to eventually decode, emulate, or even surpass the deep structures of natural design. Computer-Aided Drug Design environments are one of the many facets of the contemporary *WunderCAD/CAMmer*'s spirit—an invoked aesthetic and conceptual synthesis at the intersection of design, computational modeling, and biological systems—where traditional boundaries between Artifice and Nature, the synthetic and the organic, and the digital and the molecular progressively dissolve. These platforms function as generative laboratories of knowledge and invention, merging the formal aesthetics of design with the algorithmic precision of molecular prediction. By adopting architectures inspired even by video game logic, CADD systems do more than simulate—they intervene in, and to some extent, crack nature itself: unveiling and reprogramming the principles of biological organization. In doing so, they realize a renewed version of the *Wunderkammer*—no longer a passive *cabinet of curiosities*, but an active, computational space for discovery, synthesis, and the speculative reengineering of life. But despite their extreme sophistication, they express a *neo-mechanistic* view of matter and life. The powerful algorithms behind them, capable of predicting and simulating molecules through deterministic models and mathematical algorithms, ultimately reproduce an idea of the world as separable, sectionable, reduced to pre-determinable and predictable parts, typical of the mechanistic scientific tradition that has dominated the Western vision of the universe until today (Fagan and Maienschein 2022). However, the biological world, and particularly that of proteins and molecular interactions, is inherently complex, dynamic, and nonlinear, showing clues of some unknown form of intelligence and morphological free will that eludes us in its elusiveness, still to be discovered in its intimate quantum logic, as evidenced by cutting-edge research in the new biology of development (Rijssenbeek 2025). Hypothesizing that there is and operates some form of free will in the biological and morphological context is quite a bold, fascinating, and complex concept, and, in some ways, heretical, to the extent that it is viewed with suspicion by the global academic mainstream.[2] It involves the scientific discussion on how cells, as the fundamental units of life, can be interpreted in terms of autonomous actions determined. On one hand, there is no doubt that cells are subject to the laws that govern physics, biochemistry, and genetics, which guide their behavior, but on the other hand, morphological and adaptation processes, such as evolution, suggest that there exists a kind of decision-making field or flexibility in how cells receive and respond to external or internal stimuli. In the morphological domain, cells seem to collectively process decisions on how to organize and adapt into complex structures. For example, cellular migration, the induction of cellular differentiation, and organ formation are phenomena that seem to involve a form of response to context, which may seem analogous to free will, but is actually the result of complex interactions between genetic, chemical, and mechanical signals (Merino-Casallo et al. 2022). Traditionally, science has described organisms as systems governed by strictly deterministic biochemical and genetic laws. However, hypothesizing that cells, tissues, and entire organisms may possess some degree of decision-making autonomy opens unexpected scenarios. In this context, the concept of cellular agency comes into play, referring to the ability of cells to perceive internal and external signals and respond with adaptive choices, rather than mere automatic reactions (Su et al. 2024). This approach challenges the classical paradigm of

[2] This theme echoes a philosophical reflection previously explored in the chapter 05.0 on smartness, where we examined the epistemological tension between reductive models and emergent complexity. There, we addressed how the notion of 'smartness'—whether in materials, systems, or biological processes—resists simplistic, mechanistic interpretations, inviting instead a broader understanding of intelligence as distributed, adaptive, and context-sensitive. This current discussion builds on that foundation, reaffirming the need to move beyond deterministic frameworks in favor of more holistic, system-oriented perspectives that acknowledge the relational and processual nature of life.

molecular biology, which holds that living systems follow predefined paths dictated solely by genetic and evolutionary constraints. Speaking of free will in biology means questioning the traditional scientific reductionism and suggesting that organisms might exhibit some form of intentionality. While this does not imply consciousness or will in the human sense, it opens the possibility that organisms adopt flexible and contextual strategies, rather than being mere executors of genetic instructions. It is not surprising, then, that this perspective is often met with skepticism by the scientific community. The idea that cells can choose among different possibilities is considered heretical against the dominant biological thinking, which tends to reduce every living phenomenon to rigid chemical-physical interactions. However, recent studies in complex systems biology, epigenetics, and cellular intelligence suggest that the boundary between automatism and autonomy is less clear-cut than previously thought. Recent research conducted at the Francis Crick Institute in London reveals that endothelial cells, despite lacking a central nervous system, exhibit forms of collective *intelligence* by coordinating to form new blood vessels. Through a process known as *active perception*, these undifferentiated cells extend *filopodia* to explore their environment, and the cell gathering the most information assumes a leadership role in directing vessel formation. This behavior mirrors that of ant colonies, where decentralized yet coordinated actions lead to complex outcomes. Such findings challenge the notion that cognitive coordination is exclusive to higher organisms and offer new insights into tissue formation and potential therapeutic strategies for diseases involving impaired angiogenesis, such as cancer (Rossi Castelli 2021). Reconsidering the concept of agency in biology means opening a new field of investigation that could prove crucial in understanding how living systems organize their responses to stimuli, making decisions that cannot always be reduced to a simple cause-and-effect mechanism. If cells are capable of self-regulation and sophisticated adaptation, it may not be entirely far-fetched to hypothesize that there exists an internal logic guiding the living systems. This could have implications not only for biology but also for the philosophy of science, artificial intelligence, and bioengineering, offering new interpretations of life and its fundamental processes. Cellular free will might be viewed more as a metaphor, reflecting how organisms, at a microscopic level, possess the ability to dynamically respond to changes. However, it is probably always within the limits imposed by genetics and the environment. Likewise, in an evolutionary context, the interaction between cells and their environment could be seen as a continuous negotiation of options, leading to ever-new morphological forms driven by the needs for survival and adaptation. In bio-inspired design or biomimicry, if adopted, this principle of cellular free will could influence the design of intelligent materials and autonomous systems, where structure and response to external forces mimic, to some extent, the adaptive and decision-making capacities of cells. CADD, as a system that aims to predict and optimize molecules through linear simulations, still derives from a strictly mechanical and structural computer-based ideology, risking forms of theoretical blindness. It fails to fully grasp and embrace the fluidity and adaptability of living biology, which is not a static mechanism but an entity that evolves and adapts continuously to circumstances. Digital Imaging volving toward a more organic model, deeply inspired by Nature tech, as seen in the evolution of parametric and generative CAD, means stepping beyond strict determinism: molecule design may no longer be a search for static solutions but a process in constant transformation and interaction with the biological and chemical environment. In this context, drug designers could employ machine learning algorithms and generative technologies to create molecules that not only engage precisely with biological targets but also adapt and evolve in response to physiological and environmental changes—much like living organisms. This approach would mirror the dynamics of natural systems, in which form and function result from complex, non-linear interactions. Biomimetic approaches could serve as another powerful means to enrich computational tools, guiding them toward a more fluid and adaptive vision. If computer-aided design draws inspiration from biological principles—such as protein folding and cellular self-organization—it can evolve beyond static molecular modeling to simulate adaptive behaviors within complex biological environments. Just as proteins shift conformation to perform their functions, a molecule designed through what we might call Bio-CADD could adapt and respond dynamically to biological stimuli. This would open the door to therapeutics that do not merely act in static contexts but interact and evolve alongside physiological changes in real time (Sun et al. 2005). Such a shift could transform CADD from a conventional chemical design tool into a dynamic, systems-oriented platform. Rather than reducing biology to mechanistic laws, Bio-CADD would embrace its non-linear, emergent complexity—where biological responses are shaped by interdependent, unpredictable interactions. Drug design would no longer focus solely on structural prediction or optimization, but instead become a real-time, co-evolving process within living systems. Technologies like AlphaFold®, which already simulate protein folding with high precision, could integrate with Bio-CADD systems to predict not only structure but dynamic function, enabling the development of therapeutics conceived as embedded, responsive components within complex biological ecosystems. This convergence of computational precision and organic understanding represents a fundamental shift in the paradigm of drug development, particularly within bioengineering and predictive medicine. Viewed through the computational lens, health is no longer a static state defined by the absence of disease, but a dynamic process of systemic equilibrium. Illness, then, is not a malfunction to suppress, but a signal—a temporary disharmony within a self-organizing biological system. This view aligns with principles found in many Complementary and Alternative Medicine (CAM) practices, such as Ayurveda, acupuncture, homeopathy, energy medicine, and kinesiology—traditions often marginalized or dismissed by the academic mainstream, yet grounded in systemic and energetic models of health. With these highly predictive tools, medicine could shift from treating symptoms to fostering adaptive reconfiguration and energetic realignment of biological structures. Biomimetic technologies and advanced software might be used to support the body's innate ability to adapt, rather than overriding or suppressing its signals. In this model, the human organism is not seen as a mechanical object, but as a cybernetic, interactive, and intelligent system—capable of self-regulation and systemic repair. Artificial intelligence and computational biology could interpret physiological signals and adjust therapeutic interventions in real time, assisting the body in restoring balance without invasive techniques. Symptoms such as fever, inflammation, or pain would be reinterpreted not as failures, but as expressions of adaptive intelligence—signs that the body is actively attempting to recalibrate in response to environmental challenges. Bio-CADD thus invites a profound rethinking of medicine and design, proposing an integrated, responsive, and evolutionarily aware framework where technology collaborates with biology to support—not override—the organism's inherent wisdom. Therapeutic interventions are envisioned not as symptom suppression, but as support for the body's natural self-regulation, biofeedback, and adaptive capacity, aiming to optimize rather than normalize biological function. A key innovation lies in its potential to influence biological evolution at the individual level. By integrating tools like CRISPR with predictive molecular simulations, researchers could enable real-time adaptive changes in cells and tissues—shifting the focus from disease treatment to personalized, evolutionary well-being. This concept of controlled evolution makes the CADD an active agent in their own biological future development (Dr. Omics Labs 2024). However, it also raises profound ethical questions about the boundary between natural regulation and technological intervention (Dr. Omics Education n.d.). Maintaining a balance between the body's natural autonomy and the emerging ability to design its evolutionary trajectory is becoming increasingly critical. As biomedical and digital technologies advance, this challenge grows more significant. There is a real risk that excessive manipulation of human biology could disrupt its natural equilibrium. Nevertheless, imagining the body as an adaptive system that employs *biohacking*—the strategic use of technology to enhance biological human body self-regulation—introduces a new paradigm (Bajrektarevic and Bogdanova 2025). In this framework, biological free will—the capacity of cells to choose their responses—can integrate with technological systems in a functional and harmonious way. Bio-CADD, therefore, could redefine the very concept of health, highlighting the dynamic evolution of life and the potential to design an adaptive and ever-evolving state of well-being. Ultimately, this new perspective on the integration of organic and artificial elements could represent the meeting point between science, technology, philosophy, and creativity. Not just a design tool, but a reflection on the very nature of life, its evolution, and the interconnection between all living beings. In a world where technology and biology are increasingly merging, Bio-CADD could transform from a mere technical tool into a creative act that explores the mystery of existence, life, and our very interconnection with Nature. A step toward a design approach that is not just about seeking mechanical solutions but about embracing the beauty and intrinsic complexity of biology, incorporating variability and uncertainty as fundamental parts of the creative process. In any case, contemporary CADD—the same system that has enabled the precise determination of protein folding and the creation of vast databases containing hundreds of thousands of molecular models essential to modern pharmacology—is far more than just a digital tool accelerating techno-scientific progress in biomedicine. It represents a genuine epistemological revolution, reshaping how we understand and design therapeutic interventions. Through its predictive capabilities, CADD allows us to foresee, with growing precision, how a molecule will interact with specific biological targets—making it possible not only to optimize therapeutic efficacy but also to minimize adverse effects. In this light, CADD becomes more than a computational aid: it evolves into a *predictive oracle* of future health—a system capable of anticipating biological responses, guiding the personalization of treatments, and informing an anticipatory model of medicine where prevention, adaptation, and therapeutic foresight converge. Rather than reacting to illness, such a system empowers us to design wellness proactively, aligning medical intervention with the body's prefigurative needs and systemic behaviors.

From a technical standpoint, CADD leverages molecular modeling to construct three-dimensional representations of biological and chemical structures (Talevi 2024). These representations are generated through algorithms that process data from X-ray crystallography, NMR spectroscopy (nuclear magnetic resonance), and, more recently, from protein structure prediction technologies such as AlphaFold®. With this information, CADD software can precisely calculate how a molecule will bind to its target, what chemical interactions will occur, and what the biochemical and physiological implications will be. These scenarios gain real-world significance when considering recent technological breakthroughs—such as the announcement of Microsoft's Majorana 1®, the world's first quantum processor powered by topological qubits (Microsoft 2025). Named after the theoretical concept of Majorana fermions, this processor marks a fundamental shift in computational power and stability. From an evolutive perspective, the Bio-CADD design approach could explore how to simulate and guide adaptive biological responses, harnessing these advanced technologies to interact with cellular structures in ways never before possible. By merging artificial intelligence with the power of quantum computational models, it becomes possible to analyze biological signals in real time, supporting the organism's ability to self-regulate without resorting to invasive interventions. In this transformative vision, symptoms are no longer seen as malfunctions to suppress, but as signs of adaptive intelligence, guiding the body toward recovery. Fever, pain, and inflammation are reframed as creative responses to physiological stress—signals of the body's effort to evolve solutions in response to a shifting environment. It is through the advent of Majorana 1® that we may unlock an entirely new frontier in biological optimization. This revolutionary quantum computing platform has the potential to redefine molecular and biological simulations, bridging the capabilities of Bio-CADD with the future of predictive and adaptive medicine. By enabling simulations with atomic-level precision, Majorana 1® represents a monumental leap in our ability to understand and design biological systems.

Advances of this significance accelerate our ability to comprehend, model, and design bio-inspired drugs and materials like never before, opening new dimensions in medicine, material science, and biotechnology. The true power of Majorana 1® lies in its ability to perform advanced quantum simulations of molecular interactions, going far beyond what traditional computing could achieve. By harnessing quantum mechanics, Majorana 1® can solve problems in molecular biology with an accuracy and speed previously thought impossible. This allows us to simulate and predict the behavior of complex biological systems, from protein folding to enzyme interactions, with an unprecedented degree of certainty (Dirox 2025). Such precision holds the potential to bring about the next frontier in bio-inspired materials:

1. **Precision of Biomolecular Simulations:** Majorana 1® is at the forefront of quantum computing, enabling incredibly detailed molecular simulations. Unlike traditional systems, Majorana 1® taps into the unique properties of quantum

mechanics, dramatically reducing error margins that still afflict current computer technology and offering a quantum leap in the accuracy of biomolecular simulations. With its unparalleled ability to calculate molecular interactions at atomic precision, it can simulate how designed drugs and materials will interact with biomolecules—whether proteins, enzymes, or DNA—allowing for an understanding of these interactions that was once beyond our reach.

2. Design and Optimization of Bio-Inspired Materials: Majorana 1®'s quantum-powered simulations can enable the design of bio-inspired materials with extraordinary biocompatibility and efficiency. The computational speed and accuracy of this machine allow for simulations that take into account the complexities of biological systems in real-world contexts (Ramachandran 2024). As Bio-CADD seeks to customize materials to optimize their performance within the biological environment, Majorana 1® can provide the simulations needed to bring this concept into reality, optimizing material properties like never before.

3. Biological Customization and Predictive Approaches: With Bio-CADD integrating predictive medicine, Majorana 1® can significantly enhance the precision with which we design personalized therapies. The quantum computing capabilities of Majorana 1® allow us to model genetic variations and predict how individuals will respond to specific treatments based on their unique biological makeup. This level of personalized precision would not only accelerate the creation of tailored treatments but also allow us to model these responses in real time, paving the way for a new era in personalized medicine.

4. Analysis and Optimization of Therapeutic Effects: Traditional methods of drug design often require trial and error, but with Majorana 1, we can optimize drug therapies with quantum-enhanced simulations. By accurately modeling how drugs interact with biological systems, Majorana 1® enables us to fine-tune treatments, reducing side effects while enhancing efficacy. It optimizes pharmacokinetics and pharmacodynamics, leading to better-targeted therapies and the potential for fewer adverse effects—a leap toward the design of next-generation drugs.

5. Integration of Biology and Artificial Intelligence: Majorana 1® is not just a quantum computing machine—it is an *intelligent amplifier*. By enabling more sophisticated simulations, it also integrates seamlessly with AI-driven approaches, enhancing the optimization of bio-inspired materials and drugs. Majorana 1® works in harmony with machine learning systems, helping to analyze vast datasets and refine biological models, resulting in a synergy between AI, quantum computing, and biological innovation. This powerful integration could push Bio-CADD further into uncharted territory, advancing synthetic biology and bioengineering to new heights.

6. Discovery of New Therapeutic Targets: One of the most exciting aspects of Majorana 1® lies in its ability to explore intricate molecular interactions at an unprecedented scale. Through quantum simulations, Majorana 1® can uncover hidden therapeutic targets that would be impossible to detect using traditional methods. By delving deeper into the molecular structure of biological systems, it accelerates the discovery of new biomolecules that could lead to groundbreaking treatments for diseases once thought incurable.

From the evolving landscape of personalized medicine to the frontier of bio-inspired materials, the convergence of Majorana 1® and evolutive quantum technologies opens a gateway to innovation once confined to science fiction. In this quantum-enhanced era, computer-aided design doesn't just advance—it transforms into an imaginative engine of biological foresight: a *WunderCAD/CAMmer*, capable of weaving data, molecules, and living systems into coherent, adaptive architectures. It no longer merely assists design—it co-authors the future, sketching visions of health, sustainability, and responsiveness that echo Nature's own intelligence.

At the threshold of a post-mechanical paradigm in medicine and biology, a reassessment of tools, metaphors, and ethical frameworks becomes imperative. The forthcoming challenges extend beyond the technical; they are epistemological, ecological, and fundamentally human. While synthetic biology and quantum-enhanced biomedicine present the unprecedented capacity to shape living systems, such potential necessitates a design approach guided by discernment and responsibility. The term *futurible*, derived from the convergence of *futurum* and *possibilis*, encapsulates a fundamental tension—not simply *what will be*, but *what could be*. Embracing this concept does not entail predicting a predetermined future, but rather collaborating with a range of emerging possibilities. The aim is not to impose control over biological systems, but to participate in processes of attentive observation, simulation, and co-evolution. Within this framework, technological advancement may contribute not only to the extension of life, but to its enrichment—through systems that are both powerful and sensitive to the delicate balances that sustain living complexity.

Futurible challenges of biomedicine and synthetic biology:

- Reference to evolutionary biology perspectives on symptoms as adaptive mechanisms;
- Research on dynamic genetic therapies and self-adapting cellular models;
- Ethical discussions in synthetic biology and human enhancement (Persson and Savulescu 2012);[3]

[3] Julian Savulescu is one of the most influential philosophers and bioethicists of our time, known for his provocative approach to topics such as human enhancement, genetic ethics, and neuroscience applied to morality. Born in Australia in 1963, he is currently a professor at the University of Oxford, where he directs the Oxford Uehiro Centre for Practical Ethics.The core of his research is the concept of *enhancement*, the use of technology to improve human physical, cognitive, and moral abilities. Savulescu not only argues that it is permissible to intervene in our genetic makeup or brain functions to make us more intelligent, long-lived, or resistant to diseases, but in some cases, he considers it a moral duty. This idea is strongly reflected in the principle of *Procreative Beneficence*, which states that parents, when given access to embryo selection technologies, should choose to bring into the world children with the best possible prospects for life. This has sparked intense debates, as many see in this position the risk of a new form of eugenics, while he defends it as an extension of our ethical responsibility to provide the best for future generations. One of the most controversial aspects of his work concerns moral enhancement through biotechnology. Together with philosopher Ingmar Persson, he has argued that, given the destructive potential of human actions—ranging from nuclear weapons to the climate crisis—humanity may need biochemical assistance to develop a more refined ethical sense.

3 contd. ...

- Concept of biofeedback and biological decision-making in self-regulating systems;
- The role of computational approaches in redefining scientific methodologies;
- Predictive pharmacology and AI-driven drug design;
- The impact of computational structural biology on molecular research.

Main Application Areas of CADD:

- Structural Biology: Allows visualization and analysis of protein structures, nucleic acids, and biomolecular complexes, facilitating the understanding of their function (Wei and McCammon 2024).
- Protein Engineering: Enables the design of proteins with specific functionalities, such as more efficient enzymes or therapeutic antibodies (Abouelwafa et al. 2023). Implementation of methodologies that use advanced artificial intelligence models to read protein structures and DNA, with the aim of understanding how even small genetic variations can affect the functioning of the organism. By combining different techniques, these approaches allow for more accurate prediction of which mutations may cause diseases or disrupt gene regulation mechanisms (Schubach et al. 2024).
- Innovative Materials: Supports the design of new materials with desired properties, such as polymer-based materials, nanomaterials, or bioinspired eco-friendly artefacts (RW2 Education 2023).
- Nanotechnology: Aids in designing and simulating nanostructures and nanomaterials for applications in electronics, medicine, and energy (Keşkek Karabulut and Thurhan 2023).
- Computational Chemistry: Used for molecular dynamics simulations and quantum chemistry calculations, to predict chemical reactions and molecular properties (Domainex 2025).
- Biomaterial Development: Facilitates the design of biocompatible materials for applications in tissue engineering and regenerative medicine (Musthafa et al. 2024).
- Molecular Interaction Analysis: Enables the study of interactions between molecules, such as binding, affinity, and binding dynamics (Sarkar and Sen 2020).

...3 condt.

The idea of moral bioenhancement, illustrated in the book *Unfit for the Future: The Need for Moral Enhancement* (2012), suggests that drugs or genetic modifications could make us less aggressive, more empathetic, and more inclined to cooperate. For some, this is a dangerous manipulation of individual will; for Savulescu, however, it is a necessary reflection in the face of existential challenges humanity must confront. Beyond bioethics, he has also tackled crucial issues related to individual freedom, euthanasia, and the distribution of medical resources, always with the perspective that technological progress should be assessed not only for its risks but also for the opportunities it offers to improve the human condition. Although his positions are often controversial and sometimes criticized for their potential dystopian implications, his work continues to profoundly influence the debate on the direction we want to take in shaping our biological and ethical future.

- Molecular Sensing Design: Helps in developing molecular sensors for detecting chemicals or biomarkers (Zeng et al. 2023).
- Education and Training: Used in educational contexts to teach chemistry and biology concepts through visual simulations.

Methodological Approaches in CADD

A typical CADD workflow involves a combination of strategies, computational techniques, and tools aimed at designing and optimizing drugs via digital simulations, minimizing the initial need for laboratory testing. It represents a computational ecosystem of technologies that accelerate and enhance the precision of the drug discovery and development process.

These approaches include:

- ***Ligand-based design (LBDD)***: Based on known molecules (ligands) already active on a biological target, to design new ones with similar properties.
- ***Structure-based design (SBD)***: Uses the 3D structure of the target protein (obtained from crystallography or predictive models such as AlphaFold) to design molecules that bind effectively.
- ***Molecular Docking (MD)***: Simulates the interaction between a drug candidate and its target to estimate binding affinity and stability.
- ***Pharmacophore Modeling (PM)***: Identifies the essential chemical features responsible for biological activity.
- ***Quantitative Structure-Activity Relationship (QSAR)***: Builds mathematical models correlating chemical structure with biological activity.
- ***Molecular Dynamics Simulations (MDS)***: Analyze the dynamic behavior of molecules over time, considering the biological environment.
- ***Machine learning and AI***: Increasingly integrated to predict activity, toxicity, and pharmacokinetic properties.

Structure-Based Drug Design (SBDD) and Ligand-Based Drug Design (LBDD) represent two complementary strategies within Computer-Aided Drug Design (CADD). While distinct in methodology, they are often used in parallel to accelerate and refine the discovery of bioactive compounds. SBDD relies on detailed knowledge of the three-dimensional structure of a biological target, typically a protein. When this structure is available—through techniques such as X-ray crystallography or NMR spectroscopy—CADD tools can model the interaction between the protein and candidate ligands, evaluating parameters such as binding stability and affinity. This approach, often described as fitting a "key into a molecular lock," focuses primarily on the target structure, using ligands (either known or newly designed) to explore and optimize binding interactions. The main objective is to predict how chemical modifications affect the strength and specificity of protein–ligand binding, thus guiding rational drug design. LBDD, by contrast, is employed when the three-dimensional structure of the target protein is unavailable but ligands that bind to it are known. In this case, CADD leverages the physicochemical and biological data of these ligands to design new molecules with similar or improved properties.

This indirect yet powerful approach uses statistical and machine learning models to predict chemical–biological interactions, extrapolating from known ligand datasets to identify novel compounds with therapeutic potential.

In practice, most pharmaceutical discovery programs combine both approaches, adapting the strategy to the type of information available and the stage of development. Recent studies estimate that more than six billion small molecules are now available for biological testing against disease-relevant targets (Enamine 2025), vastly expanding the accessible chemical space beyond traditional Lipinski constraints. Public repositories such as ChEMBL (EMBL–EBI. 2025) currently catalog biological data for approximately 2.3 million compounds—still only a fraction of this chemical universe, yet essential for CADD-assisted drug discovery.

The traditional process of synthesizing and experimentally screening thousands of molecules prior to clinical evaluation has progressively shifted toward targeted *in silico* discovery. Advances in computational hardware (CPU, GPU, and storage performance) and algorithmic development now enable more accurate predictions of molecular candidates most likely to modulate disease-associated proteins, thereby streamlining early-stage drug development. Despite these advances, CADD still faces several challenges. The massive increase in available molecular data has underscored the need for efficient data integration, standardization, and automated preprocessing across multiple software platforms. Each tool typically requires specific input/output formats and specialized expertise, posing difficulties for interdisciplinary teams.

To address these limitations, DataPype® (Khan et al. 2023) has been developed as an integrated platform that unifies various CADD packages into a single automated workflow for virtual screening. It allows researchers to combine multiple analysis modules within the same pipeline—if different algorithms converge on the same set of compounds, confidence in their potential activity increases. Moreover, its ability to run on dedicated servers accelerates large-scale virtual screening and parallel computation. By integrating diverse CADD tools through a single interface, DataPype® enhances usability, reproducibility, and computational efficiency in the early stages of drug discovery.

Regarding the SBDD/LBDD split, some of the most well-known and widely used CADD suites and related molecular biology tools include:

1. **Commercial Platforms for SBDD**
   - Schrödinger Suite®, which includes: Maestro® (a modeling environment for molecular discovery), Glide® (for molecular docking), Prime® (for protein modeling), and Desmond® (a GPU-powered high-performance molecular dynamics (MD) engine). (https://www.schrodinger.com/platform/).
   - MOE® (Molecular Operating Environment), an integrated tool for docking, molecular dynamics, and computational modeling for small molecules, peptides, and biologics. (https://www.chemcomp.com/en/Products.html).
   - BIOVIA Discovery Studio®, a platform by Dassault Systèmes® for molecular modeling and simulation. (https://www.3ds.com/it/prodotti-e-servizi/biovia/).
   - OpenEye Scientific Software®, which includes tools such as FRED® (molecular docking) and ROCS®, a powerful virtual screening tool that can rapidly identify

potentially active compounds by shape comparison. (https://www.eyesopen.com/solutions).

2. **Open Source and Academic Platforms**
   - AutoDock® and AutoDock Vina®, widely used software for molecular docking in SBDD studies. (https://vina.scripps.edu/).
   - PyRx - Python Prescription Virtual Screening Tools, a user-friendly interface integrating CADD tools such as AutoDock® for molecular docking. (https://pyrx.sourceforge.io/).
   - SwissDock®, a web service based on Attracting Cavities2.0® and AutoDock Vina® for predicting drug-target interactions. (https://www.swissdock.ch/).
   - DOCK 6®, a tool developed by the University of California for flexible docking. (https://dock.compbio.ucsf.edu/DOCK_6/index.htm).
   - GNINA®, an evolution of AutoDock Vina® with deep learning for molecular docking. (https://github.com/gnina/gnina).
   - AlphaFold 2® + ColabFold®, while not specifically designed for drug design, these platforms are used to obtain high-quality protein structures relevant for SBDD. (https://colab.research.google.com/github/sokrypton/ColabFold/blob/main/AlphaFold2.ipynb).
3. **Cloud-Based and AI-Driven Platforms**
   - Insilico Medicine®, which leverages deep learning, neural networks, and generative AI software for drug discovery, scientific research, and sustainability. (https://insilico.com/).
   - Atomwise (AtomNet)®, an AI-driven platform for analyzing drug-protein interactions. (https://www.atomwise.com/how-we-do-it/).
   - AlphaFold® + UniProt®, combining structural prediction with biological databases to support SBDD. (https://www.uniprot.org/database/DB-0262).

Applications of SBDD

- Structure-based drug design enables the development of drugs when the target protein structure is known.
- Optimization of existing ligands to improve their affinity and selectivity for the protein's active site.
- Molecular docking to test the binding capability of candidate compounds with a target protein.
- Molecular dynamics simulations to analyze drug-protein interactions over time.
- Molecular interaction studies to predict binding stability and enhance drug specificity.
- Design of allosteric inhibitors that modulate protein activity by binding to non-active sites.
- Prediction of drug bioavailability and efficacy using advanced structural models.

LBDD vs. SBDD

While SBDD relies on target structure data, LBDD (Ligand-Based Drug Design) focuses on the chemical and physicochemical properties of known ligands. Some platforms, such as BIOVIA Discovery Studio®, MOE®, and FRED®, integrate both approaches, allowing researchers to work with known ligand structures (LBDD) or precise protein structural information (SBDD). These platforms provide a comprehensive analysis, leveraging the benefits of both methods to optimize drug design. In general, SBDD suites focus on protein structure modeling and molecular docking, while LBDD suites analyze known ligands to predict and design new bioactive molecules.

1. **Commercial LBDD Platforms**
   - Schrödinger Suite® uses Phase®, which specializes in pharmacophore modeling, and Canvas®, a computational chemistry toolkit that, since 2023, is now integrated into the Maestro® application and the LiveDesign® platform.[4] (https://www.schrodinger.com/platform/products/phase/) (https://www.schrodinger.com/platform/products/livedesign/)
   - BIOVIA Discovery Studio® includes tools for QSAR (Quantitative Structure-Activity Relationship), pharmacophores, and virtual screening. (https://www.3ds.com/products/biovia/discovery-studio/qsar-admet-predictive-toxicology)
   - MOE® (Molecular Operating Environment) offers molecular similarity methods, docking, QSAR, and pharmacophore modeling. (https://www.chemcomp.com/en/Products.html)
   - LigandScout®, an advanced software for pharmacophore model generation and validation. (https://docs.inteligand.com/ligandscout/)
   - OpenEye Scientific Software® provides tools such as ROCS® (molecular similarity) and FastROCS® for identifying compounds with similar activity. (https://www.eyesopen.com/fastrocs)

[4] This software is designed to identify and optimize new bioactive molecules based on known ones, without requiring detailed information about the target protein structure. The approach of Phase® focuses on building pharmacophore models, which are abstract representations of the essential chemical features that a ligand must possess to interact with a biological target. Using these models, the software can perform virtual screenings to identify new molecules with potential activity similar to that of known compounds. This methodology is typical of LBDD, which relies on the analysis and comparison of known molecular structures to predict, design, and prioritize new drug candidates with comparable or enhanced biological activity. Building on this principle, LiveDesign® provides a flexible, cloud-native working environment that connects the entire discovery team across both small- and large-molecule research. It streamlines workflows through centralized access to experimental data and *in silico* predictions, integrating advanced computational modeling tools with real-time collaborative decision-making—all within a single, unified interface.

2. **Open Source and Academic Software**
    - RDKit®, a Python toolkit for computational chemistry, used for virtual screening, QSAR, and similarity analysis. (https://www.rdkit.org/)
    - DeepChem®, an advanced machine learning platform for Ligand-Based Drug Design (LBDD), with some features supporting Structure-Based Drug Design (SBDD). (https://deepchem.io/)
    - SwissSimilarity®, an online platform for molecular similarity-based virtual screening. (https://www.expasy.org/resources/swisssimilarity)
    - Chemoinformatics Toolkits (CDK, Open Babel), used for molecular manipulation and analysis. (https://parssilico.com/blogs/112-exploring-top-open-source-toolkits-for-cheminformatics-rdkit-open-babel-and-cdk)
3. **AI-Based and Cloud-Based Platforms**
    - Insilico Medicine®, leveraging neural networks to generate new chemical structures with desired biological activity. (https://insilico.com/)
    - Atomwise (AtomNet), using deep learning to find compounds similar to existing drugs. (https://www.atomwise.com/publications/atomnet-paper/)
    - Cyclica (Ligand Express), an AI-driven approach for polypharmacology profiling. (https://www.pharmaceutical-technology.com/analyst-comment/cyclica-ai-drug-discovery-platform/)
    - IBM RXN for Chemistry, an AI platform for predicting the chemical synthesis of new ligands. (https://rxn.app.accelerate.science/rxn/sign-in

Both methods leverage computational power to analyze thousands of potential compounds within minutes, selecting the most promising candidates for laboratory testing. This has not only led to reduced development time and costs but has also opened new therapeutic opportunities in complex fields such as rare diseases and treatment-resistant cancers. Protein folding prediction is not limited to medicine; it is also applicable to the creation of new biological materials. For example, proteins can be used to develop sustainable biomaterials, with applications ranging from regenerative medicine to tissue engineering. In summary, the protein folding problem is central to life sciences and modern medicine, as it directly influences disease understanding, drug design, biotechnological innovations, and the rise of personalized medicine. The ability to accurately predict how proteins fold not only accelerates scientific research but also holds the promise of significantly improving human health and quality of life. Machine learning is undoubtedly revolutionizing CADD methodologies in a very short time, significantly improving the efficiency and accuracy of key steps such as hit identification, lead optimization, and toxicity prediction (Dr. Omics Education 2025). Unlike traditional methods, machine learning models can process vast amounts of data, learning complex patterns that determine the bioactivity, physicochemical properties, and toxicity of drug candidates. Advanced techniques such as Deep Neural Networks (DNNs), Support Vector Machines

(SVMs), and Random Forest Models (RFMs) enable a more refined prediction of molecular behaviors, making the drug design process faster and more precise.[5] As analyzed, several AI-based tools are now available, enhancing drug discovery. One crucial aspect of drug design is molecular docking, which allows researchers to predict how a drug binds to its target. Machine learning has improved this phase by enabling more accurate predictions of molecular interactions, thanks to Convolutional Neural Networks (CNNs) and Recurrent Neural Networks (RNNs).[6] These models go beyond traditional scoring functions, identifying complex molecular interactions that can determine a drug's efficacy.[7] Predictive modeling is another field where machine learning has had a significant impact. Machine learning algorithms analyze

---

[5] DNNs (Deep Neural Networks) are a type of artificial neural network that contains multiple hidden layers between the input and output. Each layer is composed of artificial neurons that take in data and process it to extract increasingly complex features. DNNs are used for complex problems such as image recognition, natural language processing, and prediction. The depth of the network refers to the number of hidden layers, which allows it to model highly intricate relationships within the data.

SVMs (Support Vector Machines) are a class of machine learning algorithms primarily used for classification and regression. The goal of an SVM is to find a hyperplane that separates data from different classes with the maximum possible margin, meaning the greatest distance between the separation boundary and the nearest points of each class, known as support vectors. This approach is highly effective for complex classification problems, especially when data is not linearly separable, thanks to the use of kernels that map the data into higher-dimensional spaces.

RFM (Random Forest Model) is an ensemble of decision trees used for classification and regression. Each decision tree in the forest is trained on a random subset of the training data and a random subset of features. The final output of a random forest is a combination of predictions from all the trees. Random forests are robust and tend to be highly accurate, reducing the risk of overfitting compared to individual decision trees. This approach is particularly useful for handling large and complex datasets.

[6] CNNs (Convolutional Neural Networks) are a type of neural network primarily designed for processing grid-structured data, such as images. CNNs use an operation called convolution, which applies a filter (or kernel) to the input data to extract local features such as edges, corners, or textures. These filters are applied repeatedly across different parts of the image, creating activation maps that highlight important characteristics. CNNs are particularly effective for image recognition, visual classification, and other computer vision applications due to their ability to capture spatial relationships between pixels.

RNNs (Recurrent Neural Networks) are a type of neural network designed to work with sequential data, such as text, speech, or time-series data. The key feature of RNNs is their recurrent connections, which allow information to be fed back into the network, influencing computations in later stages. This enables RNNs to maintain a form of memory, allowing them to capture temporal or sequential dependencies in data. RNNs are commonly used in Natural Language Processing (NLP), machine translation, text generation, and time-series analysis. However, traditional RNNs suffer from issues like vanishing gradients, which has led to the development of more advanced variants, such as LSTMs (Long Short-Term Memory) and GRUs (Gated Recurrent Units). These architectures help mitigate the vanishing gradient problem and enhance the ability to store long-term dependencies.

[7] In the specific context of drug discovery and molecular docking, scoring refers to the process of assigning a score that quantifies the *compatibility* or *affinity* of a molecule (such as a drug candidate) with its target (e.g., a protein). In practice, the score reflects the likelihood that a molecule will correctly bind to the target, with the idea that a higher score indicates a greater affinity. Scoring functions are used in molecular docking software to assess how molecules fit into the active site of a protein and to predict which molecule might be the most effective in binding and producing the desired therapeutic effect. The scoring function is thus a mathematical formula or algorithm that calculates this score based on various factors, such as binding energy, the geometry of the molecule, and other chemical interactions.

large datasets to predict the ADMET properties (absorption, distribution, metabolism, excretion, and toxicity) of drug candidates, reducing failures in the final stages of development. The introduction of generative adversarial networks (GANs) allows for the generation of novel molecular structures with optimized therapeutic potential. Machine learning also underpins virtual screening, a crucial step in drug discovery. With machine learning, virtual screening is no longer limited to physicochemical descriptors but employs ligand-based and structure-based approaches to classify and select drug candidates more efficiently. Algorithms such as QSAR models and graph-based neural networks further refine screening methods, reducing time and costs associated with experimental testing. Moreover, an increasing number of drug discovery startups are using AI to innovate pharmaceutical research. Companies such as BenevolentAI®, Insilico Medicine®, Exscientia®, and Recursion Pharmaceuticals® leverage advanced machine learning techniques to design new drugs, optimize clinical trials, and enhance precision medicine. These companies use generative models, reinforcement learning, and neural networks to create new chemical entities, pushing the boundaries of traditional pharmaceutical discovery. This sector is experiencing significant growth and expansion, with thousands of research projects and experiments being conducted worldwide and hundreds of dedicated software suites and programs (Sabe et al. 2021). It represents one of the most attractive business trends for the future of the biochemical and biomedical economy (Kasoju et al. 2023). The global terminal cover market is experiencing significant growth, with a projected market value reaching approximately $2.5 billion by 2028 (Market Competitive Research 2024).

In conclusion, here, in Table 1, that integrates the previous list of CADD suites, applications and data-driven tools in chemistry and synthetic biology:[8]

[8] Molecular Docking: Molecular docking refers to the computational technique used to predict the position and orientation of small molecules (ligands) within the active sites of target proteins. This process is pivotal in the field of drug discovery, as it helps identify how different molecules might interact with biological targets. Popular software tools for molecular docking include AutoDock®, FlexX®, RoseTTA®, and PyRx®.

Molecular Dynamics Simulations: Molecular dynamics (MD) simulations are computational methods used to simulate the atomic movements of molecules and analyze their interactions over time. These simulations provide insights into the conformational changes, stability, and energetics of molecular systems, making them essential tools in drug design and molecular research. Popular software tools for molecular dynamics simulations include AMBER®, GROMACS®, and CHARMM®.

Molecular Modeling and Virtual Screening: Software like MOE®, Sybyl®, and Discovery Studio® assist in 3D molecular modeling and selecting promising compounds through virtual screening.

ADMET Property Prediction: Software that predicts the biocompatibility of a molecule, its toxicity, absorption, and metabolism (e.g., ADMET Predictor®, QikProp®) (https://www.simulations-plus.com/software/admetpredictor/).

Biomolecular Interaction Analysis: Some tools, like ZDOCK® and STING®, focus on the analysis of interactions between proteins or between proteins and ligands; Pierce, B., Hourai, Y., & Deng, Z. (2011). Accelerating protein docking in ZDOCK using an advanced 3D convolution library. BMC Bioinformatics, 12, 348. DOI: 10.1371/journal.pone.0024657).

Table 1. List of cadd suites and their features (alphabetical order).

Software	Description	Link
ADMET Predictor®	Software for predicting the drug behavior in terms of absorption, distribution, metabolism, excretion, and toxicity (ADMET).	https://www.simulations-plus.com/software/admetpredictor/
AlphaFold®	Deep learning system developed by DeepMind for predicting protein 3D structures with high accuracy from amino acid sequences.	https://www.deepmind.com/research/highlighted-research/alphafold
AMBER®	Software for molecular dynamics simulation, used to analyze the structure and behavior of proteins and small molecules.	https://ambermd.org/
AutoDock / AutoDock Vina	Open-source docking program for predicting ligand-receptor binding.	http://vina.scripps.edu/
Biovia Pipeline Pilot®	A platform that integrates and analyzes chemical and biological data, useful in drug discovery and pharmaceutical design.	https://www.3ds.com/products/biovia/pipeline-pilot
Cerius2®	Molecular modeling software that provides tools for material design and molecular dynamics simulation.	https://www-jmg.ch.cam.ac.uk/cil/SGTL/cerius2.html
CHARMM®	Suite for molecular simulation, widely used for the study of biological macromolecules.	https://academiccharmm.org/
Chem3D®	3D molecular modeling software, useful for molecule design and visualization in the pharmaceutical field.	https://library.bath.ac.uk/chemistry-software/chem3d
ChemAxon®	Includes various tools for computational chemistry, modeling, and analysis of chemical data.	https://chemaxon.com/
ChemDraw®	Software suite for drawing chemical structures, used in molecule and drug design.	https://revvitysignals.com/products/research/chemdraw
ChemSpider®	Online chemical database offering information on chemical compounds, useful in drug design.	https://www.chemspider.com/
Cresset®	Software for drug design and analysis with a focus on 3D molecular interaction modeling.	https://cresset-group.com/
DeepChem	Open-source library for applying deep learning to drug discovery.	https://deepchem.io/
Discovery Studio – BIOVIA	Structure-based design, QSAR, ADMET prediction, and pharmacophore generation.	https://www.3ds.com/products-services/biovia/products/molecular-modeling-simulation/biovia-discovery-studio/

Table 1 contd. ...

...Table 1 contd.

Software	Description	Link
FlexX®	Molecular docking software that predicts the position and orientation of small molecules in the active sites of proteins.	https://www.biosolveit.de/wp-content/uploads/2021/01/FlexX.pdf
GOLD – CCDC	High-precision docking tool with genetic algorithm-based search.	https://www.ccdc.cam.ac.uk/solutions/csd-discovery/components/gold/
Gaussian®	Software for electronic structure modeling, useful in quantum chemical calculations.	https://gaussian.com/
GROMACS®	Open-source software for molecular dynamics simulation, widely used in studying interactions between biomolecules.	https://www.gromacs.org/
KNIME®	Open-source platform for data analysis, used in drug design to process and visualize data related to chemical compounds and bioactivity.	https://www.knime.com/
LeadFinder™	An algorithm designed to rapidly dock and score ligands using a method optimized to separate actives from inactives.	https://cresset-group.com/software/lead-finder/
LigandScout	Pharmacophore modeling, 3D alignment, and virtual screening.	https://www.inteligand.com/ligandscout/
MOE (Molecular Operating Environment)	Ligand design, QSAR, molecular modeling, docking, and protein-ligand interaction analysis.	https://www.chemcomp.com/MOE-Molecular_Operating_Environment.htm
Open Babel®	Open-source chemical toolbox suite for converting chemical file formats and manipulating chemical structures.	https://openbabel.org/
OpenEye Orion Platform	Cloud-based toolkit for virtual screening, conformer generation, and molecular property prediction.	https://www.eyesopen.com/orion
PyMOL®	Open-source software for visualizing and manipulating molecular structures.	https://www.pymol.org/
Rosetta®	Open-source suite for molecular design, including protein structure prediction and ligand design.	https://rosettacommons.org/software/
Schrödinger Suite	Comprehensive platform for molecular modeling, docking, pharmacophore modeling, and quantum mechanics.	https://www.schrodinger.com/
SimBioSys®	Visualization and diagnostic technology for delivering precision answers to critical clinical questions.	https://simbiosys.com/
UCSF Chimera®	Molecular visualization tool for manipulating and analyzing protein structures in drug design.	https://www.cgl.ucsf.edu/chimera/

References

Abouelwafa, M., Ibrahim, T. M., El-Hadidi, M. S., Mahnashi, M. H., Owaidah, A. Y., Saeedi. et al. (2023). Using CADD tools to inhibit the overexpressed genes FAP, FN1, and MMP1 by repurposing ginsenoside C and Rg1 as a treatment for oral cancer. *Frontiers in Molecular Biosciences, 10*, 1248885.

Abramson, J., Adler, J., Dunger, J., Evans, R., Green, T., Pritzel, A. et al. (2024) .Accurate structure prediction of biomolecular interactions with AlphaFold 3. Nature, 630: 493–500.

Abudayyeh, O. O., and Gootenberg, J. S. (2024). Programmable biology through artificial intelligence: From nucleic acids to proteins to cells. Nature Methods, 21: 1384–1386.

Adams, D. (1993). The Hitchhiker's Guide to the Galaxy: A Trilogy in Four Parts. Internet Archive. Available at: https://archive.org/details/hitchhikersguide0000adam_h3v6 [Accessed: 12 May 2025].

Ajjarapu, S. (2023). Computer-Aided Drug Design (CADD). Microbe Notes. Available at: https://microbenotes.com/computer-aided-drug-design-cadd/ [Accessed: 13 May 2025].

AlphaFold Protein Structure Database. (2025). Home. Available at: https://alphafold.ebi.ac.uk/ [Accessed: 13 May 2025].

Bajrektarevic, A., and Bogdanova, K. (2025). Biohacking: Key Legal, Moral and Commercial Considerations – Analysis. Eurasia Review. Available at: https://www.eurasiareview.com/24012025-biohacking-key-legal-moral-and-commercial-considerations-analysis/ [Accessed: 13 May 2025].

Bundell, S. (2019). The Hitchhiker's Guide to the Galaxy: 40 years of parody and predictions. Nature. Available at: https://www.nature.com/articles/d41586-019-02969-8 [Accessed: 12 May 2025].

Choudhry, A. (2024). RoseTTA Fold: Understanding the paper. Medium. Available at: https://medium.com/@adachoudhry26/rosetta-fold-understanding-the-paper-49d27fc387c7 [Accessed: 12 May 2025].

Dean, J., Hassabis, D., and Manyika, J. (2023) 2023: A Year of Groundbreaking Advances in AI and Computing. Google DeepMind. Available at: https://deepmind.google/discover/blog/2023-a-year-of-groundbreaking-advances-in-ai-and-computing/ [Accessed: 12 May 2025].

Desai, D., Kantliwala, S. V., Vybhavi, J., Ravi, R., Patel, H., and Patel, J. (2024). Review of AlphaFold 3: Transformative advances in drug design and therapeutics. Cureus, 16(7), e63646.

Dirox. (2025). Microsoft Unveils Majorana 1: The World's First Quantum Processor with Topological Qubits. Dirox. Available at: https://dirox.com/post/microsoft-majorana-1 [Accessed: 13 May 2025].

Domainex. (2025). Computational Chemistry | Computer Aided Drug Design. Domainex. Available at: https://www.domainex.co.uk/services/computational-chemistry [Accessed: 13 May 2025].

Dr. Omics Education. (n.d.). Ethical considerations in bioinformatics: Navigating the complex landscape. Dromicsedu. Available at: https://dromicsedu.com/blog-details/ethical%20considerations%20in%20bioinformatics%20navigating%20the%20complex%20landscape [Accessed: 13 May 2025].

Dr. Omics Labs. (2024). The Future of Computer-Aided Drug Design (CADD): Predictions and Challenges. Available at: https://dromicslabs.com/computer-aided-drug-design-cadd/ [Accessed: 13 May 2025].

Dr. Omics Education. (2025). Machine Learning in CADD: A Game Changer for Drug Design. Dr. Omics Education. Available at: https://dromicsedu.com/blog-details/machine-learning-in-cadd-a-game-changer-for-drug-design [Accessed: 13 May 2025].

EMBL-EBI. (2025). ChEMBL – A database of bioactive drug-like small molecules. European Bioinformatics Institute. Available at: https://www.ebi.ac.uk/chembl/ [Accessed: 13 May 2025].

Enamine. (2025). REAL Database – The largest enumerated database of synthetically feasible molecules. Enamine. Available at: https://enamine.net/compound-collections/real-compounds/real-database [Accessed: 13 May 2025].

Fagan, M. B., and Maienschein, J. (2022). Theories of Biological Development. Stanford Encyclopedia of Philosophy. Available at: https://plato.stanford.edu/entries/theories-biological-development/ [Accessed: 13 May 2025].

Frappier, V., and Keating, A. E. (2021). Data-driven computational protein design. Current Opinion in Structural Biology, 69: 63–69.

Gignac, G. E., and Szodorai, E. T. (2024). Defining intelligence: Bridging the gap between human and artificial perspectives. Intelligence, 104: 101832.

Google DeepMind. (n.d.). AlphaGo. Available at: https://deepmind.google/research/breakthroughs/alphago/ [Accessed: 12 May 2025].

Google DeepMind. (2024). Google DeepMind and Isomorphic Labs introduce AlphaFold 3 AI model. Google Blog. Available at: https://blog.google/technology/ai/google-deepmind-isomorphic-alphafold-3-ai-model/ [Accessed: 13 May 2025].

Google DeepMind. (2025). Google DeepMind. Available at: https://deepmind.google/ [Accessed: 12 May 2025].

Hassabis, D. (2022a). Watch now: DeepMind Founder's lecture for the Institute for Ethics in AI. University of Oxford – Humanities Division. Available at: https://www.humanities.ox.ac.uk/article/watch-now-deepmind-founders-lecture-for-the-institute-for-ethics-in-ai [Accessed: 13 May 2025].

Hassabis, D. (2022b). Demis Hassabis: Breaking the Wall to Protein Structure Prediction. YouTube. Available at: https://youtu.be/pxsCelsTIy0 [Accessed: 12 May 2025].

Kasoju, N., Remya, N. S., Sasi, R., Sujesh, S., Soman, B., Kesavadas, C. et al. (2023). Digital health: Trends, opportunities and challenges in medical devices, pharma, and biotechnology. *CSIT*, 11(1): 11–30. https://doi.org/10.1007/s40012-023-00380-3.

Kendre, P. P., Kosalaraman, K. K., Jayasree, S. S. K., Rajan, S., Jayan, A., and Muthuganapathy, R. (2023). SketchCADGAN: A generative approach for completing partially drawn query sketches of engineering shapes to enhance retrieval system performance. Computers & Graphics, 115: 55–68.

Keşkek Karabulut, Y., and Turhan, Z. Ş. (2023). Brief overview of computer-aided drug design (CADD) and commercial product examples of CADD2. pp. 1553. *In:* Yiğit, A. (ed.). Nanotechnology and organic chemistry. Özgür Publications.

Khan, M. F., Kandwal, S., and Fayne, D. (2023). DataPype: A fully automated unified software platform for computer-aided drug design. ACS Omega, 8(42): 39468–39480.

Latson, J. (2015). Did Deep Blue Beat Kasparov Because of a System Glitch? TIME. Available at: https://time.com/3705316/deep-blue-kasparov/ [Accessed: 12 May 2025].

Market Competitive Research. (2024). Market Size and Opportunity Outlook for the Computer-Aided Drug Discovery (CADD) Market: Key Segments and Emerging Markets. LinkedIn. Available at:https://www.linkedin.com/pulse/market-size-opportunity-outlook-computer-aided-592pc/ [Accessed: 12 May 2025].

Merino-Casallo, F., Gomez-Benito, M. J., Hervas-Raluy, S., and Garcia-Aznar, J. M. (2022). Unravelling cell migration: Defining movement from the cell surface. Cell Adhesion & Migration, 16(1): 25–64.

Microbe Notes. (n.d.). Computer-Aided Drug Design (CADD). Microbe Notes. Available at: https://microbenotes.com/computer-aided-drug-design-cadd/ [Accessed: 13 May 2025].

Microsoft. (2025). Microsoft unveils Majorana 1: The world's first quantum processor powered by topological qubits. Azure Blog. Available at: https://azure.microsoft.com/en-us/blog/quantum/2025/02/19/microsoft-unveils-majorana-1-the-worlds-first-quantum-processor-powered-by-topological-qubits/ [Accessed: 13 May 2025].

MIT News. (2023). AI system can generate novel proteins with specific structural design. MIT News. Available at: https://news.mit.edu/2023/ai-system-can-generate-novel-proteins-structural-design-0420 [Accessed: 13 May 2025].

Musthafa, N., Walker, H.-S., and Domagala, M. (2024). Computational modelling and simulation of scaffolds for bone tissue engineering. Computation, 12(4): 74.

Niazi, S. K., and Mariam, Z. (2023). Computer-aided drug design and drug discovery: A prospective analysis. Pharmaceuticals (Basel), 17(1): 22.

Nissinen, A. (2022). 42—How The Hitchhiker's Guide to the Galaxy (inadvertently) taught us the meaning of life. Medium. Available at: https://medium.com/@happyondemand/42-how-the-hitchhikers-guide-to-the-galaxy-inadvertently-taught-us-the-meaning-of-life-43f46c29b730 [Accessed: 12 May 2025].

Perrigo B. (2023). Demis Hassabis Believes AI Could Unlock the Secrets of the Universe. TIME. Available at: https://time.com/6246119/demis-hassabis-deepmind-interview/ [Accessed: 12 May 2025].

Persson, I., and Savulescu, J. (2012). Unfit for the future: The need for moral enhancement. Oxford University Press.

Pierce, B. G., Hourai, Y., and Weng, Z. (2011) Accelerating protein docking in ZDOCK using an advanced 3D convolution library. PLoS One, 6(9): e24657. [Epub 19 September 2011]

Prabhune, M. (2024). The Journey to a Nobel Prize: A Protein Design and Structure Research Timeline. The Scientist. Available at: https://www.the-scientist.com/the-journey-to-a-nobel-prize-a-protein-design-and-structure-research-timeline-72256 [Accessed: 12 May 2025].

Protein Structure Prediction Center. (2024). *Home*. Available at: https://predictioncenter.org/ [Accessed: 13 May 2025].

Ramachandran, R. (2024). Majorana-Based Quantum Computing: Theoretical Advances and Future Prospects. LinkedIn. Available at: https://www.linkedin.com/pulse/majorana-based-quantum-computing-theoretical-advances-ramachandran-6zsce [Accessed: 13 May 2025].

Ramadurai, K. (2023). The New Era of Programmable Biology. MedCity News. Available at: https://medcitynews.com/2023/10/the-new-era-of-programmable-biology/ [Accessed: 13 May 2025].

Rijssenbeek, J. (2025). Synthetic biology: Supporting an anti-reductionist view of life. Synthese, 205(2): 52.

Rossi Castelli, P. (2021) Cells are good at using 'collective intelligence'. IBSA Foundation. Available at: https://www.ibsafoundation.org/en/blog/cells-are-good-at-using-collective-intelligence [Accessed: 1 October 2025].

RW2 Education. (2023). CADD and Sustainability: Using Technology to Create Eco-Friendly Designs. RW2 Education. Available at: https://rw2.education/cadd-and-sustainability-using-technology-to-create-eco-friendly-designs/ [Accessed: 13 May 2025].

Sabe, V. T., Ntombela, T., Jhamba, L. A., Maguire, G. E. M., Govender, T., Naicker, T. et al. (2021). Current trends in computer-aided drug design and a highlight of drugs discovered via computational techniques: A review. European Journal of Medicinal Chemistry, 224: 113705.

Schenker, J. L. (2020). Google DeepMind unveils next generation of AI drug discovery model. The Innovator. Available at: https://theinnovator.news/google-deepmind-unveils-next-generation-of-ai-drug-discovery-model/[Accessed: 13 May 2025].

Schubach, M., Maass, T., Nazaretyan, L., Röner, S., and Kircher, M. (2024). CADD v1.7: Using protein language models, regulatory CNNs, and other nucleotide-level scores to improve genome-wide variant predictions. Nucleic Acids Research, 52(D1), D1143–D1154.

Scott, A. O. (2023). Are Fears of A.I. and Nuclear Apocalypse Keeping You Up? Blame Prometheus. The New York Times. Available at: https://www.nytimes.com/2023/09/29/books/review/benjamin-labatut-the-maniac.html [Accessed: 12 May 2025].

Su, J., Song, Y., Zhu, Z., Huang, X., Fan, J., Qiao, J. et al. (2024). Cell–cell communication: New insights and clinical implications. Signal Transduction and Targeted Therapy, 9: 196.

Talevi, A. (2024). Computer-aided drug discovery and design: Recent advances and future prospects. Methods in Molecular Biology, 2714: 1–20.

The Scientist. (2024). Insights into protein engineering: Methods and applications. The Scientist. Available at: https://www.the-scientist.com/insights-into-protein-engineering-methods-and-applications-72286 [Accessed: 13 May 2025].

uMake. (2025). Generative Adversarial Networks (GANs) in CAD. uMake Design Glossary. Available at: https://www.umake.com/glossary/generative-adversarial-networks-(gans)-in-cad [Accessed: 13 May 2025].

University of Oxford – Institute for Ethics in AI (2025.) Home. Available at: https://www.oxford-aiethics.ox.ac.uk/ [Accessed: 13 May 2025].

Wei, H., and McCammon, J. A. (2024). Structure and dynamics in drug discovery. npj Drug Discovery, 1(1).

Workman, P. (2021). Reflecting on DeepMind's AlphaFold artificial intelligence success – what's the real significance for protein folding research and drug discovery? The Institute of Cancer Research. Available at: https://www.icr.ac.uk/research-and-discoveries/cancer-blogs/detail/the-drug-discoverer/reflecting-on-deepmind-s-alphafold-artificial-intelligence-success-what-s-the-real-significance-for-protein-folding-research-and-drug-discovery [Accessed: 13 May 2025].

Zeng, L., Pernet, A., Cléroux, M., Bach, B., Froidevaux, L., and Preda, I. (2023). Development of distillation sensors for spirit beverages production monitoring based on impedance spectroscopy measurement and partial least-squares regression. ACS Omega, 8: 1–9.

10

Make a Post-Cad Transition

From Automation Toward AI Co-creation

"Fear is the main source of superstition,
and one of the main sources of cruelty.
To conquer fear is the beginning of wisdom."
Bertrand Russell

10.1 Beyond Automation: The Cognitive Turn in Design

At the heart of every human endeavor lies the shadow of unpredictability. The history of technology—like that of civilization—is marked by a persistent desire for control, tempered by the awareness that every forecast stands on shifting sands (Park et al. 2023). In the face of the exponential rise of artificial intelligence and biotechnologies, we are no longer merely choosing a destiny; rather, we are being called to recognize the full weight of responsibility in the choices we make. Given the inherent complexity of the many factors influencing the evolution of the *artificial arena*, it is exceedingly difficult—if not impossible—to make accurate predictions about how these domains will develop over the coming decades. The variables involved range from scientific and technological advancements to the economic and social dynamics of human communities, from cultural shifts to ethical implications. All of these elements are converging into a deeply interconnected and evolving organism, where each innovation can trigger unpredictable ripple effects. Throughout this publication, the different scenarios have been constantly explored, being drawn by thinkers, opinion leaders, and futurists—each one positioned within a spectrum of potential historical evolutions that range between two extremes. On one end lies utopian optimism, envisioning a harmonious future where technology substantially improves human life (Flexner 2024). On the opposite end, dystopian catastrophism gathers ominous clouds, foreseeing futures in which technological progress escapes human control and unleashes devastating social, environmental, and economic consequences

DOI: 10.1201/9781003438946-10

(Malvestio 2022). Proponents of utopian optimism tend to imagine a world in which technological artifacts, rather than threatening individuality, social balance, or the environment, instead align positively with the planet's resources—co-evolving with ecosystems and the principles of natural design. The human *ars* cited by Cicero[1] becomes the catalyst for a collective advancement toward a prosperous and interconnected humanity. In this view, technology, artificial intelligence, and biotechnology take on the role of *deus ex machina*—not only solving the major challenges of our time but also empowering individuals and societies, ushering in a new era of peace, cooperation, and sustainability (Brynjolfsson and McAfee 2014). It is a vision in which inequalities could be reduced through equitable distribution of resources and technological benefits, and where progress is seamlessly integrated with ethical values and environmental stewardship. By contrast, the prophets of dystopian catastrophism predict oppressive scenarios where technology becomes a dark and uncontrollable force—one that exacerbates inequality, undermines individual freedoms, and deepens the divide between those who wield technological power and those who are excluded from it. In these visions, often dramatized in literature and film, artificial intelligence and biotechnology are deployed irresponsibly, giving rise to omnipresent surveillance systems, unethical genetic manipulation, and an increasing dehumanization of work and daily life. In some of these scenarios, our interaction with machines could lead to the annihilation of individual autonomy and the suppression of freedom, imposed by artificial intelligence or a dominant technological elite. The most extreme epilogues predict the total eradication of humanity in a digital holocaust unleashed by machines so superintelligent and autonomous that they come to believe they can replace their very creators. These opposing visions are not unique to our time; rather, they reflect recurring technophobic psychological gradients that have surfaced cyclically throughout human history. Between these extremes lie what we might call hybrid or evolutionary scenarios, where the direction of the future remains uncertain but depends largely on our capacity to respond consciously to the challenges posed by emerging technologies. These intermediate paths, fueled by faith in human reason and resilience, hinge on maintaining a delicate balance between innovation and sustainability, between technological power and human-centered values. While technology holds the potential to dramatically improve life conditions and address pressing global problems, it also demands continuous and broadly shared reflection on its moral, political, and social consequences. The real challenge, then, is not merely to guess which scenario will prevail, but to actively manage and shape the future—to avoid sliding into dystopia while resisting the illusion that utopia can be reached uncritically and without due consideration for the inherent risks and conflicts associated with extreme technologies (Salmon et al. 2024). In this context, our responsibility—as individuals, as societies, and as creators of technology—is to navigate between these gradients, steering away from extremism and striving toward a future that acknowledges both the promise and the perils of technological evolution.

[1] "Nulla ars imitari sollertiam naturae potest"-Cicero - Rhetorica - De Natura Deorum - Liber I – 92 (See reference).

Technological progress is advancing at an unprecedented pace, with innovations intersecting in ways that were unimaginable just a few decades ago. As cited by several authors, the idea of integrating CAD/CAM systems with forms of expert knowledge and artificial intelligence has been proposed since the early 1990s (Yeslam et al. 2024). In the context of new and emerging technologies for industrial competitiveness, a conference held in October 1995 in Kuala Lumpur, Malaysia, analyzed the emergence of knowledge-based systems, capable of incorporating expert knowledge to drastically reduce design times, particularly in the phases of modification and adaptation of existing models (Ng and Poynting 1995). The ICAD system, one of the cases considered at the time, promised reductions of over 50% in development times, drastically cutting design times, particularly in the phases of modifying and adapting models, with a direct impact on corporate competitiveness. This publication foreshadowed the principles now underlying generative and predictive design, paving the way for a new paradigm of co-design between humans and intelligent systems. The process of creating a possible integration between CAD/CAM systems and AI began as early as the 1970s with the creation of LISP machines, a computing concept that, although existing for some time, seemed closer to a practical realization at that time, thanks to new technological developments at MIT (Greenblatt 1974). In the mid-1980s, during a time when artificial intelligence promised to revolutionize the world, ICAD was born: a pioneering knowledge-based engineering (KBE) system. Developed on LISP machines by a group of brilliant engineers, including Larry Rosenfeld, Avrum Belzer, and David Place, ICAD stood out for its ability to transform design knowledge into computational models. It was 1984, and the dream of automating complex design seemed within reach. ICAD allowed engineers to write semantic rules, part hierarchies, and relationships through a declarative language called IDL, based on LISP but enhanced with its own structure. Essentially, it allowed engineers to build engineering worlds made of reusable components, capable of automatically generating entire assemblies like an airplane wing, complete with thousands of rivets, each calculated according to specific constraints. Thanks to its open and flexible architecture, ICAD represented one of the most advanced expressions of *end-user computing*, giving designers unprecedented control over modeling and project generation processes. In the 1990s, companies like Boeing and Airbus made extensive use of it to design complex components in the aerospace sector. Its capabilities appeared almost *magical*: just a few lines of code were enough to generate intricate structures, validate them, and prepare them for production. As illustrated in the context of computer-aided conceptual modeling at the Department of Aeronautics and Astronautics of the Massachusetts Institute of Technology (Kolb 1990), ICAD allowed design knowledge to be represented and manipulated in a highly parametric and scalable way. But as often happens with overly promising technological dreams, ICAD also fell victim to unrealistic expectations. It became part of the wave that led to the *AI winter*—that period when the promises of artificial intelligence collided with the technical and economic reality (History of Data Science 2021). Systems like ICAD, while powerful, were expensive and difficult to integrate into less specialized environments. Companies began to doubt the returns, and enthusiasm gave way to skepticism. ICAD went through several corporate iterations: it started as ICAD Inc., became Concentra in 1995, KTI in 1998, and was finally acquired by Dassault Systèmes® in 2001. There, it met its fate: Dassault, which was pushing its CATIA V5® suite with *Knowledgeware* modules, considered ICAD redundant and discontinued its development. The system, once promising to redefine the way design is done, faded into obscurity. By 2005, version 8.3 was released, but by then its time had passed. The user communities scattered, and the code remained locked behind corporate walls. However, ICAD was not forgotten: projects like Genworks GDL (based on the open-source Gendl kernel) emerged, attempting to inherit and reinterpret its spirit (Genworks International 2019). The story of ICAD is now an emblematic example: it shows how fragile the line between visionary innovation and commercial failure can be. It's both a cautionary tale and a legacy. ICAD marked an era, contributing to defining what knowledge-based design is—and what it could still become. And while new technologies like AlphaFold® and computational design take similar concepts in a contemporary light, the echo of ICAD continues to resonate quietly in the history of computational engineering. The current convergence of artificial intelligence (AI), biotechnology, advanced robotics, nanotechnology, and quantum computing is already reshaping how we understand and design reality (Sudabathula 2025). These technologies—individually and, more crucially, in synergy—are giving rise to new forms of synthesis and hybridization that blur the boundaries between the natural and the artificial, the human and the machine. Cutting-edge domains of technoscientific evolution are generating increasingly esoteric bodies of knowledge—often inaccessible to the general public. Despite the widespread availability of data and information online, decoding it typically requires scientific and academic literacy that remains unequally distributed. And yet, mere understanding is not sufficient to guarantee democratic participation in the decision-making processes dominated by elite groups. The technocratic risk—always latent—remains an enduring challenge in steering the fate of human civilization (Dusek 2012). However, this technological evolution does not occur in a vacuum. It is constantly influenced by external factors. Political decisions, regulatory frameworks for emerging technologies, societal resistance to certain innovations, and the unequal distribution of resources all significantly shape the pace and direction of innovation. While some nations and companies push for rapid adoption and development, others may choose more conservative or regulated approaches—seeking to balance progress with ethical, social, and environmental concerns. Private enterprises, governments, and academic institutions all play pivotal yet often contradictory roles. Their interactions are shaped by economic interests, political ideologies, and ethical considerations—sometimes accelerating development, sometimes slowing or redirecting it. Globalization further amplifies this complexity. Innovations can now spread across continents instantaneously, but so can the technological inequalities and geopolitical tensions tied to control over these transformative tools. Another essential dimension is social and cultural evolution. People, as dynamic agents of change, are not mere passive spectators of technological innovation (Boker et al. 2016). Their acceptance or rejection of new technologies depends on numerous factors: education, perceived risks and benefits, and how well social norms and cultural values adapt. The fear of the unknown, the ethics of artificial intelligence, concerns about the future of work, and restrictions on personal freedoms are all fueling intense public debate.

Ethical concerns around the use of technology and biotechnology are now central to this debate. Issues such as privacy, surveillance, and the responsible design of AI—as well as environmental sustainability and social justice—are gaining increasing relevance. Decisions about how and where to direct the course of technological development are no longer purely technical; they require deep moral and philosophical reflection. In fact, artificial intelligence, biotechnology, and other emerging fields may soon raise ethical and philosophical questions that we are not yet prepared to answer—questions about control over genetic data, bioengineering, the nature of autonomy in artificial intelligences, or the implications of potentially extending human life indefinitely through biotechnological means. While technology may be moving toward greater integration between the biological and the artificial, between humans and machines, the implications of these changes on individual freedom, democracy, social equity, and sustainability remain shrouded in uncertainties and difficulties of prediction. New governance models that emerge will need to adapt swiftly to the challenges posed by emerging technologies (Hughes et al. 2025). Ethical norms must evolve and adjust rapidly to address the dilemmas raised by artificial intelligence and biotechnology, striving to ensure a balance between innovation and social responsibility. Furthermore, the most unpredictable aspect concerns humanity itself. Technological evolution cannot be separated from human behavior, which can take unpredictable trajectories in response to the changes imposed by technology itself. People's reactions to growing automation and artificial intelligence will depend on their ability to adapt to new economic and social dynamics. It is possible that a new form of collective intelligence will emerge, where the interaction between humans and AI can enhance decision-making and collective creativity. However, it cannot be ruled out that, if not properly managed, these changes could lead to greater inequality and a loss of individual autonomy, further dividing society. Social structures will need to evolve to address these challenges, adapting public policies and labor dynamics to ensure that the benefits of automation and AI are equitably distributed, creating opportunities for growth and protection for all. From the findings of this compendium, which has strived to be as exhaustive as possible in describing the evolutionary elements of the current dynamic, it is clear that in the contemporary context of computer-aided design (CAD), epochal changes are underway that are radically transforming the very conceptual essence of digital tools. The image of the *WunderCAD/CAMmer* has been created and invoked: a wordplay, a metaphor that has roots in the past, capable of evoking a vision and historical depth in a contemporary perspective, one that imbues this complex and stimulating process—inevitably transforming the evolutionary relations between the outdated categories of *naturalia* and *artificialia*—with profound meaning. The era of automation marked an extraordinary achievement for design culture (Verganti et al. 2020). The idea that every process could be made efficient, repeatable, and measurable dominated industrial and digital design in the 20th century. With the advent of CAD, design entered a realm of increasingly precise and simulable representations. Today, at the heart of the post-digital transition, something is changing: design no longer seeks to be merely correct but aims to be surprising (Valtonen 2020). Initially, CAD was limited to a simple digital version of a drafting board, but it has evolved into a sophisticated system that integrates software, hardware, and imaging tools in service of engineering design in all its possible forms. *Post-CAD* is a term that ratifies the current state of this technological evolution, where development no longer merely involves an update or a new protocol version but signifies a cognitive paradigm shift. It is no longer about *commanding* a system but engaging in a dialogue with it. AI is no longer a substitute for human intelligence but a conversational agent, a digital avatar, a presence with which to explore alternatives, generate forms, test hypotheses, and imagine worlds. The act of design becomes co-creative, hybrid, distributed between human intuition and algorithmic training. In this complex and multidimensional scenario, artificial intelligence is not merely a technological tool but an epistemic threshold: it radically changes not only what we design but also how and why we do it. The human-machine dialogue increasingly resembles a circular interaction, where the role of prompting and defining design constraints assumes strategic importance (Seymour 2018). As in Bruno Munari's design method, who in his *fantasia esatta—exact fantasy*—urged not to confuse free invention with fantasy governed by logic, in AI interaction, creativity does not emerge from chaos but from a system of well-calibrated generative rules (Hájek and Panaro 2008). The prompt, far from being merely a command, becomes a meta-linguistic design act: a grammar of intuition that defines frames within which the unexpected can manifest as a generative opportunity (Makatura et al. 2024). What makes AI truly transformative is not so much its ability to predict, but its ability to surprise within formally defined structures. When generative systems return unexpected outputs—a form, a connection, a solution not anticipated—an opening occurs, akin to creative error or lateral intuition. This is where the contemporary designer is called upon to assume a new form of responsibility, no longer tied solely to the design intent, but to the curation of the unpredictable. It is not about confining AI within reassuring limits, but creating conditions where the emergent can be questioned, contextualized, and oriented. From this perspective, design can no longer be limited to simulating reality or predicting behaviors: it increasingly approaches a generative ethics, where the meaning of the project lies also in the ability to embrace what exceeds calculation, to interpret what escapes prediction. This implies a new form of *algorithmic transparency*—not only technical but epistemological—and a renewed reflection on knowledge as a form of ethics (Bitzer et al. 2023). The designer of the future will be called not only to generate forms but to shape visions: to choose, among many possible trajectories, those that embody shared values, which take into account the complexity, fragility, and plurality of the world. The *Post-CAD* transition marks an important threshold. Crossing it means relinquishing the illusion of total control to embrace a grammar of uncertainty, emergence, and augmented intelligence. It is paradoxically a return to an archaic form of design, made of symbols, analogies, and metamorphoses, but also a leap forward into a continuously changing generative knowledge. *Post-CAD* synthesizes the heart of this transition, representing a systemic revolution in traditional digital design techniques. It is not merely about performance improvements (computational speed, stability, technological adaptability), but a radical shift in the ideation and construction processes of human artifacts. The digital industry is imagining new tools and processes that allow designers to easily translate an idea into a physical and functional form. This is giving rise to a dynamic and transversal universe, where the designer constantly interacts with artificial intelligence to explore, co-create, and adapt solutions in real time.

Post-CAD alters the traditional relationship between *problem-setting* and the flow of actions leading to *problem-solving*, enabling the management of data flows and potential variables and alternatives that were unthinkable until now. The methodological process is triggered by the ability to simultaneously assess a complex system of relationships, constraints, and stimuli that evolve throughout the process. In this new scenario, the designer is no longer an isolated author but a curator of cognitive flows who interacts with the machine, which does not merely execute commands but actively contributes to the generation, optimization, and critical evaluation of solutions. This new mode of design approaches natural principles of evolution and adaptation, where the outcome is not imposed from the outside but emerges from a continuous exchange between the forces involved. Nature does not follow a linear process but develops through the dynamic interaction of various factors, adapting to new circumstances. *Post-CAD* emulates this logic of *emergence*: design is no longer deterministic but takes on parametric and generative forms, developing and modifying in real time in response to data, relationships, and variables introduced during the process. In an evolutionary context, artificial intelligence is no longer a mere passive support, responding to algorithmic criteria, but offers contributions that open up different visions and possible innovations, assuming the role of an active actor in the design process. It no longer blindly follows commands but interacts in real time, can be invited to explore new creative solutions, co-creating with the designer to suggest input for problem sets, alternatives, and variables, adapting to changes in the project and stimulating unexpected solutions, leveraging its remarkable computational and processing capabilities. In this new paradigm, the machine is an invaluable resource for data mining, essential for research and sharing, given the vast sea of information and data flooding the network (Taranto-Vera et al. 2021). The designer increasingly finds themselves working alongside the machine, which increasingly resembles a digital *doppelgänger*,[2] developing deep learning dynamics through which the mutual intuitions of the human-machine system are valued, adding critical assessments and alternative ways of thinking and solving to optimize the flow of generated information (Yoo et al. 2021). Thus, the design process transforms into a true collaboration, where artificial and human intelligence are no longer separate entities but complementary. The challenge is no longer limited to solving predefined problems but to creating new scenarios, where the dialogue between technology and intuition leads to solutions that go beyond the individual capabilities of each actor. This dynamic relationship opens the door to a more fluid and innovative design process, which no longer relies on predetermined solutions but on the co-creation of answers to increasingly complex needs. Artificial intelligence, in this case, can become a creative element, capable of stimulating new possibilities

[2] A *doppelgänger* is a German term that refers to a non-related identical twin or look-alike of a person. Often associated with folklore and paranormal beliefs, the concept of a doppelgänger has evolved to symbolize a mysterious, sometimes ominous, reflection of oneself. In literature and popular culture, it represents themes of identity, duality, and the uncanny, often serving as a mirror to the darker sides of the self.

Table 1. Integration Between CAD/CAM Systems and AI.

Company	Software/Tool	Integrated Technologies	AI/Innovation Features	Main Applications
Altair®	HyperWorks® Inspire®	CAE, Generative Design, Simulation, Structural Optimization	Machine learning for real-time optimization and generative design	Automotive, structural engineering, architecture
Autodesk®	Fusion 360®	CAD, CAM, CAE, Generative Design	AI-driven automated design generation, real-time simulation, and optimization	Industrial design, architecture, mechanical and electrical engineering
Bentley Systems®	OpenBuildings® STAAD® AECOsim®	CAD, CAE, BIM, Infrastructure Simulation	AI-powered simulation for energy efficiency and building management	Infrastructure, architecture, civil engineering
Dassault Systèmes®	CATIA®, SolidWorks® 3DEXPERIENCE®	CAD, CAE, PDM, Simulation, Generative Design	AI for design optimization, real-time simulation, and workflow management	Automotive, aerospace, architecture, industrial engineering
Formlabs®	PreForm®	CAM, 3D Printing, Generative Design	Optimized design for 3D printing and rapid prototyping in real time	3D printing, rapid prototyping, product design
Materialize®	Magics®	CAM, 3D Printing, Optimization for Additive Manufacturing	Algorithms for 3D geometry optimization and correction for additive manufacturing	Additive manufacturing, rapid prototyping
PTC®	Creo®	CAD, CAM, Generative Design, Advanced Simulation	Generative design, real-time design optimization using AI	Mechanical design, electronics, manufacturing
Red Bull Adv.Tech.®	-	AI, Simulation, Advanced Automotive Design	Predictive simulation, real-time performance optimization with AI	Formula 1, high-performance vehicle design
Siemens®	NX® Simcenter® Teamcenter®	CAD, CAE, PLM, Generative Design	AI for design optimization, predictive simulation, data flow management	Automotive, aerospace, systems engineering
Stratasys®	GrabCAD Additive Manufacturing Solutions®	CAM, 3D Printing, Generative Design	Optimized 3D model generation for additive manufacturing, real-time design adaptation	Rapid prototyping, additive manufacturing

and actively contributing to the construction of the future (uMake 2025) (See Table 1). This change does not imply the abandonment of intentionality in design, but a reformulation of the designer's role: from a solitary author to a mediator of a shared process, where the final result is the product of continuous dialogue between the human and the artificial. In this way, *Post-CAD* deeply reconnects to the speculative and imaginative dimension of design, but in a new way that does not imply a return to the past but a leap forward. Design is no longer merely a technical act but a creative practice that explores and experiments, driven by a synergy between human imagination and computational potential. No future is written. Between utopia and dystopia, between control and chaos, what remains is our capacity to cultivate critical awareness, ethical imagination, and shared creativity. It is not about blindly believing in technological salvation nor succumbing to the panic of self-destruction, but exercising the ancient art of discernment. Because in the era of the artificial, the greatest creative act remains choosing—consciously, humbly, and courageously—which humanity we wish to become. The new *WunderCAD/CAMmer*, therefore, will not only be a container of marvelous bio-digital artifacts but a laboratory for ethical choices, where the art of design merges with the responsibility of imagining a livable future for all.

10.2 A Reflective Epilogue: The Digital *Studiolo* Towards a Generative Ethics

As discussed in the introduction of this book, the concept of the studiolo is a powerful historical matrix: an intimate, both mental and material space where knowledge is organized in an orderly, contemplative form. From Federico da Montefeltro's studiolo to the seventeenth-century wunderkammer, we've inherited the idea that knowledge arises from the combination of objects, images, fragments, and traces—whether natural or artificial (Ministero della Cultura n.d.). Today, the digital studiolo manifests as an algorithmic environment, where combinations are no longer static or finite but dynamically generated, modified, and simulated in real time. It is no longer about preserving the world but shaping it—making possible futures, design hypotheses, and ethical alternatives visible. This new context transforms our studiolo from a treasure chest into a laboratory of the possible: a space where every design, visualization, or simulation choice carries responsibility. Wonder gives way to the conscious design of the unknown. Within this new ethical workplace, authorship undergoes a radical transformation. In the Renaissance, the author was someone who constructed an order of the world through collection and interpretation; today, they become curators of processes, facilitators of complexity. The work—whether an object, an interface, an algorithm, or a material—emerges as the result of non-linear interactions between human, machine, and environment. Authorship is now diffuse: the designer is no longer a solitary demiurge but an agent among many, including non-human ones. Generative algorithms, predictive models, neural networks, and environmental sensors actively participate in the creative process. Control is no longer centralized but negotiated in real time across networked architectures. This transformation introduces a new form of computational co-creation. AIs are no longer mere tools but partially autonomous collaborators, proposing unexpected variations, novel correlations, and emergent aesthetics. The goal is not to delegate imagination to the machine, but to amplify it through relational intelligence. This challenge is not only technical but also ethical and ontological: the author of an object generated by a chain of codes, learning from billions of data points, is not easily identifiable. Responsibility for its cultural, environmental, and social implications falls on those who oversee its design. In this context, designing means taking responsibility for the cognitive infrastructure that underlies the project. The transition from the *wunderkammer* to the *digital studiolo* (DS) also changes the function of mental and perceptive space: from a container of wonders to a critical space of inquiry. The DS becomes an operational environment where the moral, epistemic, and social implications of design are questioned. It becomes an ethical chamber, an interstitial zone between art, science, and philosophy, where deep reflection occurs on what is asked of intelligent tools. In an era where algorithms generate forms, behaviors, and autonomous decisions, the ethical question is central to the project itself. This ethics is not based on external norms but emerges from within the generative processes. It is a situated, relational ethics, considering context, actors (human and non-human), and potential trajectories. It is an extended bioethics, not limited to biological life but inclusive of the synthetic, artificial, and simulated. In the DS, a pedagogy of doubt develops: every design choice, automation, and technical default becomes an opportunity to reflect on how and why choices are made, which alternatives are excluded, and which futures are favored. Thus, the design space becomes one of anticipatory responsibility, where prediction is used not only for optimization but for imagining ethical and cultural consequences. The ethical studiolo is neither nostalgic nor conservative but a laboratory of possibilities—a space for generating worlds with awareness, where co-creation is not only aesthetic or functional but deeply political. For the designer, models are not merely technical tools or graphic anticipations of the project; they become places of critical reflection and epistemology in action. Every model—parametric, predictive, generative, or simulatory—reflects intentions, cognitive limits, and cultural biases. When a model is created, an act of selection and complexity reduction occurs: decisions are made about what to include and exclude, what to represent and what to leave indeterminate. In this sense, the model is a device of visibility and invisibility, revealing as much as it omits. It thus becomes a tool of human responsibility: it forces reckoning with the unknown, unseen, and uncontrollable. It interrogates the choices made, the rules applied, and the origins of simplifications and assumptions. The model challenges, asking new questions and bringing new issues to the surface. It becomes a tool of projective reflexivity, capable of revealing technical thought's aberrations and stimulating imagination toward new formal, functional, and ethical possibilities. This reflective function is crucial in a world dominated by predictive models—whether in AlphaFold® & CO., climate simulation, smart material design, or the training of creative AI. It's not just about training intelligent machines but re-educating our gaze through them. Ultimately, the model reflects humanity, inviting reflection on itself—not just as a designer but as a situated being, a moral agent, a temporary inhabitant of an increasingly simulated and co-constructed reality. Every design act generates light, but also shadow. Every prediction, simulation, or map of possibilities creates a zone of indeterminacy, a darkroom where unforeseen implications, unexpected consequences, and systemic exclusions hide. The DS invites exploration of these liminal, marginal, and unrepresented areas. In photographic language, the darkroom is where the latent image is revealed, where light and chemistry bring out potential in the sensitive medium. Similarly, in design, critical reflection must occur in the dark—those areas not illuminated by metrics and algorithms, where ethical risks, semantic ambiguities, and unverified assumptions lurk. In these opaque spaces arise questions about who is excluded from a model, which data is missing or cannot be represented, and which forms of life remain invisible. These darkrooms stand as counterfigures to bright dashboards and hyper-optimistic visualizations, forcing us to question not what we see, but what we decide not to see. When design is limited to what can be predicted and controlled, it loses its poetic and political dimension. Darkrooms remind us that every act of modeling is an act of exclusion, and that design justice arises from the ability to reveal what is usually hidden: silenced voices, discarded materials, collateral ecologies. In the DS, we learn to give hospitality to uncertainty, designing with and despite indeterminacy. The ethical designer does not fetishize the unknown but recognizes it as an integral part of complexity. In an era where AI promises to *illuminate everything*, the task of the ethical designer is also to leave spaces of shadow, zones of silence, and margins for doubt. Only there, in the darkroom of design, can the unexpected, the possible, and the future truly emerge. Design care becomes a foundational principle, not a new concept but one requiring deep reflection in the digital age. Care extends beyond maintenance or protection of objects: it involves an ethical and environmental responsibility for people, ecosystems, and future generations. Responsible design questions the social, cultural, and environmental repercussions it brings. Care requires constant attention to context—recognizing that a design that ignores social implications fails in its highest mission. This concept intersects with sustainability, where care extends to rights, cultural differences, and social inequalities. Design must be empathetic, considering impacts on people and environments throughout its life cycle. Every design decision, whether technical or stylistic, must undergo ethical evaluation. The designer becomes one who cares for people, well-being, and all subjects involved. Design care also extends to relationships—whether within design teams or between stakeholders. Care for relationships ensures the effectiveness of a design. Finally, care expands to uncertainty and change. In the digital world, design care isn't just about creating solutions for today but about systems that adapt and evolve. Design must respond dynamically to future challenges, ecological transitions, and social changes. Design is never neutral: it is always an act of awareness, a continuous reflection on its cultural, moral, and social implications.

References

Bitzer, T., Wiener, M., and Cram, W. (2023). Algorithmic Transparency: Concepts, Antecedents, and Consequences – A Review and Research Framework. Communications of the Association for Information Systems, 52: 293–331. Available at: https://doi.org/10.17705/1CAIS.05214 [Accessed: 12 May 2025].

Boker, S., Staples, A., and Hu, Y. (2016). Dynamics of Change and Change in Dynamics. Journal for Person-Oriented Research, 2: 34–55. Available at: https://doi.org/10.17505/jpor.2016.05 [Accessed: 12 May 2025].

Brynjolfsson, E., and McAfee, A. (2014). The second machine age: Work, progress, and prosperity in a time of brilliant technologies (Reprint ed.). W. W. Norton & Company.

Cicero, M. T. (n.d.). De natura deorum libri tres. Internet Archive. Available at: https://archive.org/details/denaturadeorumli00ciceuoft/page/58/mode/2up [Accessed: 12 May 2025].

Dusek, V. (2012). Risk Management in Technocracy. *In*: Roeser, S., Hillerbrand, R., Sandin, P., and Peterson, M. (eds.). Handbook of Risk Theory. Springer. Available at: https://doi.org/10.1007/978-94-007-1433-5_46 [Accessed: 12 May 2025].

Flexner, J. L. (2024). Optimism, utopia, and blue-green futures for the archaeology of Oceania. Australian Archaeology, 90(1): 33–35.

Genworks International. (2019). Genworks GDL: General-purpose Declarative Language. Retrieved from

Greenblatt, Richard. 1974. The LISP Machine. Working Paper No. 79. Massachusetts Institute of Technology, Artificial Intelligence Laboratory.

Hájek, M., and Panaro, L. (2008). Fantasia esatta. I colori della luce di Bruno Munari. APM.

History of Data Science. (2021). AI Winter: The Highs and Lows of Artificial Intelligence. History of Data Science. Available at: https://www.historyofdatascience.com/ai-winter-the-highs-and-lows-of-artificial-intelligence/ [Accessed: 12 May 2025].

Hughes, L., Dwivedi, Y. K., Malik, T., Shawosh, M., Albashrawi, M. A., Jeon, I. et al. (2025). AI agents and agentic systems: A multi-expert analysis. Journal of Computer Information Systems, 1(1): 1–29. Available at: https://doi.org/10.1080/08874417.2025.2483832 [Accessed: 12 May 2025].

Kolb, M. A. (1990). An investigation of constraint-based component-modeling for knowledge representation in computer-aided conceptual design (Doctoral dissertation). Massachusetts Institute of Technology, Department of Aeronautics and Astronautics.

Makatura, L., Foshey, M., Wang, B., Hähnlein, F., Ma, P., Deng, B. et al. (2024). How can large language models help humans in design and manufacturing? Harvard Data Science Review. Available at: https://hdsr.mitpress.mit.edu/pub/hiii8fyn/release/2 [Accessed: 12 May 2025].

Malvestio, M. (2022). Theorizing Eco-Dystopia: Science Fiction, the Anthropocene, and the Limits of Catastrophic Imagery. European Journal of Creative Practices in Cities and Landscapes, 5(1): 24–38.

Ministero della Cultura. (n.d.). Studiolo del Duca. Movio. Available at: https://www.movio.beniculturali.it/pmmar/eccellenzegallerianazionaledellemarche/it/49/studiolo-del-duca [Accessed: 12 May 2025].

Ng, K. W., and Poynting, S. (1995). Artificial Intelligence in CAD/CAM. Presented at the 1st National CAD/CAM Conference: Emerging Technologies for Industrial Competitiveness, 10 October 1995, Hilton Hotel, Kuala Lumpur, Malaysia. Available at: https://www.researchgate.net/publication/280316366_Artificial_Intelligence_in_CADCAM [Accessed: 12 May 2025].

Park, M., Leahey, E., and Funk, R.J. (2023). Papers and patents are becoming less disruptive over time. Nature 613: 138–144.

Salmon, P. M., King, B. J., Elstak, I., McLean, S., and Read, G. J. M. (2024). Tomorrow's demons: A scoping review of the risks associated with emerging technologies. The Journal of Risk Research.

Seymour, M. (2018). Artificial Intelligence Is No Match for Human Stupidity: Ethical Reflections on Avatars and Agents. In Australasian Conference on Information Systems 2018.

Sudabathula, B. (2025). The convergence of quantum computing and AI: Reshaping the future of machine learning. International Journal of Information Technology and Management Information Systems, 16: 46–57.

Taranto-Vera, G., Galindo-Villardón, P., Merchán-Sánchez-Jara, J., Salazar, J., and Salazar-Villalva, V. (2021). Algorithms and software for data mining and machine learning: A critical comparative view from a systematic review of the literature. The Journal of Supercomputing, 77: 11481–11513.

Valtonen, A. (2020). Approaching change with and in design. She Ji: The Journal of Design, Economics, and Innovation, 6(4): 505–529. Available at: https://doi.org/10.1016/j.sheji.2020.08.004 [Accessed: 12 May 2025].

Verganti, R., Vendraminelli, L., and Iansiti, M. (2020). Innovation and design in the age of artificial intelligence. Journal of Product Innovation Management, 37(4): 360–373.

Yeslam, H. E., von Maltzahn, N. F., and Nassar, H. M. (2024, July 19). Revolutionizing CAD/CAM-based restorative dental processes and materials with artificial intelligence: A concise narrative review. Journal of Dentistry and Oral Biology.

Yoo, S., Lee, S., Kim, S., Hwang, K.-H., Park, J., and Kang, N. (2021). Integrating deep learning into CAD/CAE system: generative design and evaluation of 3D conceptual wheel. Structural and Multidisciplinary Optimization, 64(6): 2725–2747.

Index

G

H

I

L

M

N

O

P

Q

R

S

T

U

W

For Product Safety Concerns and Information please contact our
EU representative GPSR@taylorandfrancis.com Taylor & Francis
Verlag GmbH, Kaufingerstraße 24, 80331 München, Germany